American Institute of Chemical Engineers

Pharmaceutical Engineering for the 21st Century

Topical Conference at the
2007 AIChE Annual Meeting

November 4-9, 2007
Salt Lake City, Utah, USA

Printed from e-media with permission by:

Curran Associates, Inc.
57 Morehouse Lane
Red Hook, NY 12571
www.proceedings.com

ISBN: 978-1-60423-844-0

Some format issues inherent in the e-media version may also appear in this print version.

ISBN: 978-1-60423-844-0

American Institute of Chemical Engineers
Proceedings
Three Park Avenue
New York, NY 10016-5991
Phone: 212-591-8100

www.aiche.org

American Institute of Chemical Engineers

Pharmaceutical Engineering for the 21st Century
2007

TABLE OF CONTENTS

Combination of Experimental and Mathematical Techniques for Characterization of Particle Friability 1
Pavol Rajniak, Rong Fan, Luke Schenck, Kumar Dhanasekharan, Rey Chern

Analysis of Aspirin Milling in Single Ball Mill 12
R. Bertrum Diemer, Yulong Ding, Dolapo Olusanmi, Mojtaba Ghadiri, K. J. Roberts, Jon H. Olson

Optimization of Milling Using Attainable Region Theory 13
Matthew J. Metzger, Benjamin Glasser, David Glasser, Brendon Hausberger, Diane Hildebrandt, Ngangezwe Khumalo

Numerical Simulation and Optimization of Grinding in Hammer Mill 14
Bodhisattwa Chaudhuri

The Breakage Matrix Approach in Evaluating Inadvertent Breakage of Dense Particulate Mixtures 15
A. Abu Nahar, U Tüzün

Effects of Co-Milling on Crystallinity of Salbutamol Sulphate Using Adipic Acid 16
Prashant N. Balani, Wai Kiong Ng, Sui Yung Chan, Reginald B.H. Tan

Application of Box–Behnken Design and Response Surface Methodology for Modeling of Grinding Beach Sand Ore Using Ball Mill 20
Asha Immanuel Raju Chaduvula, Murali Yugandhar Nikku, Kiran Kumar Nalla, Venkata Satya Chappidi, Srinivasa Babji Josyula, Israel Yedluri

The Quality by Design Approach to Pharmaceutical Development and Manufacturing 22
Christine M.V. Moore

Challenges in Pharmaceutical Development and Manufacturing 23
Kenneth R. Morris

Optimization Under Uncertainty for the Design and Planning of Process Systems 24
Ignacio E. Grossmann

Role of PAT in Quality by Design 25
Christine M.V. Moore

Process Analytical Technology for Crystallization Processes 26
Richard D. Braatz

Measurement and Control of Product Attributes during Granulation: Challenges and Opportunities (No abstract) 27
James D. Litster

Flow Structure Mapping of Segregating Granular Mixtures Using Radioactive Particle Tracking 28
Jocelyn Doucet, Francois Bertrand, Jamal Chaouki

Mixing of Nanoparticles By Magnetically Assisted Impact Mixing Technique 29
James V. Scicolone, Alexandre Ermoline, Ryan Barrow, Rajesh Dave

Segregation of Cohesive Particles During Hopper Discharge 30
Anshu Anand, Jennifer S. Curtis, Carl R. Wassgren, Bruno C. Hancock

Eliminating Segregation In Free-Surface Flows of Particles 31
Deliang Shi, Watson Vargas, Joseph J. McCarthy

Powder Flow Induced By Rough-Surface Moving Walls: Concentration Profile of a Binary Powder Mixture ... 32
Luis Guillermo Obregon Quiñones, Carlos Velázquez Figueroa

Study of the Feedframe Effect on Particle Size Distribution ... 33
Rafael Mendez-Roman, Fernando J. Muzzio, Carlos Velazquez

The Mechanism behind Axial Segregation of Granules in a Horizontal Rotating Drum ... 34
Marleen M. H. D. Arntz, Wouter. K. Den Otter, Wim. J. Briels

Tower of Babel: Intelligent Data and Knowledge Management for Decision Support in Pharmaceutical Engineering ... 35
Venkat Venkatasubramanian

Modeling and Numerical Simulation of Pharmaceutical Processes ... 36
Hamid Arastoopour

Modelling Opportunities and Challenges in the Pharmaceutical Industry ... 37
Christopher L. Burcham

Information Management: Challenges and Opportunities in Achieving Qbd (No abstract) ... 38
Anjali Kataria

Thermodynamic Limits on Drug Loading in Nanoparticle Cores ... 39
Varun Kumar, Robert K. Prud'homme

Alternating Copolymer Structures for Targeted In Vivo Imaging and Therapy In Cancer ... 46
Michelle T. Miller, Kevin P. Brower, Jin Zhou, Robert J. Fisher, Mukesh K. Pandey, Rahul Tyagi, Arthur C. Watterson, Clark K. Colton

Implant Assisted Magnetic Drug Targeting: Ferromagnetic Nanoparticles for Enhancing the Retention of Magnetic Drug Carrier Particles ... 47
Misael O. Aviles, Jan O. Mangual, Armin D. Ebner, James A. Ritter

Free Radical Scavenging By Cerium Oxide Nanoparticles: the Effect of Adsorbed Block Copolymers and Potential-Determining Ions ... 48
Richey M. Davis, Will Miles

Drug-Delivery Nanoparticles Formed By Micellization of Polyethylene Glycol-Block-Polycaprolactone In Subcritical and Supercritical Solvents ... 49
Zachary Tyrrell, Winoto Winoto, Maciej Radosz

Silica-Based Nanoparticles for Boron Neutron Capture Therapy ... 50
Ilya Zharov

Prediction of Crystal Shape and Size Distributions Using Multidimensional Population Balances ... 51
Christian Borchert, Nandkishor K. Nere, Andreas Voigt, K. Sundmacher, Doraiswami Ramkrishna

2-D Population Balance Modelling of Gibbsite Crystallization ... 53
Andrey V. Bekker, Iztok Livk

Contin Analysis of Sals Data from Clusters of Particles beyond RDG Limit ... 63
Marco Lattuada, Lyonel Ehrl, Miroslav Soos, Massimo Morbidelli

Molecular Dynamics Simulation of Diffusion of Fractal Aggregates ... 64
Gaurav Pranami, R. Dennis Vigil, Monica Lamm

Modeling, Validation and Optimization of Antisolvent Crystallization: A New Approach In Predicting Solubility Using A Thermodynamic Model ... 65
Ombretta Foddi, Ali Abbas, Massimiliano Grosso, Jose A. Romagnoli

Modeling the Sintering of Water Soluble Amorphous and Crystalline Particles ... 67
Nicolas Descamps, Elsa Schreyer, Stefan Palzer

The Effects of Initial Polydispersity of Particle Size on the Growth Kinetics and Mechanism of Wet Granulation in a Laboratory-Scale High Shear Mixer by Means of Image Processing and Analysis ... 75
Alvaro Realpe, Carlos Velazquez

Model Predictive Control of Continuous Drum Granulation of Limestone ... 83
Thomas Glaser, Constantijn F. W. Sanders, Fu-Yang Wang, Ian T. Cameron, James D. Litster, Jonathan M. H. Poon, Rohit Ramachandran, Charles D. Immanuel, Francis J. Doyle III

Passive Acoustic Monitoring of High-Shear Wet Granulation ... 93
Albert W. Alexander

Experimental Study of Wurster Granulation for Pharmaceutically-Relevant Excipient Blends ... 94
Christopher Mancinelli, Pavol Rajniak, Frantisek Stepanek, Leon Farber, Brian Hill, Rey Chern

Suspension and Dispersion of Particles in Gels ... 95
Micaela Caramellino, Piero M. Armenante

Opportunities, Challenges, and Approaches for Modeling In the Pharmaceutical Industry ... 96
Prashant B. Kokitkar

Modeling of Breakage and Dissolution of Crystals with Varying Particle Size Distribution within the Quality by Design Framework ... 97
Pavol Rajniak, Kim Gallagher, Rey Chern, Craig Ikeda, Justin Moser

Application of Modeling in a Qbd Context ... 102
James Wertman

Coater-Dryer Models ... 103
Alexander J. Marchut, Wei Chen, Steven H. Chan, Shih-Ying Chang, Olav Lyngberg, Donald Kientzler, San Kiang

Potential of S88, Batchml and Other Standards in Streamlining Process Modeling ... 104
Alistair Gillanders

Batch Plus Modeling as a Tool To Drive R&d-Based Process Expansion ... 112
Samrat Mukherjee, Shekhar Viswanath, Jeffrey Breting

Application of Agent Technique in Bioprocess Modelling for Plant-Wide Process Improvement ... 113
Ying Gao, Katie Kipling, Jarka Glassey, Yuhong Zhou, Mark Willis, Gary Montague, Nigel Titchener-Hooker

An Integrated Quality by Design Qbd Approach for Pharmaceutical Powder Blending Unit Operation: Process Monitoring, Endpoint Determination, and Concentration Quantification ... 123
Huiquan Wu, Mobin Tawakkul, Mansoor A. Khan

Quality By Design Using an Integrated Active Pharmaceutical Ingredient-Drug Product Approach To Development ... 138
Mary T. am Ende, Vincent E. McCurdy, Thomas P. Garcia, Frank Busch, Mark R. Berry

The Use of Routine Process Capability In the Determination of Proven Acceptable Ranges (Pars) Critical Process Parameters (Cpps) and Mapping of the Design Space for an Api Process ... 139
Kevin D. Seibert, Shanthi Sethuraman, Chad Wolfe

Mathematical Modeling of the Impact of Raw Material Variabilities On Granule Particle Size Distribution During Fluidized Bed Granulation ... 140
Pavol Rajniak, Ecevit Bilgili, Rey Chern

Synchronized Approach to Developing Optimal Control and Design Spaces ... 142
Jerry D. Mitchell, Bernard McGarvey, Kumar Abhinava, Kevin Seibert, Shanthi Sethuraman, Kristi Griffiths

Correlation between Excipient Characteristics and Product Properties through Numerical Simulations of Tablet Compaction ... 143
Athanas A. Koynov, Alberto Cuitino

Characterizing the Influence of Raw Material and Process Variation on Product Quality and Yields ... 152
Chris McCready

Dynamic Impact of Water Thermodynamics On Normal Phase Chromatography ... 153
David Willcox

Application of In-Line Transmission Normal Phase Chromatography for Improved Process Understanding and Control of a Continuous Mixing Process ... 154
Gregory M. Troup, John P. Higgins, Craig A. Mckelvey, Michael Lowinger, Jeff S. Ko, Luke R. Schenck, Robert F. Meyer, Brit L. Rudeen

Solid State Form Conversion Process ... 155
Christopher L. Burcham, Jeffrey T. Vicenzi

Development and Implementation of a Drying Process for an Amorphous API Solid ... 156
Yao-En (David) Li, Alex Coles, Gerrie Liaw, and Stephen Tyler

New Spectrophotometric Method for the Determination of Ambroxol Hydrochloride Using Artificial Neural Networks ... 157
Asha Immanuel Raju Chaduvula, Kiran Babu Uppuluri

The Use of PAT in the Laboratory to Define and Optimize API Manufacturing Processes (No abstract) ... 158
Mark A. LaPack

Batch Process End Point Using PCA and DCPA Model Based On FTIR and Calorimetry Data Measurements ... 159
Shan Lin, Shekhar Viswanath, James Marek, Steve Richter, Zhi Wang, Clifford Mitchell

Numerical Bifurcation Analysis of Periodic Solution of Population Balance Models ... 160
Mykhaylo Krasnyk, Michael Mangold, Achim Kienle

Modeling Soot Formation in Turbulent Flames by Direct Quadrature Method of Moments ... 168
Federica Furcas, Alessandro Zucca, Daniele Marchisio, Antonello Barresi

A Mechanistic Model for Nucleation and Aggregation in Population Balances of Granulation: Batch Characterisation Studies and Experimental Validation ... 170
Rohit Ramachandran, Jonathan M. -H.Poon, Constantijn F. W. Sanders, Thomas Glaser, Francis J. Doyle II, James D. Litster, Frantisek Stepanek, Fu-Yang Wang, I. T. Cameron, Charles D. Immanuel

Numerical Solution of Multi-Variable Cell Population Balance Models With A Free Boundary Algorithm ... 189
Mihail E. Kavousanakis, Andreas G. Boudouvis, Nikos V. Mantzaris

Fast Monte Carlo Algorithms for Modeling and Optimization of Complex Particulate Processes ... 190
Roberto Irizarry

A New Simulation Method for Population Balances Using Constant-Number Monte Carlo With Adjustable Time Step ... 200
Samira Khalili, Antonios Armaou, Themis Matsoukas

Validation of Fcmom for Bi-Variate Population Balance Equations ... 201
Matteo Strumendo, Hamid Arastoopour

Role of Counterion Replacement on the Deactivation of Co-Salen Catalysts in the Hydrolytic Kinetic Resolution of Epichlorohydrin ... 202
Surbhi Jain, Xiaolai Zheng, Christopher W. Jones, Marcus Weck, Robert J. Davis

Experimental and Dft Study of Base- and Ligand-Free Pd/c-Catalyzed Homocoupling Reactions of Arylboronic Acids 203
Jeng-Shiou Chen, Johannes Khinast

Development of New Covalently Tethered Group 4 Metallocenes 204
Heidrun Woelfler, Georg Schitter, Rafael Eder, Claire Jeanquartier, Johannes G. Khinast

The Discovery of an Innovative and Very Green One-Step Synthesis of Tetronic Acid 205
Josef G. Schroer

Triflation Reaction Endpoint By A T R - F T I R, P L S - Based Chemometrics Model 206
James Marek, Shan Lin, Shekhar Viswanath, Clifford Mitchell

Multi-Step Hydrogenations and Process Understanding for Api Manufacturing (No Abstract) 207
Jason Mustakis, David J. Am Ende

Effect of Different Vessel Geometries on the Hydrodynamics of Partially Baffled Stirred Tank Reactor Systems Equipped with a Retreat Blade Impeller 208
Giuseppe Di Benedetto, Piero M. Armenante, Billy R. Allen

Continuous Powder Blending for Pharmaceutical Applications 209
Michael Gentzler, Ecevit Bilgili, Brit L. Rudeen, Mano Ramasamy, Steve J. Heidel

Model Predictive Control on Roller Compaction for Pharmaceutical Manufacturing 210
Shuo-Huan Hsu, Girish Joglekar, Gintaras V. Reklaitis, Venkat Venkatasubramanian

Experimental Study and Computational Modeling of Continuous Powder Mixing Processes 212
Patricia M. Portillo, Marianthi G. Ierapetritou, Fernando J. Muzzio

Determination of Granulation Growth Rates: an Evaluation Between FBRM and Sieve Analysis 214
Kevin Macias, M. Teresa Carvajal

Effect of Stirrer Design on the Homogeneity of a Pharmaceutical Powder Mixture 215
Henri Berthiaux, Cendrine Gatumel, Jean-Louis Dirion

Application of In-Line Transmission Nir for Improved Process Understanding and Control of a Continuous Mixing Process 223
Gregory M. Troup, John P. Higgins, Craig A. Mckelvey, Michael Lowinger, Jeff S. Ko, Luke R. Schenck, Robert F. Meyer, Brit L. Rudeen

Continuous Blending of Powders: Effect of Process Variables and Material Properties 224
Lakshman Pernenkil, Farzad J. Yazdi, Jason Whittaker, Charles Cooney

Crystals and Crystallization Processes: A Crystallographic Perspective 232
Qi Gao, William L. Parker

Relationship Between Self-Association of Molecules In Supersaturated Solutions and Solid State Outcome 233
Deniz Erdemir, Soma Chattopadhyay, Liang Guo, Jan Ilavsky, Heinz Amenitsch, Carlo U. Segre, Allan S. Myerson

Selective Crystallization of the Metastable Anhydrate Form in the Enantiotropic Pseudo-Dimorph System of L-Phenylalanine Using Feedback Concentration Control 234
Nicholas Kee, Paul D. Arendt, Reginald B. H. Tan, Richard D. Braatz

Growth Prediction for Molecular Crystals of Api-Complexity 242
Ryan C. Snyder, Jacob P. Sizemore, Michael F. Doherty

Solution Speciation of Impurities: A Force Field Simulation ofIts Impact On α - Glycine Crystal Habit 244
Sendhil Poornachary, Pui Shan Chow, Reginald B.H. Tan

Quantitative Analysis of H-Bonding Interactions: Relation To Crystal Shape and Size 245
Charles Acquah, Luke E. K. Achenie, Arunprakash T. Karunanithi, Shanthakumar Sithambaram, Steven Suib, Jose Gascon

Nanoparticle Formation of Highly Water-Soluble Pharmaceutical Coated with Poorly Water-Soluble Material by Using the Rapid Expansion of Water-in-Supercritical CO_2 Microemulsion 246
Worawut Temphuwapat, Masahiro Ikeda, Hideki Ichikawa, Yoshinobu Fukumori, Atsushi Tsutsumi

Processing of Superparamagnetic Iron Oxide Nanoparticles by the Dense Gas Technology 255
Un Teng Lam, Raffaella Mammucari, Neil R. Foster

Application of Supercritical Anti-Solvent To Synthesize Delivery Systems for A Sustained Release of Food Antimicrobials 256
Minfeng Jin, Qixin Zhong

Visualization and Conception of the Atomized Rapid Injection for Solvent Extraction (Arise) Process for the Production of Highly Respirable Powders 257
Roderick Sih, Neil R. Foster

Encapsulated Quercetin Particles Using a Supercritical Antisolvent Process 258
Gunilla B. Jacobson-Andrews, Elizabeth M. Szilágyi, Can Quan, Charlotta Turner, Richard N. Zare

Estimation of Nucleation Kinetics From the Kinetic Limit of Metastable Zone Width 259
Venkateswarlu Bhamidi, Sameer Talreja, Guangwen He, Paul J. A. Kenis, Charles F. Zukoski

Correlating Molecular Interactions with Solubility of Alpha-Amylase from Bacillus Licheniformis 260
Chirag M. Mehta, Edward T White, James D Litster, Matthew T. Hardin, Abraham Lenhoff

Molecular Dynamics Simulation of the Hydration and Interaction between Bovine Insulin Hexamers 269
Vishal Koparde, Kristen A. Fichthorn

Microfluidic Platforms for In-Meso Membrane Protein Crystallization 270
Griffin W. Roberts, S.L. Perry, J.D. Tice, P.J.A. Kenis

Analysis of Concentration Field within an Evaporating Hanging Droplet 271
Li May Goh, Richard D. Braatz

Crystallization of Spherical Agglomerates of a Peptide 278
Shailendra Bordawekar, Samrat Mukherjee, John Tolle

Acceleration of the Formation Rates of Metal Nano Particles with A Sonochemical Process by the Addition of Microbubbles 279
Satoshi Takeda, Hideo Tajima, Akihiro Yamasaki, Fumio Kiyono, Kazukiyo Kumagai, Yukio Yanagisawa

Particle Formation Pathways In Antisolvent Precipitation Processes 286
Jan Sefcik, Anna Jawor-Baczynska, Barry D. Moore

The Effect of Impurities on Borax Decahydrate Crystallization in MSMPR Crystallizer 287
Birgul Benli, Nusret Bulutcu

Scale-Up Investigation of Particle Production In Emulsions 297
Andreas Voigt, Michael Fricke, Kai Sundmacher

Membrane Emulsification for Polyd,L-Lactide-Co-Glycolide Acid Microparticle Production: Influencing Factors 298
Gilda Gasparini, Serguei R. Kosvintsev, Richard G. Holdich

Droplet-Based Microfluidics for Monodisperse Particle Synthesis 306
Nick Carroll, Christian Holtze, Sergio Mendez, Gabriel P. Lopez, David A. Weitz, Dimiter Petsev

Simulant Development for Savannah River Site High Level Waste 307
Michael Stone, R. E. Eibling, D. C. Koopman, D. P. Lambert, P. R. Burket

Seeding--Friend or Foe? 314
San Kiang, Chiajen Lai, Thomas LaPorte, Fernando Quiroz, Lori Spangler, Jean Tom, Zhongmin Xu

A High Intensity Emulsification Method for the Production of Nanoparticles 315
Eric Jayjock, M. Silvina Tomasone, Fernando J. Muzzio

Precipitation of Micro/nanoparticles by Enhanced Energy Dissipating Impinging Jets in Stirred-Tank Reactors 316
Giuseppe Di Benedetto, Ewa Sukcik, Piero M. Armenante, Somenath Mitra

Development of a Particle Size Analysis Method for Dry and Wet Granulation 317
Jennifer Wang, Niranjan Kottala, Robert Jerzewski, Emil Friedman

Computational Modelling of the Anderson Cascade Impactor 318
Eric Jayjock, Justin P. Lacombe, Fernando J. Muzzio

Controlled Nanoparticle Agglomeration as a Low-Density Dry Powder Aerosol Formulation Strategy 319
Lianjun Shi, Laura J. Peek, Carl Plumley, Mark Bailey, Cory Berkland

Process Engineering Aspects of Plasmid-Based Biopharmaceuticals Production: Tackling the Threatening Vaccine Shortages to Prevent Global Pandemics 320
Michael K. Danquah, Gareth M. Forde

Continuous Thermal Epimerization in High Temperature Tubular Reactor 324
Amy Jines

Continuous Newman Kwart Rearrangement In Thermal Tube Reactor 325
Martin Johnson

Controlling the Key Response Variables Through Scaling Rules During the Scale-Up of a Fluidized Bed Granulation Process 326
Ecevit Bilgili, Jeff Ko, Adam Chen, Edward Smith, Pavol Rajniak, Kyle Fliszar, Gordon Wong, Larry Rosen, Daniel Grubb, Dafni Bika

Development and Scale-Up of Efficient Api Drying Using Fluid Bed Dryer 327
Yubo Yang, Jenny Lau, Ilaria Popolla, Gerrie Liaw, Melissa Ostberg, Stephen Tyler

Quality Risk Analysis of a Tablet Coating Process 328
Steve S. Y. Wang, Steven H. Chan, San Kiang

Novel Scaling of Rotary Drum Film Coating Developed from Commercialization of Pharmaceutical Tablets 329
Alan R. Silverman, Jeffrey Givand, Brad Holstine, Michael Gentzler

Oral Absorption of Weakly Acidic Compounds 330
Monica L. Adams, James Ormes, Ajit Thakur

Oral Delivery of Insulin-Transferrin Bioconjugates Using Intelligent Complexation Hydrogels 331
Justin P. Shofner, Nicholas A. Peppas

Development of a Computer Program Predicting the Influence of Cyclodextrins On Oral Bioavailability of Insoluble Drugs: From Modeling To In Vivo Comparison 332
Ece D. Gamsiz, Lee A. Miller, Avinash G. Thombre, Rebecca L. Carrier

Using Physically-Based Dissolution Simulations to Develop Ideal Particle Distributions in Tablets 334
Dan Braido, Alberto Cuitino

A New Hypromellose Excipient for Direct Compression Controlled Release 340
Karl Jacob, Kacee Ender, Tim Cabelka, Dan Denomme, Dave Wallick

Sublingual Drug Delivery; Will It Always Give 'Fast Onset'? A Product Enhancement Perspective 341
Jeremy A. Bartlett

Conversion of Chord Length Distribution To Particle Size Distribution In Needle Systems 342
Sze Wing Wong, Christos Georgakis

Observation of Polymorphic Change Through Analysis of FBRM Data: Applications 344
Stephanie Barthe, Martha Gallivan, Ronald W. Rousseau

Relationship between Particle Size Distribution and Chord Length Distribution 345
Pui Shan Chow, Zai Qun Yu, Reginald B.H. Tan

In Situ PSD Monitoring At High Solids Concentration Using Video Imaging and Statistical Estimation 346
Paul A. Larsen, James B. Rawlings

Real-Time Particle Size and Shape Characterization Using In-Process Video Imaging Process Analytical Technology (PAT) In Dynamic Processes 347
Eric J. Hukkanen, Terry P. Redman

Automated Control of Batch Cooling Crystallization Using FBRM 348
Pui Shan Chow, Jia Wei Chew, Reginald B.H. Tan

Pharmaceutical Engineering as an Emerging Academic Field - Challenges and Opportunities 349
Piero M. Armenante

Technology Roadmap for Pharmaceutical Development and Manufacturing (No abstract) 350
Gintaras V. Reklaitis

NSF Engineering Research Center for Structured Organic Particulate Systems (No abstract) 351
Fernando J. Muzzio

Author Index

COMBINATION OF EXPERIMENTAL AND MATHEMATICAL TECHNIQUES FOR CHARACTERIZATION OF PARTICLE FRIABILITY

P. Rajniak[1], R. Fan[2], L. Schenck[1], K. Dhanasekharan[2], Rey Chern[1]
[1]Merck & Co., Inc., West Point, PA, USA
[2]Ansys-Fluent Inc., Lebanon, NH, USA

Abstract

This work involves the development of methodology for characterizing and predicting the friability of granules by combining experimental and theoretical techniques as an extension of our recent work [1].

First, experimental data of granule attrition are collected via a bench-top Sympatec laser diffraction particle size analyzer and Helos dry particle disperser. Samples are analyzed at different inlet air pressures of 0.1 to 6.0 bar. The key part of the Helos disperser (which can be considered a pneumatic conveying system) mostly responsible for attrition is the bend between the inlet and the laser diffraction measurement chamber. Increasing the inlet air pressure produce a finer PSD as a consequence of more extensive attrition.

Second, a simple maximum-gradient population balance (MG-PB) mathematical mode of breakage is developed. The model is solved using the Quadrature Method of Moments (QMOM) and used for preliminary evaluation of the breakage rate properties by fitting a model to experimental data from the SYMPATEC equipment. Different semi-empirical expressions for the breakage kernels and for the daughter distribution functions are tested. Multiple breakage distribution functions, proposed by Diemer and Olson [2], are usually needed to get satisfactory agreement with experimental data.

Finally, a model combining CFD and QMOM methodologies is developed. The combined model uses the function forms for the breakage kernel and for the daughter distribution function obtained before using the MG-PB model. However, the detailed CFD-PB model also allows us to relate the breakage kernel parameters to particular flow properties. It was found that for a given experimental set-up with diluted gas-solid flow, the breakage rate of granules is proportional to the characteristic particle size and to the square of the impact velocity between a granule and the equipment wall, and can be expressed using the form proposed by Moreno-Atanasio and Ghadiri [3].

The combined experimental and theoretical methodology was tested for evaluation of the breakage rate parameters of different granules. The methodology requires small quantities (~ 20g) of granules, which is particularly useful for comparison of different formulations during the drug development process. Moreover, physically based models calibrated by the above methodology and combining properties of the gas-solid flow with the PB models can also be employed to predict attrition in manufacture-scale pneumatic conveying systems.

2007.

[3] Paul A. Larsen and James~B. Rawlings. Maximum likelihood estimation of particle size distribution for high-aspect-ratio particles using in situ video imaging. Submitted for publication in *Technometrics*, April 2007.

Introduction

Pneumatic conveying systems are used in many industries to transport granular materials. Granule attrition, where the particles suffer wear as a result of collisions and friction, can occur during the transport of materials. Experimental and theoretical study to understand the mechanical impacts of conveying upon granules is needed so that formulations, the processing parameters and the pneumatic conveying systems can be optimized to avoid problems at the large scale. To address attrition phenomena, different experimental and theoretical approaches have been followed.

Particle size distribution (PSD) of a powder system is routinely measured by laser diffraction using bench-top equipment such as Sympatec or the Malvern Mastersizer laser diffraction particle size analyzers [1]. The powder under test is fed using a vibrating feeder and then suspended by a jet of compressed air. Increasing the air jet pressure was shown to produce a proportional increase in specific surface area [4]. This 'disadvantage' of the dry powder characterization methodology can be used as a simple technique for characterizing the friability of granules.

Theoretical

The population balance (PB) equations for size reduction of solid particles or granules by attrition (undesirable breakage) in non-homogeneous flow systems can be written in terms of volume fraction of particle size (class) i as

$$\frac{\partial}{\partial t}(f_i \alpha_s \rho_s) + \nabla.(f_i \alpha_s \rho_s \vec{u}_s) = \alpha_s \rho_s (B_i - D_i) \qquad i = 0,1,2,...,M-1 \qquad (1)$$

where α_s, ρ_s, $\vec{u}_s$ is the total volume fraction, density and velocity vector of solid particles respectively, f_i is the fraction of particles of size i, M is number of classes and B_i, D_i are the particle birth and death rates respectively. In the CFD-PB granular multiphase model the continuity equations for solid phase (1) are solved in each cell of the geometry mesh together with the momentum balance for solid phase, the continuity equation and momentum balance for the gas phase, the balance equation for the fluctuating kinetic energy (granular energy) and corresponding boundary conditions.

A one-dimensional steady state maximum-gradient population balance (MG-PB) model [5, 6] represents the simplest approximation of the flow and attrition in the investigated equipment. It is assumed that the residence time $\tau = L/\langle u_s \rangle$ is the same for all solid particles flowing with the average velocity $\langle u_s \rangle$ through the tube with the length L. The population balance equations for the MG-PB model can be written as

$$\langle u_s \rangle \frac{df_i}{dz} = B_i - D_i \qquad i = 0,1,2,...,M-1 \qquad (2)$$

The "maximum-gradient" model is also called the "plug-flow" model in other texts [5,6]. However, the first term is preferred in this manuscript to avoid confusion with the "plug flow" term used in the pneumatic conveying terminology for the dense phase flow in which the whole cross section of the pipe is filled with moving particles.

The discrete methods (DM) for solution of the PB models described by Eqs.(1) or (2) are based on representing the continuous PSD in terms of a set of discrete classes. The disadvantages of DM (also known as the classes or sectional methods) are that the classes must be defined beforehand and that a large number of classes may be required.

On the other hand the quadrature method of moments (QMOM) [7] or the direct quadrature method of moments (DQMOM) [8, 9] require only a small number of scalars (moments) to track the PSD with very small error and are also computationally affordable for use with CFD codes. The QMOM approach is based on taking the moments of the PB equations with respect to the particle size L and in place of the continuity equations for size classes (1) or (2) the continuity equations (3) or (4) for moments m_k are solved

$$\frac{\partial}{\partial t}(\rho_s m_k) + \nabla.(\rho_s \vec{u}_s m_k) = \rho_s(\overline{B}_k - \overline{D}_k) \qquad k = 0,1,2,...,2N-1 \qquad (3)$$

$$\langle u_s \rangle \frac{dm_k}{dz} = \overline{B}_k - \overline{D}_k \qquad k = 0,1,2,...,2N-1 \qquad (4)$$

The moments m_k are calculated by the quadrature approximation

$$m_k = \sum_{i=1}^{N} w_i L_i^k \qquad k = 0,1,2,...,2N-1 \qquad (5)$$

The approximation (6) of order N is defined by its N weights w_i and N abscissas L_i and can be calculated by its first $2N$ moments $m_0, m_1, ..., m_{2N-1}$. Number of scalars (moments) $2N$ for QMOM is significantly lower than number of classes M for the discrete methods. Approximations of the death and birth rate terms are in QMOM expressed as

$$\overline{D}_k = \sum_{i=1}^{N} w_i g(L_i) L_i^k \qquad (6)$$

$$\overline{B}_k = \sum_{i=1}^{N} w_i g_i \int_0^{\infty} L^k \beta(L; L_i) dL \qquad (7)$$

Many different semi-empirical expressions for the breakage kernel $g(L_i)$ and for the daughter distribution function $\beta(L;L_i)$ exist [7,10,11]. Generally used is the power-law breakage kernel depending on the size L_i

$$g(L_i) = kL_i^{\gamma} \tag{8}$$

with two empirical parameters, the breakage rate constant k and the breakage rate exponent γ. It was found [1,12,13,14] that good prediction of experimental data can be obtained using the breakage kernel

$$g(L_i) = k_v (v - v_0)^2 L_i^{5/3} \tag{9}$$

where v is the impact velocity and v_0 is the threshold impact velocity is taken into account. k_v and v_0 are empirical parameters to be evaluated by fitting the model to experimental data. The impact velocity is computed using the CFD code as the solid-phase velocity magnitude in the cells next to the wall.

An extensive discussion of the problems related to the fragment distribution function can be found in [11]. It was found [1] that the multiple breakage distribution function (10) is needed to describe the attrition of the granules

$$\theta(z) = V_i \beta(V;V_i) = p\frac{z^{q-1}(1-z)^{r-1}}{\mathrm{B}(q,r)} \qquad \text{where} \qquad r = q(p-1) \tag{10}$$

$z = V/V_i$ = self-similar daughter size, V_i = volume of parent particle, V = volume of daughter particle, p = number of daughters produced in breakage, q,r = function parameters, B = the Beta function

Experimental

The granules for attrition tests were manufactured in a bench-scale high shear granulator and on average contained 43% of Lactose Hydrous Regular Grind, 43% of Microcrystalline Cellulose Avicel, 4% of Croscarmellose Sodium, 3% of Hydroxypropyl Cellulose EXF and 7% of an Active Pharmaceutical Ingredient (API).

The particle size distribution (PSD) of the granules system was routinely measured by laser diffraction using a bench-top Sympatec laser diffraction particle size analyzer and Helos dry particle disperser. Samples are analyzed at different inlet air pressures of 0.1 to 6.0 bar. Increasing the inlet air pressure produced a finer PSD presumably as a consequence of more extensive attrition. A key part of this laboratory pneumatic conveying system is the bend between inlet and the laser diffraction measurement chamber. It is generally recognized that during pneumatic conveying the particles experience extensive impact loads at the bends because the flow direction is changing [15].

Results and discussion

It was observed that PSDs at pressures 0.1 - 0.5 are practically identical. Moments m_0, m_1, m_2, m_3 of these PSDs are evaluated using relations (11)

$$m_k = \sum_{i=1}^{31} n_i L_i^k = \sum_{i=1}^{31} \left(\frac{6x_i}{\pi L_i^3} \right) L_i^k \tag{11}$$

n_i is volumetric number of particles in class i having characteristic size (diameter) L_i and x_i is volume fraction of the class i. Number of the classes measured by the Sympatec analyzer is 31. The Sauter diameters $d_{32} = m_3 / m_2$ are compared in **Table 1.** Note that the Sauter diameter decreases with increasing pressure.

Table 1: Sauter diameters of the PSDs at different inlet pressure P_{in} [barg].

Inlet pressure P_{in}	0.1	0.5	1.0	2.0	4.0	6.0
$d_{32}.10^{+6}$ [m]	60.1	59.6	41.2	37.2	35.3	32.6

Simulation of flow without breakage using CFD model

To get better insight into the flow phenomena, a 3-dimensional (3D) CFD Eulerian granular model was developed. The CFD model solves continuity equations (1) together with the other balances in each cell of the Sympatec analyzer geometry. However, the source terms on the right side of equations (1) are zero. The model simulates impact of different inlet jet pressures on the flow properties and also evaluates average velocity and volume fraction of particles in the equipment. The model was also employed for calibration of an unknown gap size in the eductor by comparing simulation results to experimentally measured inlet and suction pressures during flow of glass particles (without breakage or attrition). The Sympatec analyzer geometry is illustrated in **Figure 1.**

Figure 1. 3-D Fluent illustration of the Sympatec PSD analyzer

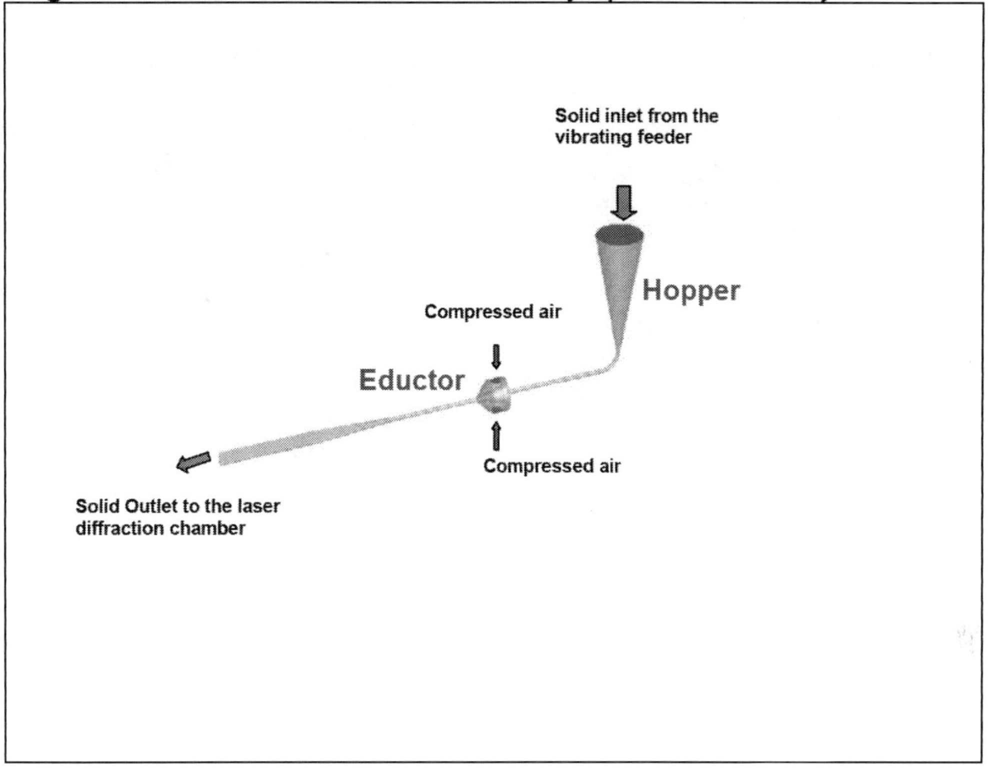

Evaluation of experiments using MG-PB model

The average solid velocities and solid volume fractions calculated using the CFD flow model were used in next step for optimization of the breakage parameters of the MG-PB model (2). The goal of optimization was to find the parameters of the breakage kernel and distribution functions which minimize the objective function (12)

$$\min\left\{F=\sqrt{\sum_{i=0,1,2}\left(\frac{m_i^{\exp}-m_i^{cal}}{m_i^{\exp}}\right)^2}\right\} \qquad (12)$$

$m_i^{\exp}$ are experimental values of the moments and m_i^{cal} are calculated values of the moments. It was assumed that there is negligible attrition of granules at inlet pressures $P_{in} \leq 0.5$ barg. The assumption is based on observation of almost identical PSDs at pressures $P_{in} \leq 0.5$ barg. Consequently the experimental PSD at $P_{in} = 0.5$ barg was employed as distribution of original unbroken granules. Very fast solution of the PBE using QMOM [9] allowed detailed mapping of the model parameter space and evaluation of the model parameters q, r, k.

Evaluation of experiments using CFD-PB model

Simultaneous modeling of flow and breakage in the 3-D mesh of the Sympatec analyzer was allowed by coupling CFD modeling of flow with the QMOM method. The combined breakage model (1) uses local impact velocities for calculation of the breakage rate (9). Values of the lumped breakage constants k from the MG-PB evaluation are used as initial approximation of the breakage constant k_v. The optimized parameters q, r of the multiple breakage distribution function (10) are employed. Then the only adjustable model parameters are the breakage constant k_v and the threshold impact velocity v_0.

The CFD-PB model simulates contours in the solid fraction, velocities of both phases, granular temperature in line with changes in particle (Sauter) diameter throughout the system. For illustration, computed values of the Sauter mean diameter along the analyzer tube are presented in **Figure 2.** The results show that the main decrease of the Sauter diameter occurs at the bend of the analyzer.

Figure 2. Computed Sauter diameter at the inlet pressure P_{in} = 4 barg

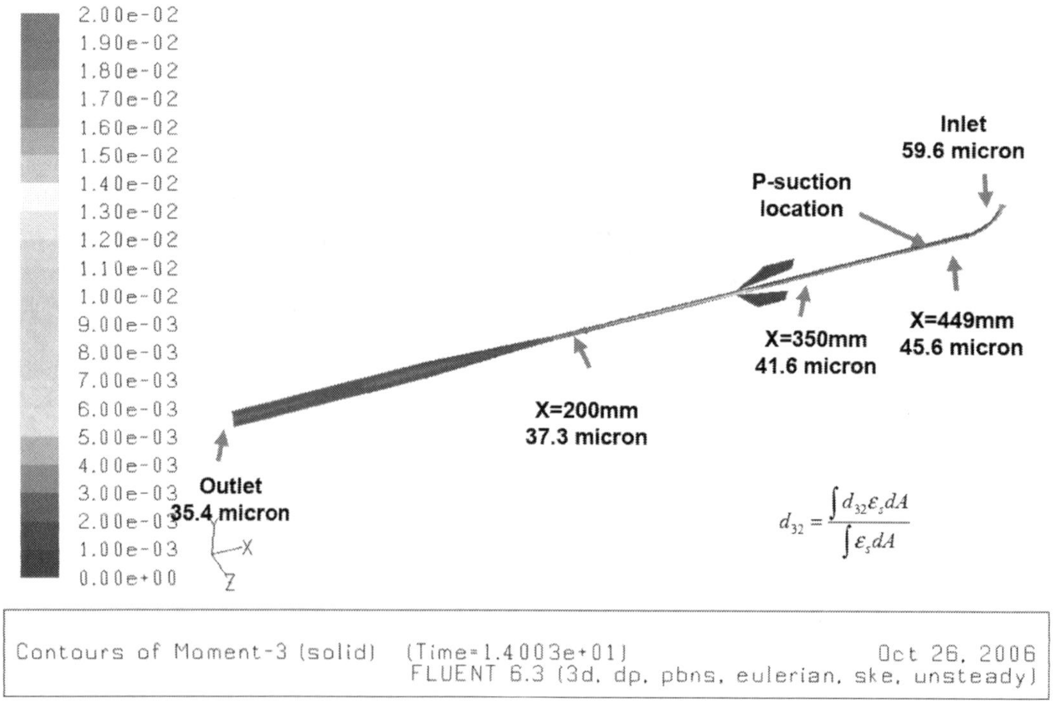

Figure 3. Prediction of experimental Sauter diameter using CFD-PB model

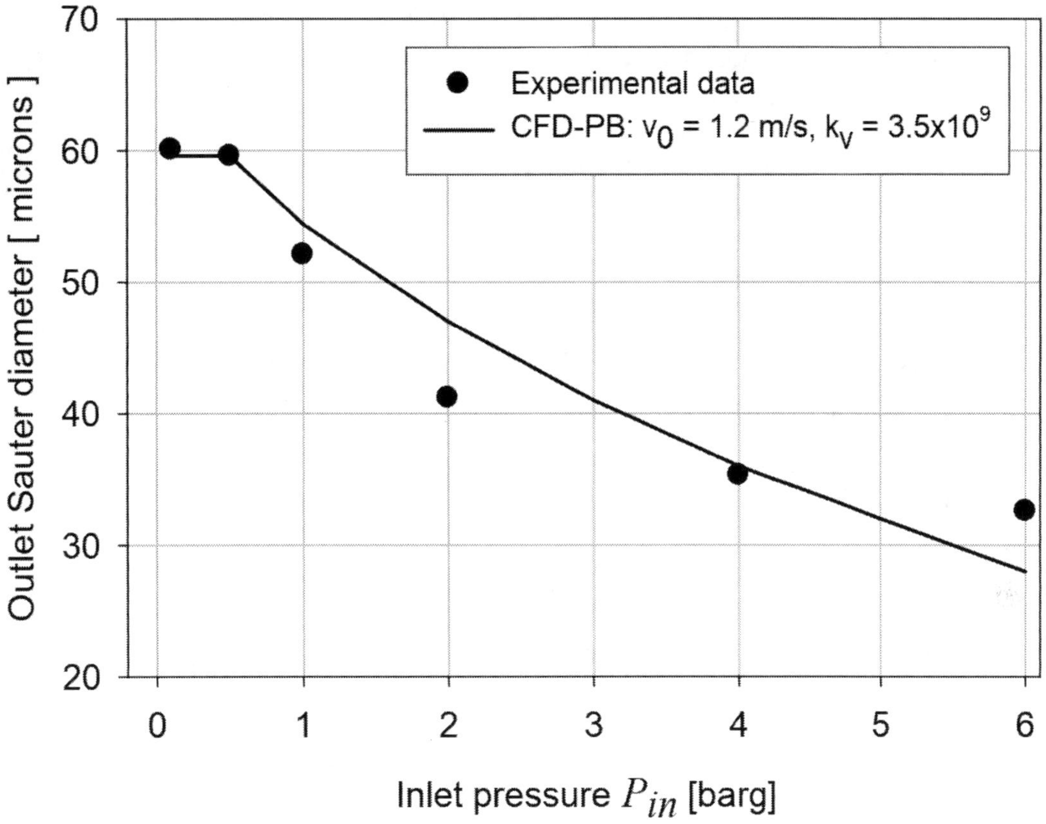

Conclusions

A combination of theoretical and experimental techniques for characterising and predicting the friability of granules in pneumatic conveying systems has been proposed and developed. A sequential procedure of the attrition/breakage data evaluation using mathematical models of increasing complexity was tested.

Firstly, a three-dimensional (2-D) computational fluid dynamics (CFD) model of the gas-solid flow within the Sympatec laser diffraction equipment was developed to simulate the impact of different inlet pressures on the flow properties. Secondly, a simple maximum-gradient population balance (MG-PB) mathematical model of attrition was developed and solved using the method of moments and used for evaluation of experimental data. The MG-PB models used the CFD model results for the average velocity and solid fraction to optimize the breakage rate and daughter distribution function parameters. Finally, a combined CFD-PB was developed and solved. The model solved continuity and momentum balance equations for both the solid and gas phase together with the population balances for the particle size reduction. The combined model used results from both the CFD and MG-PB models as inputs and initial approximation for further more detailed optimization of the breakage model parameters.

This methodology allows one to correlate the observed changes in particle size with the impact velocities within the system. Analogous physically based models combining properties of the gas-solid flow with the PB models are currently also employed for prediction of attrition and breakage in large-scale pneumatic conveying systems.

List of Symbols

B	birth rate of granules
D	death rate of granules
d_{32}	Sauter mean diameter
f	fraction of particles
g	breakage kernel
k, k_v	parameters of breakage kernels
L	characteristic particle size (diameter)
M	number of classes
m	moment of PSD
N	number of nodal points of QMOM; number of moments = $2N$
n	volumetric number of particles
P_{in}	inlet jet pressure
p	parameter of daughter distribution function (number of daughters)
q	parameter of daughter distribution function
t	time
$\vec{u}$	velocity vector
$\langle u_s \rangle$	average velocity of solid particle
V	volume of daughter particle
V_i	volume of parent particle
v	impact velocity
v_0	threshold impact velocity
w	weight of QMOM
x	volume fraction
z	distance along the tube; relative size of daughter particle

Greek symbols

α	volume fraction
B	Beta function
β	daughter distribution function
γ	exponent in breakage kernel
ρ	density
θ	self-similar daughter distribution function

Superscripts

k	order of moment of PSD

Subscripts

i size class index

in inlet

s solid

References

[1] P. Rajniak, K. Dhanasekharan, C. Sinka, N. MacPhail, R. Chern, S. Fitzpatrick, Modeling and measurement of granule attrition during pneumatic conveying in a laboratory scale system, Submitted to Powder Technology.

[2] R. B. Diemer, J. H. Olson, A moment methodology for coagulation and breakage problems: Part3 – generalized daughter distribution functions, Chem. Eng. Sci. 57 (2002) 4187 – 4198.

[3] R. Moreno-Atanasio, M. Ghadiri, Mechanistic analysis and computer simulation of impact breakage of agglomerates: Effect of surface energy, Chem. Eng. Sci. 61 (2006) 2476 – 2481.

[4] N. MacPhail, S. Fitzpatrick, P. Rajniak, C. Sinka, Measurement and modelling of granule friability, J. Pharmacy and Pharmacology 57 (2005) S64.

[5] E. Bilgili, B. Scarlett, Numerical Simulation of Open-Circuit Continuous Mills Using a Non-Linear Population Balance Framework: Incorporation of Non-First-Order Effects, Chem. Eng. Technol., 28 (2005) 153 – 159.

[6] D. M. Himmelblau, K. B. Bischoff, Process Analysis and Simulation: Deterministic Systems, John Wiley & Sons, Inc., 1968.

[7] D. L. Marchisio, R. D. Virgil, R.O. Fox, Quadrature method of moments for aggregation-breakage processes, J. Coll. Interf. Sci. 258 (2003) 322 – 334.

[8] D. L. Marchisio, R. O. Fox, Solution of population balance equations using the direct quadrature method of moments, J. Aerosol Science 36 (2005) 43 – 73.

[9] P. Rajniak, R.O.Fox, K. Dhanasekharan, F. Stepanek, R. Chern, Solution of population balance equations for wet granulation, Conference Proceedings CD Vol.2, Fifth World Congress on Particle Technology, 23-27 April 2006, Orlando, FL, USA.

[10] M. Vanni, Approximate population balance equations for aggregation-breakage processes, J. Colloid Interface Sci. 221 (2000) 143 – 160.

[11] R. B. Diemer, J. H. Olson, A moment methodology for coagulation and breakage problems: Part3 – generalized daughter distribution functions, Chem. Eng. Sci. 57 (2002) 4187 – 4198.

[12] M. Ghadiri, Z. Zhang, Impact attrition of particulate solids: Part 1. A theoretical model of chipping, Chem. Eng. Sci. 57 (2002) 3659 – 3669.

[13] R. Moreno-Atanasio, M. Ghadiri, Mechanistic analysis and computer simulation of impact breakage of agglomerates: Effect of surface energy, Chem. Eng. Sci. 61 (2006) 2476 – 2481.

[14] J. Subero. Z. Ning, M. Ghadiri, C. Thornton, Effect of interface energy on the impact strength of agglomerates, Powder technology 105 (1999) 66 – 73.

[15] H. Kalman, Attrition of powders at various bends during pneumatic conveying, Powder Technology, 112 (2000) 244 – 250.

18b Analysis of Aspirin Milling in Single Ball Mill

R. Bertrum Diemer[1], Yulong Ding[2], Dolapo Olusanmi[2], Mojtaba Ghadiri[2], K. J. Roberts[2], and Jon H. Olson[3]. (1) DuPont Engineering Research & Technology, DuPont Company, 1007 Market Street, Wilmington, DE 19898, (2) Institute of Particle Science & Engineering, University of Leeds, Clarendon Road, Leeds, LS2 9JT, United Kingdom, (3) Chemical Engineering, University of Delaware, Colburn Laboratory, Newark, DE 19716

A project on milling of organic solids is currently underway at Leeds University aimed at establishing a relationship between milling product properties and the inputs to milling, namely the feed material properties, and milling conditions. As part of this work, it was proposed to describe the milling operation using a moment-based population balance model assuming self-similar daughters, with the aim of using model parameters as one way of describing the product. This paper reports efforts to analyze Leeds' experimental data for aspirin by this approach leading to the conclusion that the assumption of self-similar daughters is not justified. An alternative model will be presented to describe what was observed.

18c Optimization of Milling Using Attainable Region Theory

Matthew J. Metzger[1], Benjamin Glasser[1], David Glasser[2], Brendon Hausberger[2], Diane Hildebrandt[2], and Ngangezwe Khumalo[2]. (1) Chemical and Biochemical Engineering, Rutgers University, 98 Brett Rd, Piscataway, NJ 08854, (2) School of Chemical and Metallurgical Engineering, University of the Witwatersrand, Johannesburg, 1 Jan Smut Avenue, Johannesburg, South Africa

Milling of particles is important to a large number of industries including the mineral processing, pharmaceutical, fertilizer and energy industries. Often it is desirable to produce particles of a certain size range while minimizing process time and/or energy usage. Production of fines can lead to powder handling problems and inefficient processes. In the pharmaceutical industry, excessive production of fines during milling of the drug crystals can increase risk of dust explosions. In mineral processing, the milling process is extremely energy intensive with large amounts of energy lost to hear or noise. There is a need to better understand milling in order to optimize milling processes and reduce energy usage.

Presented here is the application of the Attainable Region (AR) approach to optimizing the milling process. Previously, the AR approach has been used to optimize complex reactor networks and separation systems. Due to the similarities between reaction networks and the comminution process, the AR lends itself well to determining the optimal control strategy for a milling project. Though the optimization done in this work is for a ball mill and a specific material, the concepts can be applied to any system consisting of breakage and classification.

We have experimentally determined the optimal control strategy for achieving a certain size distribution with the use of the least amount of energy. The control strategy recommends a variation of speed throughout the process - fast speed at first, followed by a slightly slower speed, and then the lowest speed - to achieve the maximum amount of material of a certain size. Benefits of this approach can be realized anywhere milling is found, as a size profile over time can be predicted. Therefore, a design methodology can be constructed without complicated mathematical tools.

18d Numerical Simulation and Optimization of Grinding in Hammer Mill

Bodhisattwa Chaudhuri, Pharmaceutical Sciences, University of Connecticut, 69 N. Eagleville Road, Storrs, CT 06269

Particle size reduction of dry material by mechanical means, such as milling, is a very important operation for pharmaceutical, agricultural, food and paper industries. Size reduction is generally achieved by particle fracturing under the action of applied energy. Knowledge of particle flow, granular mixing and segregation, and size reduction in a hammer mill is thus critical to optimize the design and operation of such equipment.

We numerically modeled a pilot-plant scale hammer milling machine using Discrete Element Method (DEM). DEM is employed to study primarily the breakge and kinematics of the particle motion within the hammer mill. Moreover, we calculate the velocity, the stress field within the granular bed, breakage and the energy distribution of collisions. Algorithm of damage mechanics is incorporated to model weakening and breakage of material processed in the hammer mill. Incremental impact breakge of the particles is modeled as breakage due to single impact is certainly rare. Material properties of Cellulose and Lactose are used in the simulation.

Parametric study is performed to understand the effect of hammer speed (rotational), hammer-wall tolerance, material properties and specific exit classification conditions on the final product size distribution. Below a critical hammer tip speed, a blending action rather than comminuting is observed. Increase in hammer tip speed causes higher frequency of impact of particles per unit time and higher specific energy of impact resulting generation of much finer end product. We observe that both the specific kinetic and strain energy of the particles (colliding with hammer) increase as the impact point becomes closer to the hammer-tip. Material with lower Young Modulus(cellulose) shows higher breakage rate than Lactose at constant vessel geometry and operating condition. Our model predicts the optimal operating condition relating the size distribution of the resultant progeny with the material properties, mill geometry and mill operating conditions. The net mill power is also derived from the hammer geometry, hammer tip velocities and impact forces.

18e The Breakage Matrix Approach in Evaluating Inadvertent Breakage of Dense Particulate Mixtures

A. Abu Nahar and U Tüzün. Chemical and Process Engineering, University of Surrey, School of Engineering, Guildford, gu27xh, United Kingdom

In general, breakage matrix is an analysis technique, which is potentially useful in describing any degradation process. The application is widespread and unconstrained to model an intentional degradation process such as the study of the material breakage efficiency in a milling operation or capable of identifying which parts in a transport system contribute most to attrition problems. Breakage matrix approach is however, often limited to linear breakage assumption neglecting the nature of the many body of inter-particle interaction of individual species and the surrounding particles within mixture assembly.

Linear breakage assumption has received to open criticism for more than a century and as matter of fact, it is not necessarily held true in many degradation scenarios. Broadbent and Calcott, 1956 a&b observed linear or independent breakage for their case of coal transporting system even tough they remarked earlier, "..Breakage in an assembly is partly due to interaction between particles in the assembly". Therefore, it can be reasoned at small extent, inter-particle interaction may result in abnormal breakage effects of the output size distribution. The successfulness of breakage matrix as a prediction tool is no doubt for self-governing condition, in fact, Campbell and Webb, 2001 in their work essentially suppose the breakage matrix approach is worthless for significant interacting mixture. Non-interacting conditions are sensible for some degradation scenarios for example during roller milling process in which each particle passing through the two rotating rolls and break independently, or in dilute phase pneumatic conveyor where the particles impact on pipe bends one at a time. Consider an example of dense phase condition where dense assemblies are subjected under compression loads/ shear, the breakage of individual particles within the mixture obviously far from negligible and are supposedly influenced by the presence of particles in other size classes.

This paper attempts to demonstrate the usefulness of breakage matrix as a more versatile tool in predicting degradation than it has been appreciated to date with additional feature of coefficient of interaction as a function of micro-structural limit fines composition and size ratio to account for the role of inter-particle interaction in influencing the binary mixture output size distribution after degradation. The prediction capability of this proposed more flexible breakage matrix approach is verified with some degradation data set obtained from a compression test bed of dense particulate mixtures.

References:

1-) Broadbent, S.R., Callcott, T.G.,(1956-a) J. Inst Fuel., pp: 524-528 2-) Broadbent, S.R., Callcott, T.G., (1956-b) J. Inst Fuel., pp: 528-539 3-) Campbell G.M., Webb C., (2001), J. Powder Technol., 115, pp: 243-255 4-) Tuzun U., Artega P., (1992), Chem. Eng. Sci., 47, pp:1619-1630

Effects of co-milling on crystallinity of Salbutamol Sulphate using Adipic Acid

Particle size in pharmaceuticals is important in defining drug delivery and bioavailability in an increasing range of pharmaceutical applications. This formulation aspect affects performance of a drug product. It governs for instance, sedimentation and flocculation rates in suspensions, therapeutic effect of dry powder inhalations (DPI) and good powder homogeneity in tablets. Comminution is commonly used for this purpose and also to reduce particle size as it is difficult to achieve a narrow particle size distribution (PSD) via crystallization alone. However, milling also induces disorder in the crystal structure leading to the formation of amorphous regions, which are concentrated particularly at the surfaces of the particles (1-4). During post-milling storage, these amorphous regions are unstable and may transform back to the crystalline state. It is therefore important to monitor and control these process induced changes, as these also affect powder properties such as chemical and physical reactivity, dissolution and finally the drug product performance (5).

Salbutamol sulphate (SS), a pharmaceutical active used in DPI, typically transforms from the crystalline to the amorphous form during milling. Amorphous SS does not remain stable and tends to undergo agglomeration under normal storage conditions for pharmaceutical powders. Pfeiffer and co-workers (6, 7) suggested that the increase in particle size of micronized SS was caused by re-crystallization of amorphous regions produced during micronisation. This uncontrolled recrystallization may occur at any time following micronisation, during the storage of the milled material, formulation of the powder for dry powder inhalers, or even as late as in the hands of the patient. This can have serious consequences with respect to the effectiveness of powders for inhalation, because particle growth may generate fractions of the particles with diameters outside of the respiratory range (between 1 and 10 μm).The purpose of this study is to stabilise the crystalline form of milled SS by co-milling with a pharmaceutically acceptable excipient - adipic acid (AA).

Different mass ratios (1:1, 1:2, 1:3, 2:1, 3:1) of SS and AA were co-milled at 300 rpm in a planetary ball mill for different durations. The duration was optimized and set at a period of 1 hr. The co-milled mixtures were characterized and compared with the individual components and physically blended components using Powder X-Ray Diffraction (PXRD).PXRD patterns of co-milled mixtures of SS and AA (Figure 1) showed that the characteristic crystalline peaks of SS were retained on co-milling. This indicated that co-milling stabilized the crystalline form of SS. It was also observed that all AA:SS ratios indicated similar intensity of peaks for SS suggesting that a minimum ratio of AA:SS is useful in maintaining the stable crystalline form of SS on co-milling. This was different from the amorphous halo observed in the diffractogram of SS after it was ball-milled for 1 hr. The prominent crystalline peaks characteristic of SS were also absent from the physically blended mixtures of SS and AA (Figure 2).The amorphous content, surface and thermal properties of these co-milled mixtures are further analyzed using Dynamic Vapor Sorption (DVS), Differential Scanning Calorimetry (DSC) and Thermogravimetric Analysis (TGA).

Figure 1

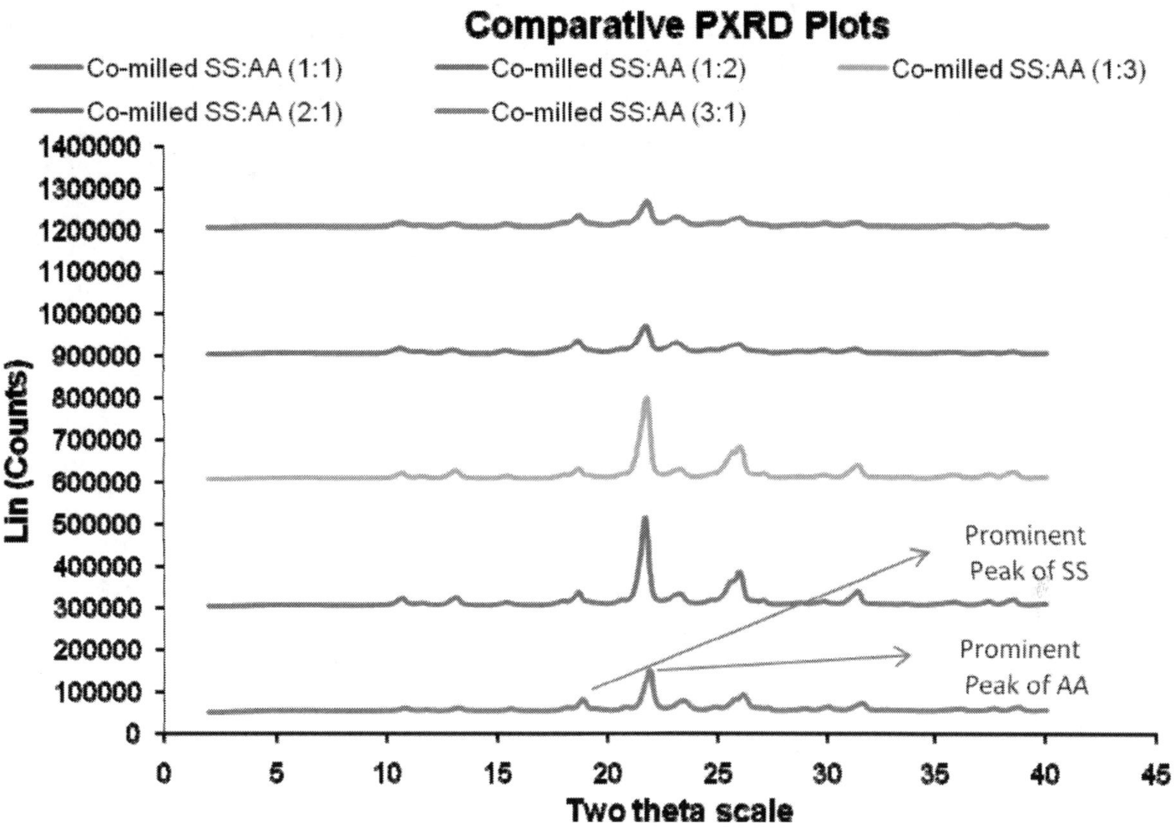
Comparative PXRD Plots
Co-milled SS:AA (1:1)
Co-milled SS:AA (1:2)
Co-milled SS:AA (1:3)
Co-milled SS:AA (2:1)
Co-milled SS:AA (3:1)
Lin (Counts)
1400000
1300000
1200000
1100000
1000000
900000
800000
700000
600000
500000
400000
300000
200000
100000
0
Prominent Peak of SS
Prominent Peak of AA
0
5
10
15
20
25
30
35
40
45
Two theta scale

Figure 2

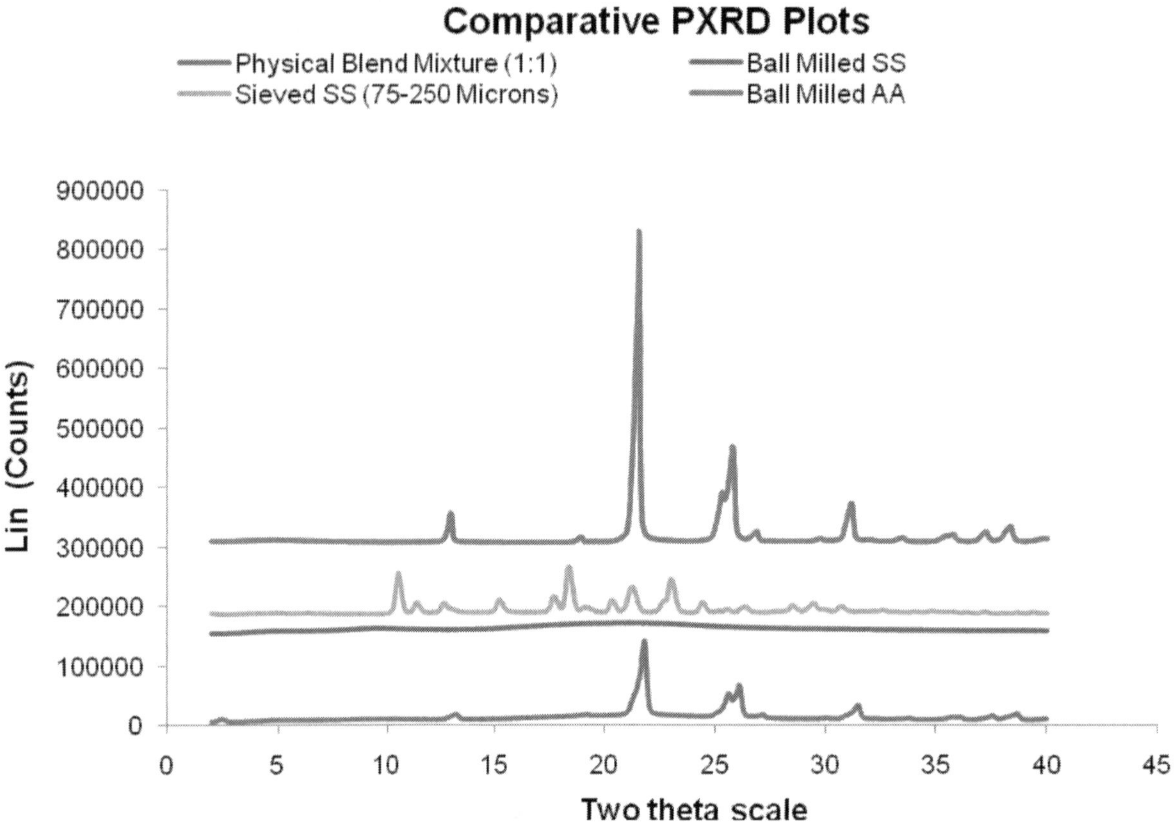

References:

1. Chikhalia, V., R.T. Forbes, R.A. Storey and M. Ticehurst. The effect of crystal morphology and mill type on milling induced crystal disorder. European J. Pharm. Sci. 27: 19-26 (2006).

2. Mackin, L., S. Sartnurak, I. Thomas and S. Moore. The impact of low levels of amorphous material (< 5%) on the blending characteristics of a direct compression formulation, Int. J. Pharm. 231: 213-226 (2002).

3. Mackin, L., R. Zanon, J.M. Park, K. Foster, H. Opalenik and M. Demonte. Quantification of low levels (<10%) of amorphous content in micronised active batches using dynamic vapour sorption and isothermal microcalorimetry. Int. J. Pharm. 231: 227-236 (2002).

4. Begat, P., P.M. Young, S. Edge, J.S. Kaerger and R. Price. The effect of mechanical processing on surface stability of pharmaceutical powders: Visualization by atomic force microscopy. J. Pharm. Sci. 92: 611-619 (2003).

5. S,Guinot and F. Leveiller. The use of MTDSC to assess the amorphous content of a micronized drug substance. Int. J. Pharm. 192: 63-75 (1999).

6. Pfeiffer, K.B., H. Haeusler, P. Grass and P. Langguth. Conditioning following powder micronization: Influence on particle growth of salbutamol sulfate. Drug Dev. Ind. Pharm., 29: 1077-1084 (2003).

7. Pfeiffer, K.B., P. Langguth, P. Grass and H. Haeusler. Influence of mechanical activation on the physical stability of salbutamol sulphate. Eur. J. Pharm. Biopharm. 56: 393-400 (2003).

18g Application of Box–Behnken Design and Response Surface Methodology for Modeling of Grinding Beach Sand Ore Using Ball Mill

Asha Immanuel Raju Chaduvula[1], Murali Yugandhar Nikku[2], Kiran Kumar Nalla[2], Venkata Satya Chappidi[2], Srinivasa babji Josyula[2], and Israel Yedluri[3]. (1) Chemical Engineering, Andhra University, Andhra University, Visakhapatnam - 530 003, AP, INDIA, Visakhapatnam, 530003, India, (2) Chemical Engg Dept, Andhra University, Andhra University, Visakhapatnam - 530 003, AP, INDIA, Visakhapatnam, 530003, India, (3) Andhra University, Andhra University, Andhra University, Visakhapatnam - 530 003, AP, INDIA, Visakhapatnam, 530003, India

The aim of our research paper was to apply Box-Behnken experimental design and response surface methodology for modeling of grinding beach sand ore using ball mill. The Box-Behnken experimental design was used to provide data for modeling and the variables of model are time of milling, speed of the mill, ball size, number of balls, density of material, feed size and feed quantity. Maximum specific surface is obtained at higher speed of the mill (85 RPM in our case). This also correspondence to the speed at which we have minimum energy consumption per unit surface area generated. Specific surface area created increases with increased in cylinder size. The increase is rapid for larger cylinder sizes. Specific energy consumption is almost same for all cylinder sizes. Specific surface area decreases with increase in feed quantity. Energy required for the unit surface generated has a minimum value at 400 g of feed. Specific energy consumption also decreases with feed quantity at all ball sizes. The specific surface area generated increases with additive quantity for a cylinder size of 0.875 inches and above. Energy savings, expressed as a percentage of the energy consumption without additive, increase with an increase in the additive quantity. The increase is rapid between 1.25 and 2.5 % of feed quantity. Using these sets of experimental data obtained by mathematical software package (MATLAB 7.0), mathematical models were then developed to show the effect of each parameter and their interactions on product. Predicted values and obtained values using model equations were in good agreement with the experimental values. This study proved that Box-Behnken design and response surface methodology could efficiently be applied for modeling of grinding of beach sand ore using ball mill.

The Study yielded the following Equation:

RANGE OF BALL MILL DESIGN FACTORS STUDIED ARE

Variables Minimum Maximum Max/ Min Time of milling, min 5 60 12 Speed of the mill 24 85 3.53 Ball size 1/2 1 2 Number of 1„c/nballs 10 150 15 Number of 3/4„c/nballs 50 237 4.74 Number of 1/2 „c/nballs 50 800 16 Feed size, ƒÝ 231 1731 7.5 Feed quantity, gm 200 1500 7.5

NOMENCLATURE

Bs = diameter of the ball, m Dm = average diameter of the particle, m Fs = average diameter of the feed, m g = acceleration due to gravity, m/s2 N = number of revolutions per minute, rpm Ps = average product diameter, m Q = feed quantity, kg S = surface area of the product, m2/kg t = time of milling, s V = linear velocity of the mill at the outer surface, m/s Ws = mass of the balls, kg Y1 = correlation factor

Keywords: Box-Behnken design; Response surface methodology; Modeling, Ball mill

The Quality by Design Approach to Pharmaceutical Development and Manufacturing

Christine M. V. Moore, Branch Chief - Manufacturing Sciences, FDA/CDER/ONDQA

Abstract

In FDA's view, Quality by Design (QbD) is a systematic approach to pharmaceutical development and manufacturing where: (1) the target product profile is defined to meet patient requirements and the critical quality attributes (CQAs) are identified; (2) the formulation and process are designed to consistently meet product CQAs; (3) the impact of formulation components and process parameters on product quality is understood; (4) critical sources of process variability are identified and controlled; and (5) the process is continually monitored and updated to assure consistent quality over time.

Many of the tools to implement QbD concepts are discussed in recent ICH quality guidelines. ICH *Q8 - Pharmaceutical Development* introduces the concepts of design space and flexible regulatory approaches. ICH *Q9 – Quality Risk Management* and ICH *Q10 – Pharmaceutical Quality System (Step 2)* discuss other essential elements of implementing a QbD approach.

In 2005, FDA's Office of New Drug Quality Assessment (ONDQA) began a CMC Pilot Program to explore ways of implementing QbD concepts in regulatory submission and review. This presentation will provide an overview of QbD concepts and recent ICH guidelines, and discuss experience gained from the CMC Pilot Program.

45b Challenges in Pharmaceutical Development and Manufacturing

Kenneth R. Morris, Industrial and Physical Pharmacy, Purdue University, 575 W. Stadium Mall Drive, West Lafayette, IN 47907-2088

The pharmaceutical industry is facing significant challenges to maintain high quality and competitiveness. In the U.S. the FDA has lead the drive towards more science based practices and regulations. This is captured in the concepts of Quality by Design (QbD). Quality by design in product development and its use in scale up and manufacturing is really no more than the application of accepted scientific and engineering principles to design for function. The gaps in our knowledge and ability in implementing QbD concepts in a comprehensive manner fall largely into three categories: 1) Materials Science gaps, 2) Process Modeling gaps, and 3) Information management and informatics. These concepts are captured in measure by the ICH CTD and guidelines in the form of the development rationale for review and in the design space for flexible control limits.

The "business case" for such activities is in some ways more complicated than the scientific case. Comparing the potential gains in efficiency, the decrease in OOS investigations, and the lower hurdle for incorporating new technology against the start-up resources required and the sustainable benefit is challenging, yet there are some data to help in the analysis. The analysis for generic and OTC products in emerging markets such as China may be more favorable towards modernization due to the reduced margins, but the start up resources must still be assessed.

Concepts in combining the scientific, business, and regulatory aspects of modernization will be discussed and examples will hopefully help provide some insight into the challenges.

Optimization Under Uncertainty for the Design and Planning of Process Systems

Ignacio E. Grossmann, Dept of Chemical Engineering, Carnegie Mellon University, 5000 Forbes Ave, Pittsburgh, PA 15213

This presentation will give an overview on major formulations and strategies for incorporating effects of uncertainties in the design, planning and scheduling of process systems. We first present general approaches and formulations that rely on deterministic and probabilistic representations. The former lead to problems related to flexibility and robustness, while the latter lead to problems that either work with scenarios in one or several decision stages, or with probabilities for meeting constraints. We illustarte the application of these techniques on several applications outside the pharmaceutical domain that include optimal design of heat exchanger networks, optimal batch scheduling with uncertain demands, optimal design of offshore platforms, and optimal planning of sustainable supply chains with uncertain emissions. Finally, we close with a specific application in the pharmaceutical industry that deals with simultaneous optimization of new product development and design of batch facilities. As will be shown throughout these examples the handling of uncertainty gives rise to very challenging optimization problems, which however, provide important benefits compared to ad-hoc approaches that use deterministic methods.

Role of PAT in Quality by Design

Christine M. V. Moore, Branch Chief – Manufacturing Sciences, FDA/CDER/ONDQA

Abstract

In September 2004, the FDA published the guidance, *PAT – A Framework for Innovative Pharmaceutical Development, Manufacturing, and Quality Assurance*. An underlying principle in this guidance is that quality cannot be tested into products, but should be built-in, or should be by design. This "Quality by Design" approach is carried through in recent ICH documents as well as through recent activities in CDER's Office of New Drug Quality Assessment (ONDQA). The recent ICH *Q8 - Pharmaceutical Development* guidance embraces similar concepts about quality and provides opportunities for flexible regulatory approaches. ICH *Q9 - Quality Risk Management* and ICH *Q10 - Pharmaceutical Quality Systems (Step 2)* give guidance for implementing robust, reliable and flexible pharmaceutical manufacturing. This presentation will discuss the role of PAT in the Quality by Design approach to pharmaceutical development and share some recent ONDQA experiences with reviewing process analytical systems in pharmaceutical manufacturing.

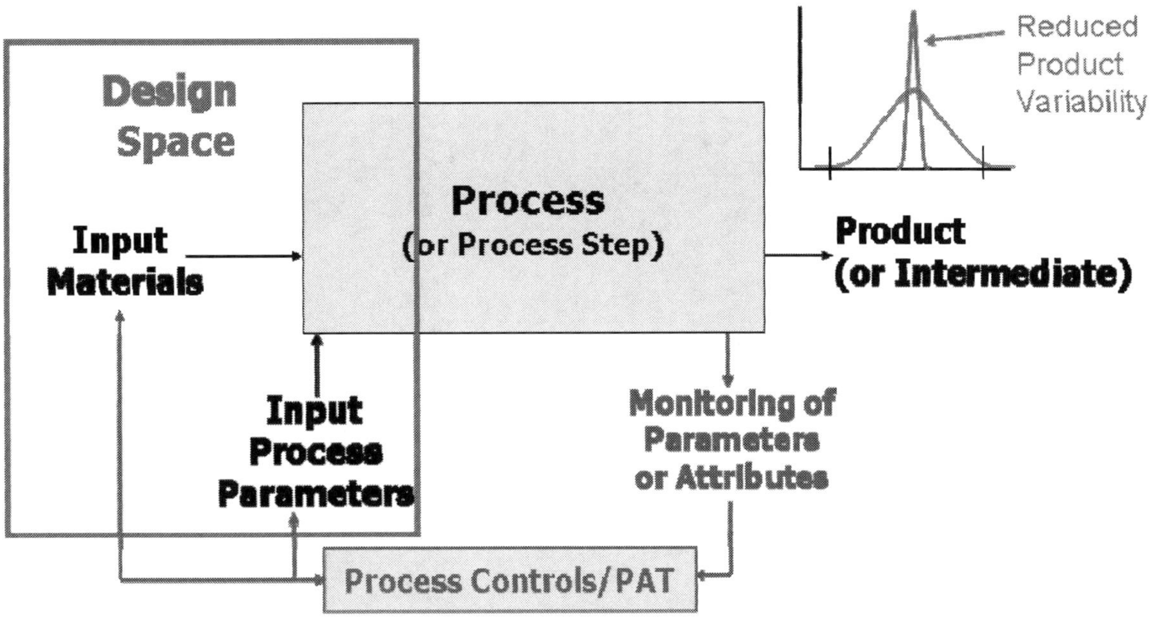

Reducing process variability through implementation of PAT for process control.

103b Process Analytical Technology for Crystallization Processes

Richard D. Braatz, University of Illinois, 600 South Mathews Avenue, 93 Roger Adams Laboratory, Box C-3, Urbana, IL 61801-3602

This presentation provides an overview of the state-of-the-art in the application of process analytical technology (PAT) to crystallization processes. A description is provided of in-situ sensors including ATR-FTIR and Raman spectroscopies, laser backscattering, and process video microscopy as well as the use of the sensors and automation for the design of the operating procedures for batch and semibatch crystallization processes to produce crystals of desired size and polymorphic form. Industrial applications to antisolvent crystallization are described, as well as later applications to various polymorphic and pseudo-polymorphic crystallizations. The presentation summarizes techniques that have developed to simulate the effects of nonideal mixing for large-scale crystallizers, to guide the modification of bench-scale crystallization recipes to increase process reliability during scaleup. The talk ends with a description of recent innovations in using high-throughput microfluidic platforms for identifying the kinetic parameters needed in the simulation of nonideal mixing in crystallization.

Web Page: brahms.scs.uiuc.edu

103c Measurement and Control of Product Attributes during Granulation: Challenges and Opportunities

James D. Litster, Chemical Engineering and Industrial Pharmacy, Purdue University, 575 Stadium Mall Drive, West Lafayette, IN 47907-2091

To be entered by Jim Litster

129a Flow Structure Mapping Of Segregating Granular Mixtures Using Radioactive Particle Tracking

Jocelyn Doucet, Francois Bertrand, and Jamal Chaouki. Department of Chemical Engineering, Ecole Polytechnique de Montreal, University of Montreal Campus, 2920 ch. de la Tour, Montreal, QC H3T1J7, Canada

This work presents experimental flow measurements of segregating binary mixtures of granular material in a cylindrical tumbler using radioactive particle tracking (RPT). The particles, made of spheronized lactose/avicel with carboxymethyl cellulose binder, have a mean diameter of 200, 400 and 750 microns. The use of RPT enabled the tracking of one radioactive particle flowing with other particles for a long period of time and in a non-intrusive manner. Assuming ergodicity, it was then possible to obtain a complete map of the flow in a cylindrical tumbler mixing system.

The RPT technique used in this work will first be described. The results obtained will next be analyzed. In particular, important characteristics of the flow will be presented, which includes the rate of mixing, the mean velocity profile, the Lyapunov exponents and the entropy. Finally, based on these metrics, the structures of monodisperse and bidisperse particle mixtures will be compared to enlighten some of the important features at the basis of the segregation mechanisms in a cylindrical tumbler.

129b Mixing Of Nanoparticles By Magnetically Assisted Impact Mixing Technique

James V. Scicolone, Alexandre Ermoline, Ryan Barrow, and Rajesh Dave. New Jersey Center for Engineered Particulates, New Jersey Institute of Technology, Newark, NJ 07102

Mixtures of nanoparticles and nanocomposites, manufactured using nanoparticles as building blocks, have unique properties owing to small particle/grain size, and large interface area between individual nano-sized constituents. These properties have a great potential to improve performance of drugs, biomaterials, catalysts, energetic materials etc. However, to fully employ this potential, nanoparticles should be mixed at nanoscale level, which is a difficult task because they tend to form aggregates on the order of microns. Therefore, a major challenge in utilizing nanoparticles is to achieve their mixing at nano-scale level.

The objective of this study is to explore the homogeneity of nanoparticle mixtures obtained using a dry mixing technique, Magnetically Assisted Impact Mixing (MAIM). The MAIM process is an environmentally friendly dry mixing process, which eliminates current downstream processing (drying) time associated with the wet chemistry mixing methods. In this technique nanopowders are placed along with sub-millimeter sized magnets in a container inside a coil producing an oscillating magnetic field. The magnets undergo rotational and translational motion and promote rearrangements of nano-particles between agglomerates of different materials, mixing them at scales below the aggregate size. Multiple parameters were studied to optimize the MAIM process: magnet size, magnet to sample weight ratio, mixing time, and electromagnetic field strength. The characterization of the resulting nanoparticle mixtures were conducted using Scanning Electron Microscope (SEM) equipped with an EDS (energy dispersive X-ray spectroscopy).

129c Segregation Of Cohesive Particles During Hopper Discharge

Anshu Anand[1], Jennifer S. Curtis[1], Carl R. Wassgren[2], and Bruno C. Hancock[3]. (1) Chemical Engineering, University of Florida, Gainesville, FL 32611, (2) Mechanical Engineering, Purdue University, 575 Stadium Mall Drive, West Lafayette, IN 47907-2088, (3) Pfizer Inc., Groton, CT 06340

Granular materials may readily segregate due to differences in particle properties such as size, shape, and density. This segregation may occur in many industrial processes involving transport or other handling of granular materials and will occur even after a material has been uniformly blended. The discharge of granular material from a hopper is a typical example of such a process that may cause segregation. The present work investigates the causes and extent of segregation of cohesive granular materials during flow from a hopper. In the system considered, the most important source of cohesion is pendular liquid bridges existing between particles. The liquid bridge force is strongly dependent on the geometry of the liquid bridge and is a result of surface tension and the pressure differential created in the liquid.

The cohesive granular material is modeled as bi-disperse inelastic, frictional spheres via the discrete element method (DEM). Regression expressions are implemented into the DEM code to describe the liquid bridge force as an explicit function of separation distance between the particles, contact angle, surface tension and liquid volume. Segregation and discharge rate data are obtained from the discharge of quasi-three-dimensional wedge-shaped hoppers. The extent of the discharge segregation over time is determined as a function of the hopper parameters and particle properties. In addition, simulations for a monodispersed cohesive granular system show an increase in residence time and significant differences in the shape of the hopper free surface when compared to the cohesionless granular system. For the cohesive granular system, sticking of particles to the hopper wall and clumping of particles as they flow out of the hopper are observed.

129d Eliminating Segregation In Free-Surface Flows Of Particles

Deliang Shi[1], Watson Vargas, and **Joseph J. McCarthy**[2]. (1) Chemical Engineering, University of PIttsburgh, Pittsburgh, PA 15116, (2) Chemical and Petroleum Engineering, University of Pittsburgh, 1238 Benedum Hall, Pittsburgh, PA 15261

Segregation, or un-mixing, of particles has been a topic of intense research and industrial frustration for many decades. When particles differ in almost any mechanical property, processing typically leads to pattern formation, layering, or complete separation of the materials and this non-homogeneity can cause dramatic revenue loss and product failure in industries such as pharmaceuticals, ceramics, and agriculture, to name but a few. In particular, particles of different size and density may segregate quite strongly in free-surface flows; the larger (lighter) particles often rise to the top, while the smaller (denser) particles sink to the bottom. In this paper, we show both experimentally and computationally that by introducing periodic flow inversions that occur at a frequency above some critical value, we can effectively eliminate both density and size segregation. The critical frequency is related to the inverse of the characteristic time of segregation and is shown to scale with the shear rate of the particle flow. This observation could lead to new designs for a vast array of particle processing applications and suggests a new way for researchers to think about segregation problems. We demonstrate this approach in several industrially-relevant devices, including chute flows and tumbling mixers.

129e Powder Flow Induced By Rough-Surface Moving Walls: Concentration Profile Of A Binary Powder Mixture

Luis Guillermo Obregon Quiñones, Chemical Engineering, University of Puerto Rico, P.O.Box 5997, Mayaguez P.R. 00681-5997, Mayaguez, PR 00681 and Carlos Velázquez Figueroa, Chemical Engineering, ERC, Center for Structured Organic Composites, University of Puerto Rico, Mayaguez Campus, P.O.Box 5997, Mayaguez P.R. 00681-5997, Mayaguez, PR 00681.

The interest in understanding the behavior of granular materials has been growing as they are extensively used in industry in many engineering processes such as mixing, which is the focus of this study. It is necessary to understand the flow behavior of particles to then understand the phenomenon of powder mixing. This can be done using concepts of fluid flow that include mass and momentum balances with constitutive equations containing material characteristics such as friction angle, repose angle, particle density, etc. The purpose of this research is to study the phenomenon of powder flow induced by rough surface moving walls and to simulate the concentration profile of the combination of two powders to predict the degree of mixing. To accomplish the latter, a new term in the constitutive equation will be used, the interparticle force between the particles of the different components, along with either the theory of mixture or theory of interacting continua and the governing equations. These simulation results will be compared to those obtained from simulation with the Discrete Element Method and experimental data obtained when mixing two different powders in the rough surface moving walls apparatus developed in the laboratory. Preliminary results show that the particle size strongly affects the flow of the bulk causing different degrees of mixing at the end of each experiment. The phenomena of dilatation and consolidation were remarkable

129f Study of the Feedframe Effect on Particle Size Distribution

Rafael Mendez-Roman, Chemical Engineering, Engineering Research Center for Organic Particulate System, University of Puerto Rico Mayaguez, PO Box 21272, Mayaguez, PR 00680-9007, Fernando J. Muzzio, Chemical & Biochemical Engineering, Rutgers University, 98 Brett Road, Piscataway, NJ 08854, and Carlos Velazquez, Chemical Engineering, ERC on Structured Organic Composites, University of Puerto Rico, Road 108, Km. 3, Mayaguez, PR 00680.

In pharmaceutical industry, the tablet characteristics are an important parameter. Some of the variables that can affect the final properties of a tablet may be present when feeding the die. Feedframe is a common way used to feed the powder to the tablet press dies. There exist different designs of feedframes between manufacturers. It has different paddle diameters and number of paddles, heights, speed, shapes. These equipment variations can make a difference in the powder flow into a die. The filling of the dies also is an essential part to control the final properties of tablets. If the rate is too slow, the dies are starved, and if it is fast the powder can dearate or densify in an inconsistent manner [1]. Because of the complexity of powder flow through the feedframe there is no actual model or principle to describe it. The interest on granular material behavior has been growing in the last years especially for understanding and prediction of powder flow, mixing, and segregation and compaction. The advance obtained from these processes let us focus in one problem that the pharmaceutical industry is facing. This is the mechanical properties of tablets and their final composition. To understand the causal effect, this project focuses on studying the effect of feedframe on the particle size distribution, segregation and flow variability of organic granular materials. These variables have a strong effect on the compression of powders and thus in the final properties of tablets.

129g The Mechanism behind Axial Segregation of Granules in a Horizontal Rotating Drum

Marleen M. H. D. Arntz[1], Wouter. K. Den Otter[2], and Wim. J. Briels[2]. (1) Food and Bioprocess Engineering Group, Wageningen University, Bomenweg 2, Wageningen, 6703 HD, Netherlands, (2) Computational Biophysics, University of Twente, P.O. Box 217, Enschede, 7500 AE, Netherlands

Mixing of granules with different size, density, or shape is a common and important unit operation in the chemical, food and pharmaceutical industries. However, the phenomena at the basis of the mixing process are poorly understood. Operations intended for mixing may even result in segregation. Probably, the most general known segregation phenomena are the radial and axial segregation in a horizontal rotating drum. Since axial segregation is always preceded by radial segregation, knowledge about axial segregation will provide insight in radial segregation. We therefore studied the mechanism behind axial segregation in a horizontal rotating drum with various particle-based numerical simulations, covering a wide range of system parameters. Several existing theories were tested against the simulation results and found to be only partially satisfactory. A new hypothesis to explain the axial segregation is proposed that describes the broad range of observed processes in literature and in our simulations.

Tower of Babel: Intelligent Data and Knowledge Management for Decision Support in Pharmaceutical Engineering

Venkat Venkatasubramanian
Laboratory for Intelligent Process Systems
School of Chemical Engineering
Purdue University
West Lafayette, IN 47906

Abstract

The compelling drivers for the pharmaceutical industry are speed to market and getting the manufacturing process right the first time. However, getting there is a complex challenge, particularly with respect to information management. Getting there involves going through the following stages after the viability of a newly discovered molecule is established: laboratory scale, pilot plant scale and commercial scale manufacturing. During this a *staggering* amount of information of *different types, ranging from raw data to lab reports to sophisticated math models,* is shared and revised by computational tools in each stage. Subsequent to process development, technical specifications and reports must be developed to satisfy regulatory requirements. However, due to the *Tower of Babel* like incompatibility among such diverse types of data, information, and knowledge, an appropriate automated decision support environment to address these needs is very much lacking. Additionally, at present most of the information and knowledge are generated and processed directly by humans. With the onset of information and knowledge explosion, it is clear that we need intelligent software systems to effectively manage and access information for efficient decision making. Proper informatics support to process development is crucial in order to achieve speed to market, and getting the process right the first time.

To date the industrial response to all these challenges, however, has been sub-optimal. Even with rapid progress on information integration and sharing in business functions (such as ERP systems) and on plant floor (such as MES systems), the area of process/product development has been largely neglected. Several individual islands of automation exist, but a comprehensive, integrated decision support environment that integrates these islands does not exist. Therefore, practitioners must make do with a limited computer-based assistance to acquire, manage, analyze and interpret complex product and processing information with enormous amounts of human intervention. This increases the inefficiencies, uncertainties, costs, delays, and product quality concerns in all stages of product development. This also hampers the interaction between process development and business or manufacturing functions.

In this paper, we present an overview of our approach to address these challenges. In our work, we start from modeling the data/information/knowledge as well as their flows during decision-making in the entire process development. An ontological informatics infrastructure is developed to support such decision making spanning the entire process, including product portfolio selection, capacity allocation decisions, pilot plant operation, drug product formulation design, process simulation, production planning and scheduling, process safety analysis, and supply chain management for API process as well as drug product development.

183b Modeling and Numerical Simulation of Pharmaceutical Processes

Hamid Arastoopour, department of chemical and environmental engineering, Illinois Institute of Technology, Chicago, IL 60616

Fluid/particle flow systems, including fluidized beds, are an essential part of many pharmaceuitcal processes. the optimum design, productivity enhancement and scale-up of these processes require a thorough understanding of fluid/particle flow patterns. Achievement of this understanding involves the numberical simulation of pharmaceutical processes using a Computational Fluid Dynamics (CFD) approach along with modeling of particle variation in size and other properties. Furthermore, numerical simulation contributes to a decrease in the time required to move from the laboratory-scale to the commercial-scale process.

This presentation provides an overview of the types of computational methods and recent contributions of numerical simulations that are used in pharmaceutical processes, including particle mixing, size reduction and growth, and particle deposition. The detailed numerical simulation of anti-solvent crystallization processes will be presented at different process conditions such as feed rate, feed position and seed amount. The simulation uses a CFD code coupled with the population balance equation (PBE) to describe the flow behavior of the solution (water, ethanol, dissolved glycerine) and the crystals to predict mixing and crystal formation and growth.

The presentation will be concluded with a discussion of the limitations and opportunities for research and development in this area.

183c Modelling Opportunities and Challenges in the Pharmaceutical Industry

Christopher L. Burcham, Chemical Product Research and Development; Process Definition, Eli Lilly and Company, Lilly Research Laboratories, Indianapolis, IN 46285

The development of pharmaceuticals involves the development of the process and chemistry to produce the active pharmaceutical ingredient as well as the development of the process and formulation to produce the final drug product. There are multiple opportunities to incorporate modeling into the process, yet adoption of simulations has been slow in the industry. Examples of the opprotunities and impediments to utilization will be discussed.

183d Information Management: Challenges and Opportunities in Achieving Qbd

Anjali Kataria, Conformia Software, Inc., 150 Mathilda Place, Suite 202, Sunnyvale, CA 94086

Text Not Available.

Thermodynamic Limits on Drug loading in Nanoparticle Cores

Varun Kumar and Robert K. Prud'homme, Dept. of Chemical Engineering, Princeton University, Princeton, NJ.

Abstract

Recently, biodegradable nanoparticles based on block copolymers have attracted attention as effective drug delivery vehicles. Maximizing the amount of drug loaded into particle is the desired goal, but experimental results find loading only between 3 to about 25 wt% drug (for paclitaxel). The reasons for the low loading and variability in loading have not been fully explained. In this study, we present a model that quantitatively explains the observed phenomena. The thermodynamic model of drug loading is based on the molar free energy of the drug, which depends on the block copolymers size (entropic term), the interaction parameter between the drug and the hydrophobic core (enthalpic term), and the pressure-volume work to load the particle. The pressure-volume work, related directly to the interfacial tension between the core and the corona region, has not been previously considered with respect to drug loading. To validate the model, calculations were compared with experimental results for organic solutes, including paclitaxel, loaded into poly(ethylene glycol)-*b*-poly(ε-caprolactone), PEG-*b*-PCL block copolymer micelles. The model developed was found to predict the loading values in close agreement with experiments reported in literature.

Keywords: paclitaxel; drug delivery; nanoparticles; interfacial energy; aggregation number; micelle; blob size; particle size; interaction parameter

1. Introduction

We start with the Flory-Huggins theory and consider an additional term that originates from the interface between the nanoparticles cores and the aqueous external phase. To understand the interfacial term, the hydrophilic polymer conformation at the interface is analyzed. In the later sections, we discuss how the nature of polymer blocks, their size and the type of solutes control the loading. The present model is based on the equilibrium solubilization of drugs in nanoparticle cores.

2. Flory Huggins theory

The phenomenon of equilibrium partitioning of a hydrophobic bulk polymer and a hydrophobic solute can be represented by the Flory-Huggins expression. However, for polymeric nanoparticles we need to consider an additional interfacial free energy term. The interfacial free energy between the hydrophobic core and the hydrophilic aqueous phase i.e. the Laplace pressure associated with the high curvature of the interface is the restraining force which counterbalances the swelling force.

Subscript **1** denotes for the drug solute and **2** the hydrophobic core polymer. The partial molar free energy of the drug component represents the free energy change in the system upon the addition of a molecule of solute (component 1), with the number of polymer molecules fixed. It is given by[1,2]

$$\Delta \overline{F}_1 = \Re T\underbrace{(\ln\phi_1 + (1-\frac{1}{J})\phi_2}_{\text{Mixing entropy}} + \underbrace{\phi_2^2 \cdot \chi)}_{\text{Enthalpy}} + \underbrace{\frac{2\gamma}{R} V_{1M}}_{\text{Pressure-volume work}} \qquad (1)$$

The equilibrium drug loading is then evaluated by applying the equilibrium condition,

$$\Delta \overline{F}_1 = 0 \qquad (2)$$

3. Role of interfacial forces

As a working hypothesis, the surface tension can be decomposed into two terms, a dispersive and a polar surface energy parameters[3]

$$\gamma_i = \gamma_i^p + \gamma_i^d \qquad (3)$$

The interfacial tension between two materials γ_{ij} can be estimated from the known dispersive and polar surface energy parameters by using the geometric mean (GM) or harmonic mean (HM) expressions. The harmonic mean is given by

$$\gamma_{ij} = \gamma_i + \gamma_j - 4 \cdot \frac{\gamma_i^d \cdot \gamma_j^d}{\gamma_i^d + \gamma_j^d} - 4 \cdot \frac{\gamma_i^p \cdot \gamma_j^p}{\gamma_i^p + \gamma_j^p} \qquad (4)$$

Nagarajan *et al.* estimated the interfacial tension between a polymer and solvent from the expression for the Flory chi parameter,[4]

$$\gamma_{ij} = (kT/l^2) \cdot (\chi_{ij}/6)^{\frac{1}{2}} \qquad (5)$$

where, the characteristic segment length, l, depends on the volume of the solvent molecule, v_w, as $l = v_w^{1/3}$.

The effective interfacial tension between the core/solution is equal to the area weighted interaction between core/water and core/S. The effective interfacial tension is expressed as

$$\gamma_{eff} = f \cdot \gamma_{C,S} + (1-f) \cdot \gamma_{C,W} \qquad (6)$$

where, $\gamma_{C,S}$ = interfacial tension between core, C, and the soluble blocks, S, of the micelles

$\gamma_{C,W}$ = interfacial tension between the core and water.

f = fraction of area covered by the S block.

The Overlapping Coil Regime[5] is obtained when blob size is much shorter than the end-to-end distance of the free soluble chain in solution: $\xi <$ d. In this regime, the fractional area (grafting fraction) of the core covered by the hydrophilic chain is given by[5]

$$f = \frac{b^2}{\xi^2} \tag{7}$$

Eqn.7 is valid when the size of first correlation blob near the surface, ξ, is considerably smaller than the core radius. We will be discussing two approaches to estimate the blob size which then gives fractional coverage and thus the effective interfacial tension, γ_{eff} .

A1. Curvature Induced Effect: The structure of corona region in spherical micelles has been extensively studied.[6,7] Daoud and Cotton[7] proposed a model which accounts for the curvature induced effect on the polymer brushes and takes into account the radial variation of the monomer concentration. The size of these blobs, $\xi(r)$, is radial dependent. Russel and co-workers show that the first or smallest blob size is given by

$$\xi_{\min} = 4R/(p^{\frac{1}{2}} - 2) \tag{8}$$

and the last or largest is $\xi_{\max} = 4(R+d)/(p^{\frac{1}{2}} + 2)$.[6] Here *p* is the aggregation number of micelle, R is the core radius and *d* is the corona thickness.

A2. Flat Plate Approximation: Another approach for the estimation of blob size is based on the fact that the chain at an interface will try to minimize its free energy. The free energy of a single chain at an interface is given by[8],

$$F = \gamma_{eff} \cdot \pi \frac{\xi^2}{4} + \frac{3}{2} kT \frac{d^2}{b^2 N} \tag{9}$$

where, the second term is the entropic contribution to the free energy of a real chain, which is the energy required to stretch an ideal chain to end-to-end distance *d*. We however will focus on a very simple case where *d* is a stack of blobs of uniform size, ξ. If N is the number of monomers in the chain and g the number of monomers in the blob, then number of blobs in a chain, n = N/g.

$$d = \xi \cdot n = \xi \frac{N}{g} \tag{10}$$

For the blob[8], $$\xi = bg^{\nu} \tag{11}$$
where, b = Kuhn monomer length; for an ideal chain, $\nu = 1/2$ and for a real chain, $\nu = 3/5$. Using Eqns.10 and 11, the free energy Eqn.9 becomes

$$F = \gamma_{eff} \cdot \pi \frac{\xi^2}{4} + \frac{3}{2} kT \frac{N b^{2(1-\nu)/\nu}}{\xi^{2(1-\nu)/\nu}} \tag{12}$$

Minimizing the free energy with respect to blob size (for a real chain), we get the expression for ξ as

$$\xi = \left(\frac{4.kT.N.b^{4/3}}{\gamma_{eff}.\pi} \right)^{\frac{3}{10}} \tag{13}$$

Combining Eqns.6, 7 and 13, an implicit expression for the fractional coverage is obtained as

$$f = b^2 \cdot \left(\frac{4kTNb^{\frac{4}{3}}}{\pi[\gamma_{C,S} \cdot f + (1-f).\gamma_{C,W}]} \right)^{\frac{-3}{5}} \tag{14}$$

4. Comparison with published results for loading of polymeric micelles

We present calculations of predicted loading of paclitaxel in polyethylene glycol-*b*-polycaprolactone (PEG-*b*-PCL). A representative calculation for **PEG-5K** and **PCL-5.7K** has been carried out.

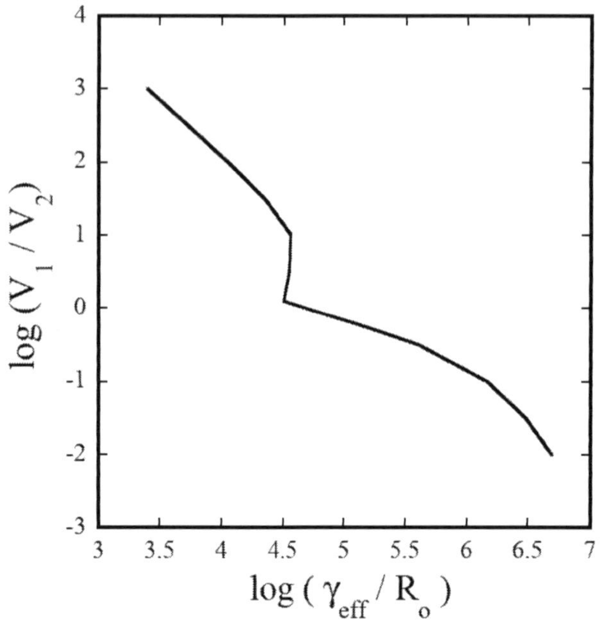

Figure 1. Calculated equilibrium curve for the loading of paclitaxel in PEG-*b*-PCL micelles as a function of the Laplace pressure, γ_{eff} / R_o.

Table 1. Values of paclitaxel loading for various sizes of the PEG-*b*-PCL block copolymer. **FPA** stands for "flat plate approximation" and **CIE** for "curvature induced effect". Loading values shown are for the FPA case.

PEG (Mw)	PCL (Mw)	J	$\log(\gamma_{eff} / R_o)$ by FPA	$\log(\gamma_{eff} / R_o)$ by CIE	V_1 / V_2	DLC, wt%
2K	2.7K	3.3	6.51	6.39	0.036	1.98
5K	5.7K	7.0	6.32	6.24	0.066	3.60
5K	18.0K	22.0	6.08	5.93	0.112	5.88

The polymer poly(D, L-lactide) (PDLLA) has higher degree of compatibility (i.e. lower χ=0.44[9,11]) with paclitaxel, so we predict the loading of paclitaxel into the PEG-*b*-PDLLA system.

Table 2. Values of the drug loading content (DLC) for paclitaxel with PEG-*b*-PLA as micellar system.

PEG (Mw)	PLA (Mw)	J	γ_{eff} (mN/m)	V_1/V_2	DLC *predicted*	DLC *experimental*[12]
2K	2K	2.26	23.1	0.218	9.5	10
5K	2.14 K	2.42	30.27	0.128	3.6	5

The validity of the model is extended further to other organic solutes by considering the system poly(propylene oxide) studied by Nagarajan *et al.*.[4,13]

Table 3. Values of the loading content for organic solutes and comparison with the experimental results[4]

Organic solutes	J	χ_{12}	V_{1M}	$V_1/(V_1+V_2)$ *predicted*	$V_1/(V_1+V_2)$ *experimental*
Benzene	42.36	0.0015	87.9	0.57	0.51
Toluene	35	0.028	106	0.52	0.40

5. Conclusions

We have presented a model that predicts the loading of drug into block copolymer cores which accounts for enthalpic and entropic interactions between the drug and the polymer core and the effect of interfacial energy or work to insert drug molecules into the core. This work term has not been included in previous analyses of drug loading. The effect is significant for drug-loaded particle in the 10-200 nm size range. The results predict the trends and levels of drug loadings that have been experimentally observed for the loading of paclitaxel into nanoparticles form. This work presents a methodology for *a priori* prediction of drug or solute loading in polymer micelles using only data available on the individual components in the system. It provides a starting point to optimize drug nanoparticle compositions by tuning block copolymer structure and composition.

Acknowledgements

We would like to acknowledge financial support from the NSF NIRT award to Princeton University.

References

1. Ugelstad J, Mork PC, Kaggerud KH, Ellingsen T, Berge A. 1980. Swelling of oligomer-polymer particles. New methods of preparation. Advances in Colloid and Interface Science 13(1-2):101-140.
2. Morton M, Kaizerman S, Altier MW. 1954. Swelling of Latex particles. J Colloid Sci 9:300.
3. Wu S. 1982. Polymer Interface and Adhesion.
4. Nagarajan R, Ganesh K. 1989. Block copolymer self-assembly in selective solvents: theory of solubilization in spherical micelles. Macromolecules 22(11):4312-4325.
5. de Gennes PG. 1980. Conformations of Polymers Attached to an Interface. Macromolecules 13(5):1069-1075.
6. Biver C, Hariharan R, Mays J, Russel WB. 1997. Neutral and Charged Polymer Brushes: A Model Unifying Curvature Effects from Micelles to Flat Surfaces. Macromolecules 30(6):1787-1792.
7. Daoud M, Cotton JP. 1982. Star shaped polymers: a model for the conformation and its concentration dependence. J Phys II 43:531-538.
8. Rubinstein M, Colby R. 2003. Polymer Physics.
9. Sheihet L, Piotrowska K, Dubin RA, Kohn J, Devore D. 2007. Effect of tyrosine-derived triblock copolymer compositions on nanosphere self-assembly and drug delivery. Biomacromolecules 8(3):998-1003.
10. Bailey AI, Salem DK, Walsh DJ, Zeytountsian A. 1979. The interfacial tension of poly (ethylene oxide) and poly (propylene oxide) oligomers. Colloid and Polymer Sci 257:948-952.
11. Liu J, Xiao Y, Allen C. 2004. Polymer-drug compatibility: A guide to the development of delivery systems for the anticancer agent, ellipticine. Journal of Pharmaceutical Sciences 93(1):132-143.

12. Burt HM, Zhang X, Toleikis P, Embree L, Hunter WL. 1999. Development of copolymers of poly (D,L- lactide) and methoxypolyethylene glycol as micellar carriers of paclitaxel. Colloids and Surfaces B: Biointerfaces 16:161-171.
13. Nagarajan R, Barry M, Ruckenstein E. 1986. Unusual selectivity in solubilization by block copolymer micelles. Langmuir 2(2):210-215.

184b Alternating Copolymer Structures For Targeted In Vivo Imaging And Therapy In Cancer

Michelle T. Miller[1], Kevin P. Brower[1], Jin Zhou[1], Robert J. Fisher[1], Mukesh K. Pandey[2], Rahul Tyagi[2], Arthur C. Watterson[2], and Clark K. Colton[1]. (1) Department of Chemical Engineering, Massachusetts Institute of Technology, 25 Ames St., Rm. 66-457, Cambridge, MA 02139, (2) Department of Chemistry, University of Massachusetts, Lowell, MA 01854

The biggest single problem that prevents a dramatic reduction in the mortality due to cancer is the limitation on current medical imaging techniques, computed tomography (CT) and magnetic resonance imaging (MRI), that provide detailed anatomical snapshots of the body but fail to provide accurate, basic information necessary to manage the patient's disease optimally. The limitations are manifested in several ways: (1) Small primary tumors go undetected. (2) Metastatic disease is grossly underdiagnosed. (3) Treatment response to therapy is poorly measured. A related problem in cancer therapy is the lack of selectivity of chemotherapeutic agents that are toxic to proliferating cells. The resulting side effects limit dosing and prevent use altogether.

A solution to this problem is to selectively carry contrast agents and drugs into cancer cells so as to enhance uptake and selectivity. We have developed a highly adaptable amphiphilic alternating copolymer system that self-assembles into micelles for in vivo imaging agent and therapeutic delivery applications in cancer. These unique alternating copolymer micelle nanoparticles were designed as delivery vehicles targeted to human cancer cells expressing the underglycosylated mucin-1 antigen, which is found on almost all epithelial cell adenocarcinomas, by use of the peptide EPPT. The synthetic scheme includes the enzymatic polymerization of multifunctional linker molecules (dimethyl 5-hydroxyisopthalate) with poly(ethylene glycol). This chemo-enzymatic synthesis is much faster and more convenient than an entirely chemical synthesis. Polymerization studies are underway to control molecular weight and scale up the enzymatic reaction while trying to better understand the importance of mass transfer limitations during this reaction. Subsequent synthetic steps have been developed to attach peptides (for targeting), perfluorocarbons (19F MR imaging), fluorescent dyes (NIRF imaging), and radioiodine (nuclear imaging and radioimmunotherapy) to the backbone. Attachment of hydrocarbon or perfluorocarbon sidechains provides amphiphilicity to produce the multimodal self-assembling micelles. Additionally, an encapsulation procedure for the chemotherapeutic agent, doxorubicin, has been established. Development of the synthetic schemes has been coupled with in vitro toxicity experiments using various cell viability assays to minimize the toxic effect of these copolymer structures. Other in vitro studies are underway that include the investigation of cellular uptake by I125 radioactive analysis and fluorescence confocal microscopy and the exploration of the treatment of cancer cells by delivery of encapsulated doxorubicin. Theoretical models of polyvalent binding are employed to predict the results of the in vitro uptake experiments.

184c Implant Assisted Magnetic Drug Targeting: Ferromagnetic Nanoparticles For Enhancing The Retention Of Magnetic Drug Carrier Particles

Misael O. Aviles, Jan O. Mangual, Armin D. Ebner, and James A. Ritter. Department of Chemical Engineering, University of South Carolina, Columbia, SC 29208

One of the key problems associated with drug administration is the difficulty to target specific areas or sites in the body, like cancerous tumors or arterial blockage. Typically in these cases, exceedingly large doses of a drug are needed to ensure that some of the drug reaches a specific site, a fact which unavoidably imposes substantial toxic side effects at non-targeted organs. One way to achieve drug targeting in the body is to incorporate magnetic particles into drug carriers and then to retain them at the site using an externally applied magnetic field. This process is referred to as magnetic drug targeting (MDT). However, the main limitation of MDT is that under a given set of conditions an externally applied magnetic field alone may not be able to retain a sufficient number of magnetic drug carrier particles (MDCPs) to justify its use. Such a limitation may not exist in implant-assisted (IA) MDT. When a ferromagnetic element (e.g., an implant) is placed in a magnetic field, it becomes magnetically energized creating a very strong but localized magnetic field that is far more capable of concentrating magnetic particles at the site of the implant compared to the magnetic field alone.

An IA-MDT system based on the use of seed nanoparticles to increase the capture of MDCPs in capillary tissue was studied in vitro. This IA-MDT system uses ferromagnetic particles as seeds (i.e., the implant) for collecting the MDCPs at a desired site. Two model systems were designed for in vitro flow experiments to mimic capillary tissue. They both utilized a cylindrical polyethylene porous polymer prepared using a compression melt molding and salt leaching method. In one case the magnetite nanoparticle seeds were embedded within the matrix during the melt molding process. In the second case, the seeds were captured magnetically prior to capturing the MDCP surrogates, by flowing them through the matrix just like the MDCPs. The second case is similar to how the seeds would be captured in vivo. In both cases 0.87 ƒÝm polydivinylbenzene particles embedded with magnetite nanoparticles were used as the MDCPs surrogates. In all cases a permanent magnet was used as the external magnetic field and the flow conditions were similar to those found in capillaries.

This presentation will disclose the effects of several variables on the performance of the proposed seed concept as an IA-MDT system in terms of the MDCP capture efficiency (CE). The fluid velocity, the distance to the magnet, the applied magnetic field, the seed nanoparticle size and concentration, and the MDCP concentration were studied. The results showed a significant increase in the capture of the MDCPs surrogates when the seeds were present. They also revealed considerable insight to the design of such an IA-MDT system.

184d Free Radical Scavenging By Cerium Oxide Nanoparticles: The Effect Of Adsorbed Block Copolymers And Potential-Determining Ions

Richey M. Davis, Virginia Tech, Department of Chemical Engineering, Blacksburg, VA 24061 and Will Miles, Chemical Engineering, Virginia Tech, Department of Chemical Engineering, Blacksburg, VA 24061.

Cerium oxide nanoparticles have been shown to have profound biological effects in cell culture and animal experiments. These same effects have not been seen in experiments with micron-sized ceria. [1-2] Nanosize ceria has extended the lifespan of mixed brain cell cultures by six fold, and the lives of Drosophila significantly. Results so far suggest that the increased lifespan is related to the mitigation of free radical damage due to UV-light, H2O2, and excitotoxicity (overactive signaling) by the ceria nanoparticles. In-vitro experiments show that the ceria nanoparticles are acting as free radical scavengers. The size, zeta potential, and BET surface area of ceria particles from two different sources were measured and related to the free radical scavenging rates for selected model radical species. The effects of coating the ceria with soluble polymer were investigated. These polymers include Pluronics PEO-PPO triblock copolymers and PEO chains anchored to the ceria using ion-containing anchor blocks. Finally, the scavenging rates were related qualitatively to the ratio of Ce+3/Ce+4 which is believed to be one of the factors controlling the scavenging mechanism.

1. B.A. Rzigalinski, D. Bailey, L. Chow, S.C. Kuiry, S. Patil, S. Merchant, and S. Seal. Faseb Journal, 2003, 17, pp A606-A606.

2.E.T. Strawn, C.A. Cohen, and B.A. Rzigalinski. Faseb Journal, 2006, 20, pp A1356-A1356.

184e Drug-Delivery Nanoparticles Formed By Micellization Of Polyethylene Glycol-Block-Polycaprolactone In Subcritical And Supercritical Solvents

Zachary Tyrrell, Winoto Winoto, and Maciej Radosz. Chemical & Petroleum Engineering, University of Wyoming, 1000 E University Ave, Laramie, WY 82071

Block copolymer micelles used for drug delivery are normally formed and processed in aqueous solutions. A major challenge in this area is how to increase the concentration of hydrophobic drugs in the nanoparticle core. Other issues are the lengthy dialysis procedure and the inefficient freeze dry procedure. We address these issues by processing the polymeric nanoparticles in highly compressible non-aqueous solvents, such as subcritical and supercritical polar fluids. Toward this end, one needs phase equilibrium data, such as cloud-point and micellization pressures as a function of temperature. These data will be presented for model diblocks, such as polyethylene-block-polycaprolactone, in near critical fluids.

184f Silica-Based Nanoparticles For Boron Neutron Capture Therapy

Ilya Zharov, Chemistry, University of Utah, 315 So. 1400 E., Salt Lake City, UT 84102

Boron neutron capture therapy (BNCT) is a binary type of cancer therapy that requires a substantial number of boron atoms to be delivered specifically to cancer cells. Preparation, characterization and cell targeting studies of novel fluorescent silica nanospheres incorporating a large number of boron atoms through a surface polymerization or a sol-gel process will be described.

225a Prediction Of Crystal Shape And Size Distributions Using Multidimensional Population Balances

Christian Borchert, University of Magdeburg, Magdeburg, Germany, **Nandkishor K. Nere**, Purdue University, 1283, FRNY, 480 Stadium Mall Dr., West Lafayette, IN 47907, Andreas Voigt, MPI Magdeburg, Sandtorstr. 1, Magdeburg, 39106, Germany, K. Sundmacher, Process Systems Engineering, Otto-von-Guericke-Univ. Magdeburg, Universitaetsplatz 2, Magdeburg, 39016, Germany, and Doraiswami Ramkrishna, School of Chemical Engineering, Purdue University, West Lafayette, IN 47907-1283.

Solution crystallization produces materials with specific crystal shapes and size distributions. Both the crystal size and shape play an important role in deciding the processing rate of material in various downstream operations, namely filtration, washing, and drying. These also govern the storage and handling characteristics, the flow ability, and the extent of dust formation, dispersibility, and stability. Furthermore, crystals of different shapes have different bioavailability and compressibility, the properties that are important in tablet manufacturing. Thus, the significant impact of crystal size and shape demands for its tighter control. Hence the accurate prediction of crystal shape along with its distribution becomes essential.

The relative face specific growth rates have recently been used to predict the shape evolution of a growing convex crystal (Zhang et al., 2006). These predictions are obtained assuming constant relative face growth rates. Obviously, the assumption makes the population of the crystals to be described by one dimensional population balance model (Zhang and Doherty, 2004). It has been observed that there exists significantly different crystal shapes in a crystallizer (Weigler, 2005 and Yang et al., 2006). Furthermore, it has been reported in the literature (Joshi and Paul, 1974 and Mullin and Whiting, 1980) that the face specific growth rates have different dependencies on the supersaturation leading to relative face growth rates strongly varying with supersaturation. Thus the population of crystals at any instant will have widely varying crystal shapes and sizes depending upon the initial crystal shape and size distribution, face specific growth dependence on supersaturation as well as the supersaturation profile. Eventually these observations demand for the complex consideration of varying relative growth rates in accordance with the supersaturation while predicting the crystal shape and size distribution. The additional complexity can be well treated with the multidimensional population balance modeling (MDPBM) approach coupled with the supersaturation dynamics. It is this approach, which will be presented to show some interesting predictions of the evolution of the crystal size and shape distributions in industrially relevant batch and continuous crystallizers.

The proposed multidimensional population framework has the potential to model, optimize and control the evolution of desired crystal form based on fundamental understanding of crystal growth including effects of supersaturation and crystal-solvent interactions.

References

1. Joshi, M. S. and Paul, B. K. "Effect of Supersaturation and Fluid Shear on the Habit and Homogeneity of Potassium Dihydrogen Phosphate Crystals", J. Cryst. Growth, 22, 321-327, 1974.

2. Mullin, J. W. and Whiting, M. J. L. "Succinic Acid Crystal Growth Rates in Aqueous Solution," Ind. Eng. Chem. Fundam., 19, 117-121, 1980.

3. Weigler, F., Process investigations for Gold colloid production, Study Thesis, Magdeburg 2005.

4. Yang, G., Kubota, N. Sha, Z., Louhi-Kultanen, M. and Wang, J. "Crystal Shape Control by Manipulating Supersaturation in Batch Cooling Crystallization", Crystal Growth and Design, 6, 2799 – 2803, 2006.

5. Zhang, Y. and Doherty, M. F. "Simultaneous Prediction of Crystal Shape and Size for Solution Crystallization", AIChE J., 50, 2101-2112, 2004.

6. Zhang, Y., Sizemore, J. P. and Doherty, M. F. "Shape Evolution of 3-Dimensional Faceted Crystals". AIChE J., 52, 1906-1915, 2006.

2-D POPULATION BALANCE MODELING OF GIBBSITE CRYSTALLIZATION

Andrey V. Bekker and Iztok Livk
Parker CRC for Integrated Hydrometallurgy Solutions / CSIRO Light Metals Flagship (Division of Minerals)
Karawara WA 6152, Australia

A population balance model for a gibbsite crystallization process was solved using numerical and analytical techniques. The crystallization model connects crystal size distribution with simplified gibbsite crystallization kinetics, such as secondary nucleation and crystal growth. The solution technique was studied considering two equivalent model formulations: (i) 1-D dynamic model, and (ii) 2-D steady-state model. Numerical solutions were compared to an analytical solution derived for the case of constant crystallization kinetics. A Newton's method-based finite element PDE solver has been adapted to solve both crystallization models. Evolutions of gibbsite crystal size distribution in respective batch and pipe crystallizers were predicted numerically and compared to the analytical solution. It is shown that for the numerical algorithm applied here, the stationary 2-D statement of the problem compares favourably to the equivalent 1-D dynamic statement in terms of the accuracy achieved and computational time required.

Introduction

Crystallization from solution represents a separation and purification technique with a spectrum of applications ranging from specialty chemicals to bulk chemicals manufacturing. In most cases, crystal size distribution (CSD) significantly affects the product quality and process productivity. The population balance equation (PBE) has been widely accepted as an appropriate mathematical framework for modeling CSD in such systems (see Randolph & Larson (1988), and Ramkrishna (2000)).

In this work, we study population balance modeling of gibbsite crystallization from solution, the operation that represents a critical step in the production of alumina, via the Bayer process. First principle mathematical models, such as gibbsite crystallization model, can be used for improving design, optimization, and control of the process. In most crystallization systems, there are many different crystallization mechanisms taking place simultaneously requiring relatively complicated nonlinear crystallization kinetics models. This can lead to very complex crystallization process models. Commonly, these cannot be solved analytically and numerical solution is usually sought. At present, a wide range of numerical methods for direct solution of population balance crystallization models have been developed and reported in the open literature. They include techniques such as discretized population balance (Hounslow *et al.*, 1988), pivotal schemes (Kumar & Ramkrishna, 1996a), and finite element techniques (Wulkow *et al.*, 2001). In the work of Nicmanis & Hounslow (1998), for example, a Galerkin finite element method was used to solve a one dimensional steady-state population balance problem. In most of these cases, a customized computer code was written in order to obtain the solution to a given problem. With ever increasing complexity of population balance models, which can now involve multiple internal and external coordinates, there is a strong incentive to develop modeling methodologies which can be directly integrated into available commercial partial differential equation (PDE) solvers. Developing a modeling framework suitable for solving multi-dimensional population balance problems in crystallization is also the aim of the present work.

Specifically, the objective of this work is to investigate the potential of a general Newton's method-based finite element algorithm to solve a nonlinear population and mass balance model of a gibbsite crystallization process. The suitability of Newton's method for solving nonlinear discrete population balance equations for steady-state crystallization problems was demonstrated in the work of Hounslow *et al.* (2005). In this work, a Newton's method-based FEM PDEs solver has been adopted to solve a population balance model of a gibbsite crystallization process, as previously described by Li *et al.* (2001). For this purpose, a simplified crystallization model that only includes secondary nucleation and crystal growth rates was formulated. The same crystallization problem was modeled using two different PBE formulations: (i) 1-D dynamic PBE, and (ii) 2-D steady-state PBE.

In the first part of the paper, a general statement of the 1-D dynamic problem is given in terms of the crystal size distribution (CSD) and solution supersaturation. Solving the 1-D model, we address numerical difficulties, such as numerical diffusion and spurious oscillations, and propose a method to minimize their impact. In the second part, an equivalent 2-D steady-state problem is formulated and solved using the same numerical approach. It is demonstrated that the solution of the 2-D steady-state approach is superior to the 1-D dynamic one. Advantages of the former for solving the chosen crystallization problem are also discussed.

1-D Dynamic Crystallization Model

A 1-D dynamic crystallization model can be formulated to describe batch crystallization from solution, where the main process parameters, such as CSD and supersaturation, are all functions of time. For gibbsite crystallization, empirical power-law kinetic expressions can be used to describe gibbsite secondary nucleation and crystal growth rates, as shown, for example, by Li *et al.* (2000, 2001, and 2003). The dynamic model that describes CSD and supersaturation balance in this system can be stated as:

$$\begin{cases} \dfrac{\partial n}{\partial t} + G\dfrac{\partial n}{\partial L} = 0\,, \quad L \in (L_{\min}, L_{\max}),\ t \in (0, t_{end}]; \\ \dfrac{\partial \sigma}{\partial t} = -aB_u - bG\mu_2\,, \\ G = k_g\sigma^2,\ B_u = k_n\sigma^4 k_a\mu_2,\ \mu_2 = \displaystyle\int_{L_{\min}}^{L_{\max}} L^2 n\, dL, \\ n\big|_{L=L_{\min}} = \dfrac{B_u}{G}, \qquad \dfrac{\partial n}{\partial L}\bigg|_{L=L_{\max}} = 0, \\ n\big|_{t=0} = n_0(L), \qquad \sigma\big|_{t=0} = \sigma_0. \end{cases} \tag{1}$$

were n is the number density, σ is the relative supersaturation, t is time, L is crystal size, n_0 is the initial number density of seed, σ_0 is the initial relative supersaturation, L_{min} is the smallest size, L_{max} is the largest size, G is the crystal growth rate, B_u is the secondary nucleation rate, k_g and k_n are growth and nucleation kinetic constants, μ_2 is the second moment of crystal size distribution, and k_a is the crystal surface shape factor (π for spheres). a and b are the normalization coefficients, defined as V_n/k_M, and $k_a/(2\ k_M)$, respectively, with $V_n = k_V\ L_{min}^{\ 3}$, and $k_M = 2\ \varepsilon\ M[Al(OH)_3]/(\rho_s\ M[Al_2O_3])$. k_V is the volume shape factor ($\pi/6$ for spheres), ε is the ratio of liquid volume to total volume, $M[Al(OH)_3]$ and $M[Al_2O_3]$ are molecular weights for aluminium trihydroxide and aluminium oxide, and ρ_s is the solids density.

The above 1-D dynamic crystallization model describes the distribution of gibbsite crystals by crystal size dimension in an isothermal homogenously mixed tank. Due to the *n(L)* and *dn(L)/dL* discontinuities, which exist in crystallization systems with fines destruction or product classification (Rawling *et al.*, 1993), the numerical solution of a dynamical PBE crystallization model often exhibits artificial effects, such as oscillations and numerical diffusion, which are not originally included in model equations. Numerical dispersion can be removed from the PBE solution by employing the method of characteristics (Kumar & Ramkrishna, 1996a, 1996b). However, this method can be difficult to implement in cases where nonlinear or multiple crystal growth rates are present, or where problems are large, such as PBEs coupled with other PDEs (Lim *et al.*, 2002). Upwind first-order finite difference approximation methods can also be used to solve the above crystallization model, but in this case mesh sizing needs to be considered with care to reduce possible numerical diffusion (Sotowa *et al.*, 2000). On the other hand, popular higher order approximation methods give spurious oscillations, which can result in the population density of crystals having negative values. In addition, weighted essentially non-oscillatory (WENO) approximation schemes with an artificial compensation method exist (Lim *et al.*, 2002). These can handle large gradients, while reducing numerical diffusion and spurious oscillations, and achieving more than first-order accuracy. However, this technique requires a tuning parameter and cannot satisfactorily solve problems with discontinuities in several variables.

To solve the 1-D dynamic model in Eq.1, we adopted a Newton's method-based finite element PDE solver. In this case, a fully implicit finite-element approximation was used to solve the model consisting of a set of nonlinear integro-partial differential equations. With this approach, the model solution is obtained iteratively using Newton's method and adaptive non-uniform mesh strategy. To validate numerical results, an analytical solution was derived for the constant crystallization kinetics case.

Constant Crystallization Kinetics Case

Implementation of a numerical algorithm usually has to balance the accuracy and required computational time (CPU time). Numerical errors depend not only on stability of the numerical algorithm, but also on the nature of the problem. Discontinuities, which can be present in initial and boundary conditions, and complex crystallization kinetics can both lead to a stiff system response. For the case of constant crystallization kinetics, $G = G_0 = \text{const}$, $B_u = B_0 = \text{constant}$, the statement in Eq. 1 can be reformulated in order to improve the behavior of the problem. For this reason, a diffusion term is added to the PBE:

$$\begin{cases} \dfrac{\partial n}{\partial t} + G_0 \dfrac{\partial n}{\partial L} = D\dfrac{\partial^2 n}{\partial L^2}, & L \in (L_{\min}, L_{\max}),\ t \in (0, t_{end}]; \\ \dfrac{\partial \sigma}{\partial t} = -aB_0 - bG_0\mu_2, & \mu_2 = \displaystyle\int_{L_{\min}}^{L_{\max}} L^2 n \, dL, \\ n\big|_{L=L_{\min}} = \dfrac{B_0}{G_0}, & \dfrac{\partial n}{\partial L}\bigg|_{L=L_{\max}} = 0, \\ n\big|_{t=0} = n_0 \exp\left(-\dfrac{L - L_{\min}}{l_0}\right), & \sigma\big|_{t=0} = \sigma_0. \end{cases} \tag{2}$$

The additional D-term on the RHS of the first equation merely introduces an artificial diffusion and has no other physical meaning in this case. Note also that the initial CSD condition in this case assumes an exponential shape. The constant kinetics model represents a hypothetical crystallization system with a very small change in the level of supersaturation.

The initial condition parameters, $n_0 = 5\cdot10^{16}$ m^{-4} and $l_0 = 10^{-6}$ m, correspond to the seed with density $\rho_s = 2400$ kg·m^{-3} and mass loading of $1.4\cdot10^{-3}$ g·L^{-1} $Al(OH)_3$. A very small initial amount is necessary to maintain a positive supersaturation balance for longer time periods, which are considered in our simulations. The following model parameters, also valid for gibbsite crystallization, were used: $L_{\min} = 1.4\cdot10^{-6}$ m; $L_{\max} = 200\cdot10^{-6}$ m; $G_0 = 5\cdot10^{-10}$ m·s^{-1}; $B_0 = 0.05n_0G_0 = 1.25\cdot10^{6}$ m^{-3}·s^{-1}; $\sigma_0 = 1$; $V_n = (\pi/6)\cdot L_{\min}^3 = 1.4\cdot10^{-18}$ m^3; $M[Al(OH)_3] = 78$ g·mol^{-1}; $M[Al_2O_3] = 102$ g·mol^{-1}; $k_M = 4.46\cdot10^{-2}$; $a = V_n/k_M = 3.1\cdot10^{-17}$ m^3; $b = \pi/(2\,k_M) = 35.2$.

Spurious oscillations test

For $D = 0$ in the first equation of Eq. 2 a simple convective equation is obtained. The solution of this system now only depends on the initial and boundary conditions, but not on supersaturation. The first equation can therefore be integrated independently with given initial and boundary conditions. For the case of constant kinetics, the PBE can be integrated analytically, and the expression obtained can then be used to also integrate analytically the supersaturation balance equation. In that manner full a analytical solution of the model can be obtained for this case. The analytical solution is presented in detail in Bekker & Livk (2007).

$$\begin{cases} L_{mid} = L_{\min} + G_0 t; \\ n(L,t) = \begin{cases} L < L_{mid}: & \dfrac{B_0}{G_0}, \\ L >= L_{mid}: & n_0 \exp\left(-\dfrac{(L - L_{\min} - G_0 t)}{l_0}\right); \end{cases} \\ \sigma(t) = \sigma_0 - aB_0 t + \dfrac{b}{3} B_0 L_{\min}^3 t - \dfrac{b}{12}\dfrac{B_0}{G_0} L_{mid}^4 - b n_0 l_0^4 \left\{ \left[\dfrac{1}{3}\left(\dfrac{L_{mid}}{l_0}\right)^3 + \left(\dfrac{L_{mid}}{l_0}\right)^2 + 2\left(\dfrac{L_{mid}}{l_0}\right) \right] - \right. \\ \left. - \left[\dfrac{1}{3}\left(\dfrac{L_{\min}}{l_0}\right)^3 + \left(\dfrac{L_{\min}}{l_0}\right)^2 + 2\left(\dfrac{L_{\min}}{l_0}\right) \right] - \left[\left(\dfrac{L_{\max}}{l_0}\right)^2 + 2\dfrac{L_{\max}}{l_0} + 2 \right] \exp\left(-\dfrac{L_{\max}}{L_0}\right) \left[\exp\left(\dfrac{L_{mid}}{l_0}\right) - \exp\left(\dfrac{L_{\min}}{l_0}\right) \right] \right\}. \end{cases} \tag{3}$$

The exponential initial condition in Eq. 2 exhibits a large gradient on one side of the CSD, which together with given initial and boundary conditions can lead to a discontinued shock wave. These features offer an opportunity to investigate the effect of artificial spurious oscillations, which is inherent to higher order approximation methods.

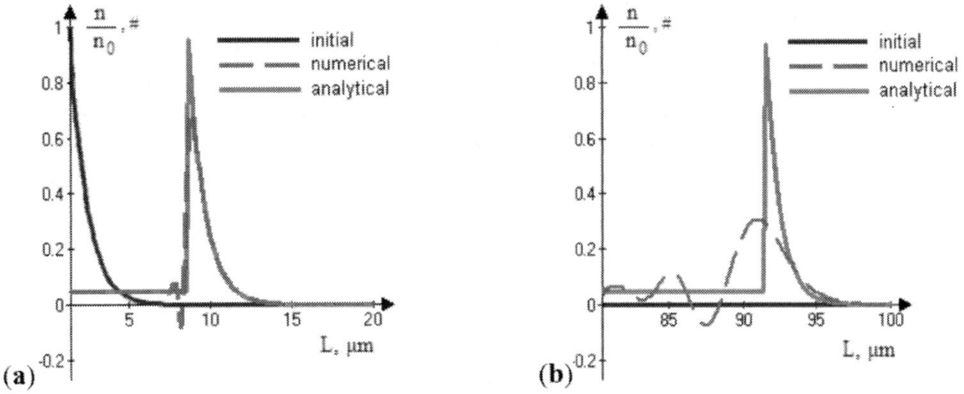

Fig.1. Analytical and numerical CSDs: (**a**) after 4 hours, and (**b**) after 50 hours. Analytical results are solid red lines; numerical solutions are doted green lines. The initial CSD is presented by solid blue lines.

In Fig. 1, numerical solution of the model in Eq.2 is compared to corresponding analytical evolutions of gibbsite CSDs. In the numerical solution $D = 10^{-34}\ m^2 \cdot s^{-1}$ is taken into account. The results show that the difference between the analytical and numerical solution increases with time. The increased difference is mainly due to spurious oscillations, which result in negative values for crystal number density. This would further negatively affect the solids mass balance and supersaturation balance. A way to reduce spurious oscillations is to increase the value of artificial diffusion D. To identify the appropriate value for the artificial diffusion, a separate test was conducted.

Artificial diffusion test

To test the appropriate value for the artificial diffusion, D, the statement in Eq. 2 has been modified by making n_0 equal to zero. If D is also set to zero then that statement has the following simple analytical solution:

$$\begin{cases} L_{mid} = L_{\min} + G_0 t; \\ n(L,t) = \begin{cases} L < L_{mid} : & \dfrac{B_0}{G_0}, \\ L >= L_{mid} : & 0; \end{cases} \\ \sigma(t) = \sigma_0 - aB_0 t + \dfrac{b}{3} B_0 L_{\min}{}^3 t - \dfrac{b}{12} \dfrac{B_0}{G_0} L_{mid}{}^4 . \end{cases} \tag{4}$$

The initial condition $n(L)_{t=0} = 0$ describes pure liquid without solid seeds. But due to high supersaturation, the crystallization system generates fine crystals of the size $L_{\min}$ at a nucleation rate equal to B_0. This artificial diffusion test can also help investigate the effect of numerical diffusion.

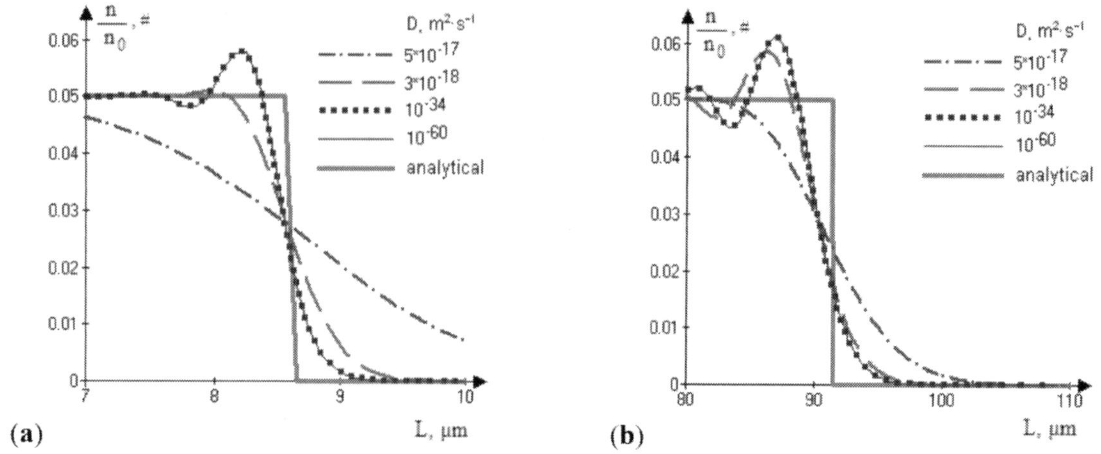

Fig.2. Analytical and numerical CSDs: (**a**) after 4 hours, and (**b**) after 50 hours. Numerical solutions are presented for different values of *D*.

A comparison between the analytical and numerical solution for two different times is shown in Fig. 2. The differences notable between the analytical and numerical results are due to spurious oscillations and diffusion. By adjusting the value of *D*, one can control the balance between the two. Increasing *D* reduces spurious oscillations, but increases artificial diffusion. For example, $D = 5 \cdot 10^{-17}$ $m^2 \cdot s^{-1}$ can totally damp spurious oscillations, even for the 50-hour time, but results in a significant overestimation of number of large particles at shorter times. For the time of 4 hours, the value of *D* of $3 \cdot 10^{-18}$ $m^2 \cdot s^{-1}$ represents a good trade-off between the two effects, but it cannot dampen the oscillations at 50 hours. In order to improve the solution at longer residence times additional measures need to be taken, as described in the next section.

Newton's relative error test

The model solution for the 50-hour residence time can be improved by tightening the tolerance applied for each iteration of Newton's method. As shown in Table 1, for a 1.6 GHz processor and second order polynomial finite elements with 1000 computational cells, a decrease in Newton's relative error (N.r.e.) results in a significant increase in required CPU time.

Table 1. CPU time after the Newton's relative error

CPU time	Newton's relative error
60 s	$5 \cdot 10^{-3}$
2400 s	$5 \cdot 10^{-5}$

Due to the accumulation of numerical error with simulation time, the acceptable relative error of Newton's method for the simulation time of 50 hours is two orders of magnitude lower than the appropriate relative error applied for the 4-hour simulation. As indicated in Table 1, a decrease in the relative error comes with a considerable cost in terms of the required computational time.

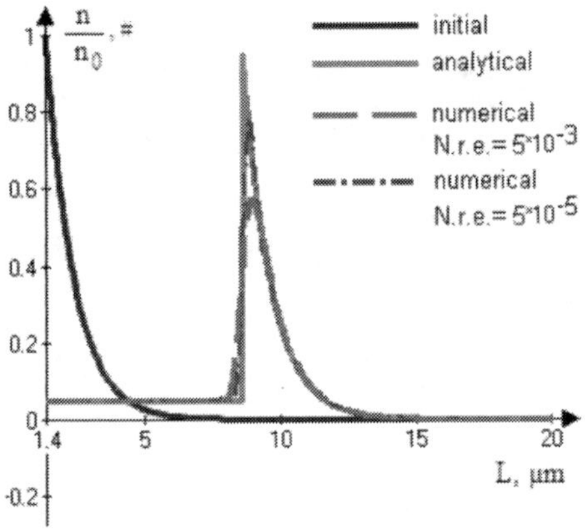

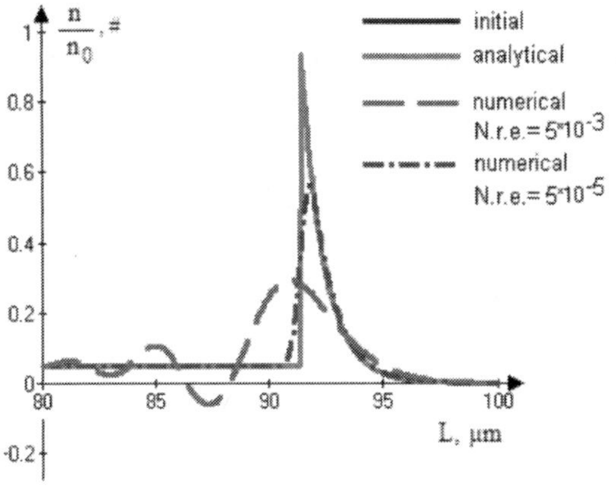

Fig.3. Analytical and numerical CSDs for $D = 3\cdot10^{-19}$ $m^2\cdot s^{-1}$: (**a**) after 4 hours, and (**b**) after 50 hours. The solid and dotted line are analytical and numerical solution, respectively.

2-D Gibbsite Crystallization Modelling

2-D Steady-state model

As demonstrated previously by Heath & Livk (2006), a 1-D dynamic gibbsite crystallization model can be transformed into an equivalent 2-D steady-state model. The time dimension of the 1-D dynamic model is in the 2-D steady-state model simply replaced by the space dimension, interpreted as the length of a plug-flow tubular crystallizer. The equivalent 2-D steady-state model can thus be represented as

$$\begin{cases} \Omega = \{(X,L);\ X\in(0,X_{\max}),\ L\in(L_{\min},L_{\max})\ \}\subset\mathbb{R}^2\ \ t\in(0,t_{end}] \\ v\dfrac{\partial n}{\partial X}+G_0\dfrac{\partial n}{\partial L}=D\left(\dfrac{\partial^2 n}{\partial L^2}+\dfrac{\partial^2 n}{\partial X^2}\right), \\ v\dfrac{\partial \sigma}{\partial X}=-aB_0-bG_0\mu_2, \qquad \mu_2=\displaystyle\int_{L_{\min}}^{L_{\max}} L^2 n\,dL, \\ n\big|_{L=L_{\min}}=\dfrac{B_0}{G_0}, \qquad \dfrac{\partial\sigma}{\partial L}\Big|_{L=L_{\min}}=0, \quad \dfrac{\partial n}{\partial L}\Big|_{L=L_{\max}}=0, \quad \dfrac{\partial\sigma}{\partial L}\Big|_{L=L_{\max}}=0, \\ n\big|_{X=0}=n_0\exp\left(-\dfrac{L-L_{\min}}{l_0}\right), \quad \sigma\big|_{X=0}=\sigma_0, \quad \dfrac{\partial n}{\partial L}\Big|_{X=X_{\max}}=0, \quad \dfrac{\partial\sigma}{\partial L}\Big|_{X=X_{\max}}=0, \\ n\big|_{t=0}=n_0\exp\left(-\dfrac{L-L_{\min}}{l_0}\right), \quad \sigma\big|_{t=0}=\sigma_0. \end{cases} \tag{5}$$

Mathematically, the 2-D steady-state model is obtained by replacing $\dfrac{\partial}{\partial t}$ in Eq. 2 with $v\dfrac{\partial}{\partial X}$, where v is the velocity of the plug flow in the X direction. The 2-D model predicted $n(L, X)$ is equivalent to the $n(L, t= X/v)$ for the 1-D dynamic model. The applied value of v is such that the diffusion in X-direction is made negligible. Analogy between the both systems is further illustrated in Table 2 and Fig.4.

Table 2. Batch residence time and corresponding lengths in a plug-flow pipe reactor.

Batch, t	Plug flow, X
4 hour	0.08 X_{max}
50 hour	X_{max}

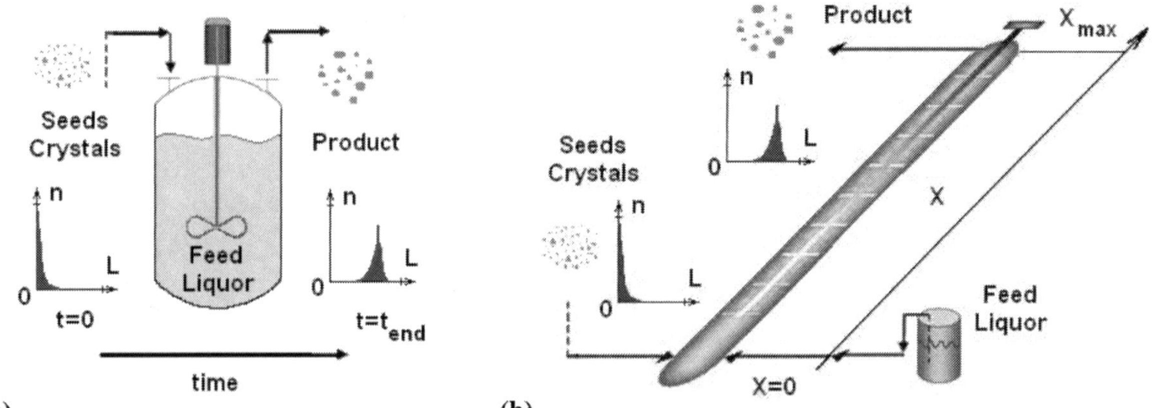

(a) (b)

Fig. 4. A well-mixed batch crystallizer **(a)**, and an equivalent tubular crystallizer **(b)**.

Similar to the 1-D dynamic case, we used a Newton's method-based finite element PDE solver to solve the 2-D steady-state model in Eq.5. Steady-state finite-element approximations were used to solve the set of nonlinear integro-partial differential equations. The model solution was obtained iteratively using Newton's method and adaptive non-uniform mesh strategy. The solution obtained for the constant kinetics case was also validated against the analytical solution (Bekker & Livk, 2007).

Constant kinetics results

The solution of the 2-D problem (Eq. 5) also requires implementation of the elliptic regularization (Lions & Stampacchia, 1967), which is introduced via the addition of the artificial diffusion term on the RHS of the PBE, alike to the 1-D problem (Eq. 2). For small values of D spurious oscillations are present also for the 2-D statement. In contrast to the 1-D case, adaptive mesh strategy is used simultaneously to solve the 2-D model in both dimensions, L and X. This strategy leads to a global convergence for the 2-D case. In the 1-D case, where adaptive mesh is applied at each time-step, in contrast only partial convergence is achieved.

Results obtained for the 2-D case are presented in Fig. 5. Shown in Fig.5 (a), is the evolution of CSD as a function of crystal size and pipe length. It is evident that the peak in CSD distribution shifts to larger crystal sizes with increasing pipe length. At the same time, the peak in CSD decreases with pipe length, which is attributed to the artificial diffusion. Fig. 5 (b) shows a comparison between the analytical solution and numerical solution for the 1-D and 2-D model. The plot shows a good agreement between the three solutions. Note that the difference between the both numerical solutions is mainly due to different level of numerical diffusion at the same value for D being used in both cases.

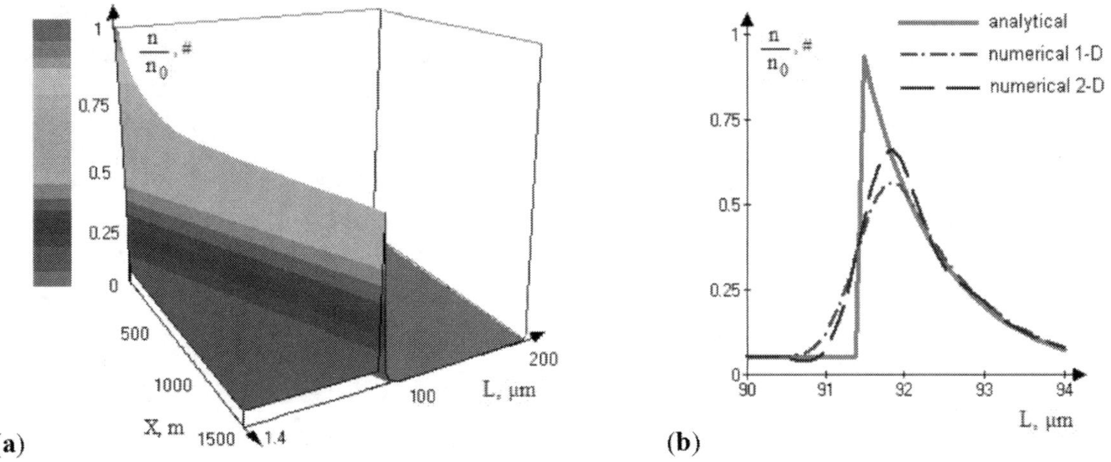

Fig.5. CSD as a function of pipe length for the 2-D problem: (**a**) complete distribution, and (**b**) at X_{max}.

Table 3 compares the CPU time required to solve the 1-D and 2-D statement using a 1.6GHz processor. It is clear that for a comparable accuracy, a much lesser computational effort is required to solve the 2-D model.

Table 3. Required CPU time for the two equivalent statements.

CPU time	Statement
60 s	2-D
2400 s	1-D

The regularization method, as implemented here for the 2-D problem, is expected to lead to a superior performance in all cases where larger gradients in CSD, multi-modal CSD, or stiffness in space (time) are present.

Conclusions

Solution of a simplified gibbsite crystallization model was studied using a Newton's method-based finite element PDE solver. Fully implicit and steady-state finite-element approximations were used to solve the respective sets of nonlinear integro-partial differential equations arising in a 1-D dynamic and 2-D steady-state problem. In both cases, the model solution is obtained iteratively using Newton's method and adaptive non-uniform mesh strategy. Numerical results for the case with constant crystallization kinetics were validated against an analytical solution. A comparison between numerical and analytical results showed a difference owing to numerical diffusion and spurious oscillations.

It was shown that spurious oscillations and numerical diffusion can be successfully controlled by the addition of an artificial-diffusion term. For longer simulation times however, a reduction of the relative tolerance of Newton's method is also required to dampen the oscillations.

It was shown that for longer residence times the stationary 2-D statement of the problem compares favourably to the equivalent 1-D dynamic statement in terms of the accuracy and computational cost. A simultaneous 2-D adaptive mesh technique and elliptic regularization used to solve the 2-D statement leads to a more efficient global convergence in size-space domain than a separate 'adaptive size-mesh and adaptive time-step' strategies used to solve the 1-D statement.

References

1. Bekker A.V. & Livk I. (2007). PBE Modelling of Gibbsite Crystallization with Constant and Nonlinear Kinetics. *In preparation.*
2. Heath A. & Livk I. (2006). Coupled Population Balance and CFD Model for Continuous Gibbsite Crystallizer. *Fifth International Conference on CFD in the Process Industries CSIRO*, Melbourne, Australia 13-15 December 2006.
3. Hounslow M.J., Lewis A.E., Sanders S.J. & Bondy R. (2005). Generic crystalliser model: I. A model framework for a well-mixed compartment. *A.I.Ch.E. Journal*, **51**, 11, 2942–2954.
4. Hounslow M.J., Ryall R.L. & Marshall V.R. (1988). A discretized population balance for nucleation, growth and aggregation. *A.I.Ch.E. Journal*, **34**, 11, 1821–1832.
5. Kumar S. & Ramkrishna D. (1996a). On the solution of population balance equations by discretisation I. A fixed pivot technique. *Chemical Engineering Science*, **51**, 8, 1311–1332.
6. Kumar S. & Ramkrishna D. (1996b). On the solution of population balance equations by discretisation II. A moving pivot technique. *Chemical Engineering Science*, **51**, 8, 1333–1342.
7. Li J., Prestidge C.A. & Addai-Mensah J. (2000). The Influence of Alkali Metal Ions on Homogeneous Nucleation of $Al(OH)_3$ Crystals from Supersaturated Caustic Aluminate Solutions. *Journal of Colloid and Interface Science*, 224, 317–324.
8. Li T.S., Livk I. & Ilievski D. (2001). The influence of crystalliser configuration on the accuracy and precision of gibbsite crystallisation kinetics estimates, *Chemical Engineering Science*, **56**, 7, 2511–2519.
9. Li T.S., Livk I. & Ilievski D. (2003). Supersaturation and Temperature Dependency of Gibbsite Growth in Laminar and Turbulent Flows, *Journal of Crystal Growth*, 258, 409–419.
10. Lim Y.I., Le Lann J.-M., Meyer X.M., Joulia X., Lee G. & Yoon E.S. (2002). On the solution of population balance equations (PBE) with accurate front tracking methods in practical crystallization processes. *Chemical Engineering Science,* 57, 3715–3732.
11. Nicmanis M. & Hounslow M.J. (1998). Finite-Element Method for Steady-State Population Balance Equations. *A.I.Ch.E. Journal*, **44**, 10, 2258–2272.
12. Ramkrishna D. (2000). Population Balances: Theory and Applications to Particulate Processes in Engineering. *Academic Press*, New York, NY.
13. Randolph A.D. & Larson M.A. (1988). Theory of Particulate Processes: Analysis and Techniques of Continuous Crystallization, 2nd ed., *Academic Press*, Toronto.
14. Rawling J.B., Miller S.M. & Witkowski W.R. (1993). Model Identification and Control of Solution Crystallization Processes: A review. *Industrial Engineering and Chemistry Research*, 32, 1275–1296.
15. Sotowa K., Naito K., Kano M., Hasebe S. & Hashimoto I. (2000). Application of the method of characteristics to crystallizer simulation and comparison with finite difference for controller performance evaluation. *Journal of Process Control*, 10, 203–208.
16. Wulkow M., Gerstlauer A. & Nieken U. (2001). Modeling and simulation of crystallization processes using Parsival. *Chemical Engineering Science,* 56, 2575–2588.

225c Contin Analysis of Sals Data from Clusters of Particles beyond Rdg Limit

Marco Lattuada, Lyonel Ehrl, Miroslav Soos, and Massimo Morbidelli. Institute for Chemical and Bioengineering, Department of Chemistry and Applied Biosciences, ETH Zurich, Wolfgang-Pauli-Str. 10, HCI F 133, Zurich, 8093, Switzerland

Static light scattering (SALS) is a widely used technique to gather information about properties of colloidal dispersions, because it can provide useful information about size distribution of particles and clusters together with information about structural arrangement of particles within clusters. However, the analysis of light scattering data is straightforward only in the case of spherical particles thanks to Lorenz-Mie theory. When particles and clusters are not spherical, as it is the case when colloidal suspensions undergo aggregation, resulting in the formation of fractal clusters, the application of Lorenz-Mie theory can provide misleading information. In this work, we present a strategy which overcomes these limitations and allows one to extract reliable information about the cluster mass distribution from the analysis of scattering data. The proposed methodology relies on three main steps. First of all, it is necessary to gather structural information about clusters like those commonly encountered in coagulation processes. We have accomplished this by using a tunable fractal dimension algorithm, which allows generating clusters with a prescribed fractal dimension, up to values close to three. Consequently, the structure of these clusters has been analyzed in terms of particle-particle correlation functions. The second step of this procedure is the calculation of the scattering properties of clusters. Since shear aggregation processes usually lead to the formation of highly compact (i.e. high fractal dimension) aggregates, often obtained using large primary particles, conventional RDG theory cannot be applied. We have therefore used a mean-field version of the T-Matrix theory, which allows one to efficiently compute scattering properties of clusters with both high fractal dimension and large primary particle size given their particle-particle correlation function. The third step of this procedure is to use these individual cluster scattering properties in combination with CONTIN algorithm to perform parsimonious fitting of light scattering data to extract cluster mass distributions. This procedure is applied to SALS data from shear-induced aggregation of polystyrene particles with various sizes.

225d Molecular Dynamics Simulation Of Diffusion Of Fractal Aggregates

Gaurav Pranami[1], **R. Dennis Vigil**[2], and Monica Lamm[1]. (1) Chemical and Biological Engineering, Iowa State University, 2114 Sweeney Hall, Ames, IA 50011-2230, (2) Chemical Engineering, Iowa State University, Ames, IA 50010

The synthesis of structured nanoparticles for use in a variety of applications, particularly pharmaceutical materials, has led to a need for improved models for aggregation of colloidal particles. Typically such aggregates are smaller than the Kolmogorov scale and the aggregation is therefore governed by Brownian motion rather than by fluid shear. In such cases, the calculation of Brownian aggregation rate kernels depends crucially on the particle morphology, which influences both the collision cross-section and the particle diffusivity. In this work, we use a robust method to construct rigid fractal aggregates with any desired fractal dimension and particle mass. These fractal clusters are then imported into molecular dynamics simulations and diffusion studies are carried out in the infinite dilution limit. The resulting diffusivity data are used to deduce the relationship between diffusivity (both translational and rotational), fractal dimension, particle mass, and these expressions are compared with predictions developed by other investigators that are based upon estimation of equivalent hydrodynamic radii.

225e Modeling, Validation And Optimization Of Antisolvent Crystallization: A New Approach In Predicting Solubility Using A Thermodynamic Model

Ombretta Foddi[1], Ali Abbas[2], Massimiliano Grosso[1], and Jose A. Romagnoli[3]. (1) Dipartimento di Ingegneria Chimica e Materiali, Università degli Studi di Cagliari, Piazza d'Armi, Cagliari, Italy, (2) School of Chemical & Biomedical Engineering, Nanyang Technological University, Blk 1, Unit 100, 637722,, Innovation centre, 16 Nanyang Drive, Singapore, (3) Chemical Engineering, Louisiana State Univeristy, Cain Department of Chemical Engineering, South Stadium Road, Baton Rouge, LA 70803

Crystallization is a consolidated unit operation widely used in the process industry to obtain high purity particulate products at low costs. It is one of the more commonly used techniques in the separation of products of great worth or where purity is of significant importance, such as in the pharmaceutical industry. The fundamental driving force for the crystallization process is the supersaturation, which is the difference between the solute concentration in the solution and its saturation concentration. Supersaturation is typically generated by cooling, evaporation and/or antisolvent addition.

In the last decades the salting-out technique has drawn more attention and has been more frequently seen as a valid alternative to cooling and evaporation. The antisolvent addition reduces the solubility of the product compound in the original solvent thereby facilitating supersaturation generation. This method is attractive since it can lead to significant savings in energy consumption and operation costs in comparison to the conventional techniques. Furthermore it can be seen as an alternative methodology whereby the limited temperature stability of the solid product precludes the use of evaporation, as in the case of pharmaceuticals and biochemicals, or when, because of the weak temperature dependence of the solute solubility, it is not possible to use the cooling techniques, such as for sodium chloride in water.

Currently most of the industrial crystallization processes are still designed and tested by means of time consuming and expensive trial-and-error experimentations. The use of model based techniques is extremely useful in improving the implementation of these technologies and in understanding the complex mechanism involved in the process at different scale. The main control objectives in a crystallization process are usually defined in terms of crystal characteristics, such as size, morphology and purity. In the case of antisolvent crystallization, the antisolvent feedrate trajectory is crucial to control the supersaturation level and therefore to achieve the desired crystal size distribution (CSD). Thus, optimization of feed flow rate profiles is of great importance, though one of the major obstacles for designing the optimal feed profiles is the accuracy of the mathematical models.

In this work a new approach to predict solubility is applied to the antisolvent crystallization modeling. The thermodynamic model proposed by Kolker et al. [1] is used here to describe the ternary equilibrium and to calculate the supersaturation for the water-NaCl-ethanol system. In contrast to previous empirical approaches, in which experimental data need to be derived and validated, this model is more general and requires only standard thermodynamic properties for the pure components and for the solutes at infinite dilution in each solvent. The thermodynamic model is then blended into an overall mechanistic crystallization model. The complete mathematical model for the crystallization process is developed based on the conservation of mass, energy and population moments. In particular, population balance equations are numerically solved to predict end product properties expressed in terms of particles' mean size and CSD.

Alternative kinetic models for growth and nucleation are also investigated and compared. The kinetic parameter estimation is performed experimentally by means of maximum likelihood method. In this step data collected from a number of experiments under different antisolvent feeding conditions are used. Once the kinetic model is identified, model validation is carried out using experimental data obtained for new antisolvent feed profiles. The results are in accordance with the experimental data, and show that the dynamic behavior of the antisolvent crystallization process can be successfully modeled and predicted.

The overall model described here is proposed as a useful tool to understand the basic phenomena involved in the process and is used to find optimal operational procedures under different objectives. In particular, a dynamic optimization was conducted in order to find the optimal time-varying trajectory for the antisolvent flow rate. The results presented here have been obtained from different final size constraints. These results show that model-based approaches are powerful tools which can be implemented for the design of optimal operation policies in the area of particle size control.

[1] Kolker, A., De Pablo, J. Thermodynamic modeling of the Solubility of salt in mixed aqueous-organic solvents. Ind. Eng. Chem. Res. 1996, 35, 228-233

Modeling the sintering of water soluble amorphous and crystalline particles

N. Descamps, E. Schreyer

Nestlé Product Technology Centre,
Lange Straße 21, 78224 Singen, Germany,
Email: nicolas.descamps@rdsi.nestle.com; elsa.schreyer@rdsi.nestle.com

ABSTRACT

Plasticization of water soluble amorphous carbohydrates by moisture sorption is a ubiquitous phenomenon influencing the stability and functionality of pharmaceutical excipients and food products. Indeed, because of its very low molecular weight, water enhances molecular mobility and is thus a highly efficient plasticizer. As a consequence, the glass transition temperature of an amorphous matrix decreases strongly as a function of the water content. Thus when a glassy carbohydrate is placed at high relative humidity, its glass transition temperature may reach a lower value than the one of the product itself. A dramatic decrease of the viscosity of the material happens, leading to the development of viscous flow. Viscous flow is responsible for the formation of viscous bridges between the particles also called sintering. Amorphous water-soluble particles are sintering together depending on time, temperature, stress and water content. In case the particles are in equilibrium with the surrounding air the sinter kinetics can be predicted by combining the known sinter equations with the theory of glass transition. However in practice the particles are often not in their equilibrium state and moisture migration and heat transfer are influencing the sintering kinetics.

Based on the sinter equation from Rumpf et al.[1] and the theory of glass transition the growth of the sinter bridge between two amorphous particles was modeled as a function of time, viscosity and surface tension. In addition amorphous maltodextrin particles were exposed under a microscope in controlled climate conditions. The development of the sinter-bridge between two particles was monitored depending on time, temperature and relative humidity of the surrounding air. The obtained time dependent distribution of the sinter-bridge diameter was compared with the calculated bridge diameter using Rumpf's model[1]. The same experimental approach was used to measure bridge formation between crystalline (NaCl) and amorphous (maltodextrin DE21) particles.

It appears from this study that Rumpf et al.'s model[1] well describes the sintering kinetics of amorphous particles in a qualitative way. However, its use for a quantitative prediction is limited. Indeed, the experimental sintering kinetics of amorphous particles is slower than the theoretical calculations. Concerning the bridge formation between crystalline salt and amorphous maltodextrin components, the measured kinetics is faster than predicted by the theory.

Keywords: Sintering, amorphous, crystalline, modeling

1. INTRODUCTION

Powder agglomeration is a process for the pharmaceutical, chemical and food industry. Agglomeration can be either desired (fluid bed agglomeration, tabletting) or undesired (caking and stickiness). In the case of amorphous products, agglomeration is based on sintering. The sintering process is commonly described as the coalescence of two single particles. Such phenomenon is associated with the formation of a bridge between the two particles. The strength of the bridge depends on its diameter compared to the particle diameter and on the tensile strength of the material. The formation of this bridge is linked to the physical state of the material and its response to external disturbances such as temperature and humidity, in term of viscosity and elasticity. This in turn influences the occurrence and kinetics of the powder agglomeration phenomenon.

Many food powders contain amorphous water soluble components, such as low molecular carbohydrates. At low temperature and water content, such products are in the glassy state where they behave like solids[2]. However, when an amorphous component is exposed to sufficiently high temperature and/or relative humidity, the molecules start to move generating a viscous solution. The reducing viscosity enables viscous flow driven by surface tension. This leads to the formation of interparticle bridges inducing sintering. Such phenomenon may occur between amorphous particles and also between crystalline and amorphous particles.

In the first part of this paper, the properties of amorphous products influencing the sintering process are briefly described. Then, experimental data obtained for the sintering of:

- two amorphous particles
- one amorphous and one crystalline particle

were compared to the predicted values calculated according to Rumpf et al.'s model[1].

2. INFLUENCE OF THE MATERIAL PROPERTIES ON THE SINTERING PROCESS

Sintering is the formation of viscous bridges between the particles, whose growth is driven by surface tension. The process kinetics depends on the viscosity (η) of the material. All glass forming amorphous compounds have a viscosity of about 10^{12} Pa·s at their glass transition temperature (T_g)[3]. However, a dramatic decrease of viscosity is observed when the temperature is increased above T_g. This drop in viscosity enables the material to flow. The amplitude of the drop is reflected by the amplitude of the activation energy E_a. E_a is visualized in an Arrhenius plot by the variation of η as a function of temperature i.e. the slop of log(η) vs. 1/T. The non-Arrhenius temperature behaviour of the viscosity is represented by several equivalent phenomenological laws, including the Williams-Landel-Ferry (WLF) equation[4]:

$$\log\frac{\eta}{\eta_g}=\frac{-C_1(T-T_g)}{C_2+(T-T_g)} \qquad (1)$$

were η_g is the viscosity value at T_g, T is the temperature and C_1 and C_2 are constants.

For a glassy water soluble substance, the relative humidity will have a strong influence on the physical properties of an amorphous powder. According to the Gordon and Taylor equation (G-T)[5], the T_g of binary blends of glass forming compound miscible in water is given by:

$$Tg=\frac{kw_w T_{g,w}+w_s T_{g,s}}{kw_w+w_s} \qquad (2)$$

where $T_{g,S}$ and W_S are respectively the glass transition temperature and the mass ratio of the substance S. $T_{g,w}$ and W_w are the glass transition temperature and the mass ratio of water. K is a fitting parameter.

T_{gw} is expected to be about -135°C[6]. From the Gordon-Taylor equation, it appears that the effect of water will be to decrease the T_g of the blend. It acts as a plasticizer and contributes to increase the molecular mobility. In consequence it strongly decreases the temperature difference T-T_g, with regard to the working temperature T. From what has been described above, the viscosity of an amorphous material can be decreased equivalently by increasing the temperature at a given humidity, or by increasing the moisture level at a given temperature. This enables the material to build viscous bridges between the particles by viscous flow[7], which is reflected to as sintering.

The parameter which characterizes the geometry of the sinter bridge is the ratio x/a (bridge radius/particle radius) as illustrated in Figure 2 ad 5. Kinetics in the viscous regime is expected to change with the particle size, so that the evolution in time (t) of the ratio follows a general law[8] such as:

$$\left(\frac{x}{a}\right)^n=K\cdot t \qquad (3)$$

where n and K are constants. The value of exponent n depends on the sintering mechanism (evaporation/sublimation, diffusion, viscous flow…ref). For viscous flow, it has been predicted by Frenkel[9] that n=2. The value of K depends, according to Frenkel, on the surface tension (γ) and the viscosity (η). Thus, the Frenkel's equation for sintering can be written as:

$$\left(\frac{x}{a}\right)^2=\frac{3\gamma}{2a}\frac{t}{\eta} \qquad (4)$$

Rumpf et al.[1] derived an expression which includes the sintering effect of an applied load, F on sintering. The constant of proportionality for the first term was found to be 8/5 instead of the 3/2 found by Frenkel:

$$\left(\frac{x}{a}\right)^2=\left(\frac{8}{5}\frac{\gamma}{a}+\frac{2F}{5\pi a^2}\right)\frac{t}{\eta} \qquad (5)$$

The pressure factor introduced by Rumpf is very important since it allow modelling powder agglomeration in silos or big storage bags.

3. MATERIAL AND METHOD

Amorphous spray dried maltodextrin DE21 (Glucidex® 21) was purchased from Roquette and crystalline salt NaCl was purchased from Salins du Midi-Marine Salt Trading Company. Sorption isotherms of maltodextrin and salt were performed on IgaSorp® Moisture Sorption Analyser from Hiden Isochema. The G-T diagram of maltodextrin was established by measuring Tg of samples with known water content by differential scanning calorimetry (DSC). DSC experiments were performed using a Mettler-Toledo 822e. The instrument was calibrated for temperature using the melting temperature of indium and n-octane and for heat flow using the heat of fusion of indium. The analyses were carried out with typically 10mg of sample hermetically sealed in pre-weighed aluminium pans. After each experiment, the tightness of the pan was systematically checked by weighing. The DSC oven was flushed with pure nitrogen gas. All experiments have been performed with a heating rate of 5°C/min and a cooling rate of 20°C/min. The glass transition temperature of a sample is defined as the onset of the glass transition jump.

To evaluate the sintering kinetics, the evolution of the sinter bridge diameter was measured for different couples of particles of maltodextrin DE21 and for different couples of maltodextrin DE21 and NaCl particles. Experiments were performed at room temperature (25°C). Each particle was placed under a light microscope in controlled relative humidity (RH) of the surrounding air. The RH was controlled using salt solutions, diluted if necessary to allow a range from 53%RH to 80%RH, using $Mg(NO_3)_2$, NH_4NO_3 and NaCl salts from Fluka Chemie AG. The RH was constantly checked during the measurements with a Hygroscop DT from Rotronic® and could fluctuate for 2% maximum around the target value.

The particles were first placed at the controlled RH until they were at equilibrium with the surrounding air (the surface of the maltodextrin DE21 particles gets brittle). The contact was then induced manually.

The slide on which the particles are placed is coated with a hydrophobic material (dimethyldicholorosilan), to provide spherical droplet once the particle is significantly plasticized. Finally, pictures were taken regularly to evaluate the growth of the bridge (see figures 1 and 4).

The water content of the amorphous maltodextrin was evaluated according to its sorption isotherm. The glass transition temperature was thus deduced from the G-T equation. From the T-Tg value, the viscosity of the maltodextrin could be calculated using WLF model (equation 1) with the "universal values" of C_1 and C_2 (C_1=-17,4 and C_2=51,6°C).

To compare the experimental results with the theory, the parameter K of equation (3) must be calculated, according to Rumpf et al.'s definition[1]. The surface tension (γ) was measured for diluted solutions of maltodextrin. The results were then extrapolated for higher concentrations. The value thus found is: γ=60mN.m^{-1}.

The force applied on the particles (F) was assumed to be zero.

4. RESULTS AND DISCUSSION

4.1. Sintering of two maltodextrin particles

A series of pictures showing the evolution in time of the sintering is presented in figure 1.

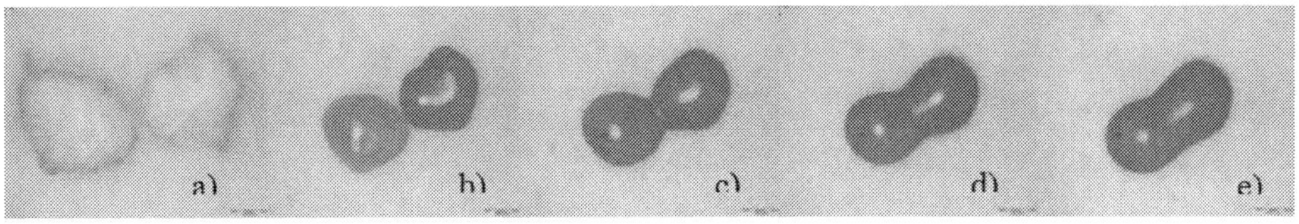

Figure 1: Two dextrose syrup particles at 25°C and 76% RH. a) t_1=0 min, b) t_2=30 min (initiation of contact), c) t_3=60 min, d) t_4=90 min, e) t_5=120 min.

On picture a), the particles are in their initial state and exhibit a clear roughness. On picture b), after only 30 minutes in humid conditions, the particles are already in the rubbery state. Their surface becomes smoother, and adopts a spherical shape. The contact is then initiated manually and the sintering process starts.

The evolution of the size of the sinter bridge can then be measured and experimental results compared with the theory.

Figure 2 presents the evolution of the ratio x/a measured and calculated for several couples of particles placed in different humidity conditions. It shows clearly the influence of the relative humidity on the sintering kinetics of amorphous particles. Indeed, changes in the relative humidity imply changes in the difference T-Tg and thus in the viscosity of the material (see Eq. 1). This trend is well reproduced by Rumpf et al.'s model, but with a faster kinetics: 4 times quicker for RH=75-76% and 16 times quicker for RH=80%.

It thus seems that differences exist between theory and experiment in the coalescence kinetics of two amorphous maltodextrin particles. However, before drawing any conclusion on the capacity of Rumpf et al.'s[1] model to represent such phenomenon, it is interesting to know whether the assumption made in equations 4 and 5 on the nature of the sintering regime is correct. Indeed, as seen before (cf. §2.), sintering can be driven by different mechanisms. The value of the parameter n in equation 3 reflects the nature of the mechanism. In order to analyse the character of the sintering process occurring here, n must be evaluated. Such evaluation can be done by plotting experimental results with logarithmic scales: Log(x/a) vs. Log(t), and determining the slope of the linear regression as shown on figure 3.

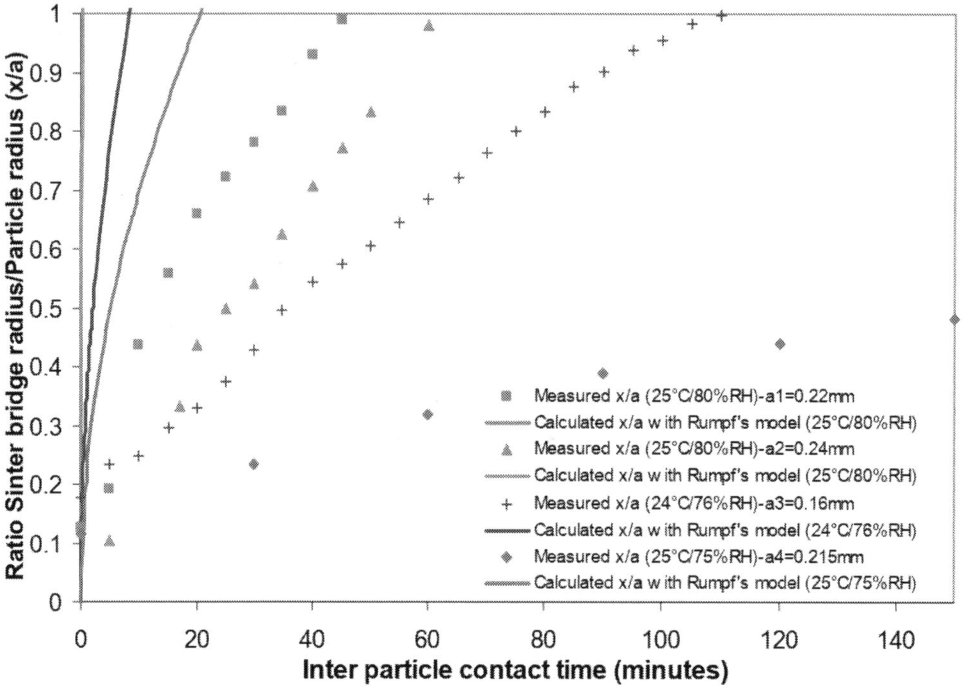

Figure 2: Measured and calculated ratio x/a (sinter bridge radius/particle radius) depending on the contact time between the particles. The dots represent the experimental points measured at room temperature and several RH. The curves correspond to the theoretical sintering kinetics calculated with Rumpf et al.'s[1] model for each experimental condition.

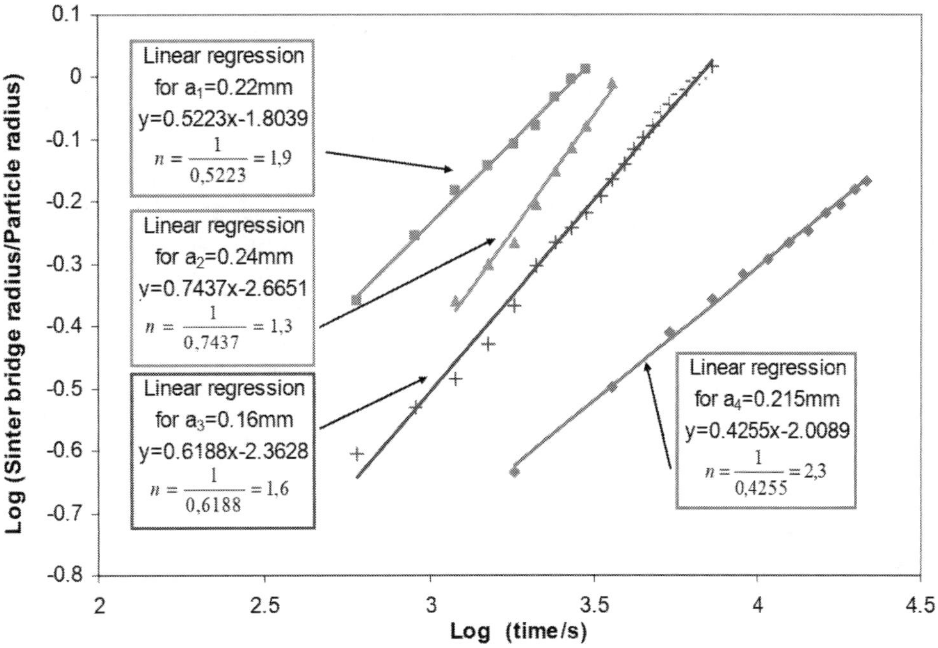

Figure 3: Example of logarithmic plot of the evolution of the relative bridge diameter (x/a) as a function of time. In the insets are given the formula of the linear regression and the value of n for each series.

Plots presented on figure 3 show a linear evolution of Log(x/a) as a function of Log(t). This demonstrates the validity of the law presented in equation 3 in the case of maltodextrin. Linear regressions were then made on each series in order to estimate the value of the parameter n and check the applicability of the Frenkel-Rumpf equation which predicts a value of 2 for the exponent. Exponents n are calculated from the slope of each regression. The mean of all the calculated values of n is: n=1,8, which is close to the one used by Frenkel and Rumpf and thus asserts that sintering of amorphous maltodextrin is driven by viscous flow of matter. The Rumpf et al.'s equation[1] is preferred to model experimental results since it can take into account the applied force on the particles, which will be useful in future work to model the sintering kinetics of powder during storage for instance.
The main conclusion we can draw from this comparison is that Rumpf et al.'s model[1] predicts qualitatively well the global evolution of the sinter bridge diameter. However, a quantitative use of the theoretical data is more doubtful. Indeed, a gap between theory and experiments is observed. This gap may be due either to the fact that particles are not totally equilibrated in humidity during the experiment, or to a wrong estimation of the viscosity value used in the model. Indeed, the theoretical values of x/a are influenced a lot by the viscosity value and the WLF equation with "universal" constants is not trustful, as showed by Peleg[10].

4.2. Sintering of one maltodextrin and one NaCl particle

A series of pictures showing the evolution of the sintering process is shown in figure 4 below.

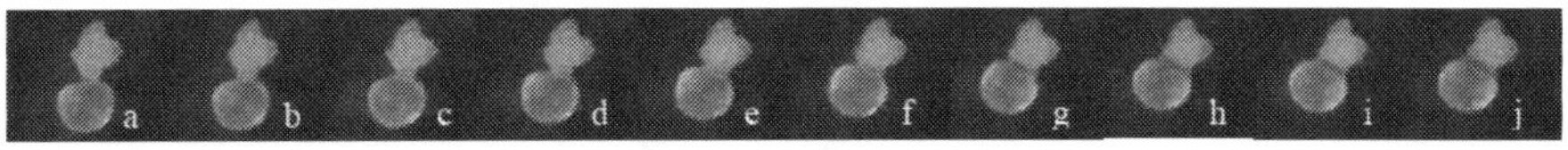

Figure 4: one maltodextrin DE21 particle (top) and one NaCl particle (bottom) at 25°C and 66%RH, pictures taken every 12.5 min, from 0 min (contact initiated manually) to 125 min.

The RH humidity of 66% gives a water content of 11.7% wet basis corresponding to a viscosity of $10^{10.5}$Pa.s. This very high value of viscosity explains why no smoothing of the surface can be observed, opposite to figure 1. But paradoxically sintering can be observed and measured.
As shown in the previous analysis on sintering of two maltodextrin particles, the particle size has a great influence on the kinetics of the sintering process. Contrary to the previous analysis section, in this experiment, particles of very different sizes were studied in different humidity conditions (from 53% to 72% RH). Therefore it was necessary to normalise their size in order to compare the kinetics at different

relative humidity. By using the Rumpf et al.'s model (equation 5), we can easily normalise the x/a ratio, as follow:

$$a_1 = Ba_2$$
$$\sqrt{B} \cdot x/a_1 = \sqrt{(8 \cdot \gamma \cdot t)/(5 \cdot a_2 \cdot \eta)} \qquad (6)$$

where a_2 is the reference particle radius. On figure 5 the normalised measured ratio x/a and the calculated ratio x/a from the model are plotted as a function of time. Figure 5 shows clearly that the ratio x/a does not change if the relative humidity stays below 63% RH, several measurements done at 53% to 60% RH did not show any sintering (the value x/a=0.1 on the graph is due to the manually induced contact between the particles). 63% RH corresponds to Tg=25°C for maltodextrin DE21. This would mean that it is needed to reach or exceed Tg to observe sintering.

To verify that the relative humidity was not the only one responsible of the process, but that excess of Tg was necessary, experiments with NaCl particles and maltodextrin DE12 particles were conducted at high relative humidity (from 60% to 72%RH). Maltodextrin DE12 exhibits a higher Tg than maltodextrin DE21 (40°C versus 25°C at 62%RH). In the chosen conditions, the difference T-Tg was negative, so the measurements were done below Tg, although the relative humidity is as high as in the experiment with the maltodextrin DE21. No sintering happened. This shows that a combination of high relative humidity and excess of Tg is necessary to induce sintering.

Once the Tg is exceeded, it is clear that the sintering process follows a faster kinetics than predicted by the model. The lines representing the model are below the measured data. Moreover, the kinetics of the process appears to be independent from the relative humidity at which the particles were placed.

Thus it can be concluded that as soon as the Tg is exceeded, a process starts, which is equivalent at any relative humidity, and as a consequence at any T-Tg or viscosity of the maltodextrin particles, and that is faster than the prediction from equation 5.

In order to assess the mechanism responsible for the sintering process in this experiment, experimental results were plotted again on a logarithmic scale as shown in figure 6. To validate Frenkel and Rumpf's assumption as explained in the previous experiment, linear regressions were made on each serial to estimate the value of the parameter n from the experimental data. The exponent n was calculated from the slope of the linear regression and gave a average value of 2.34. This confirms that viscous flow is the mechanism driving the sintering between maltodextrin DE21 and NaCl particles. It confirms that there is a flow of matter between the particles while coalescence. But this phenomenon is not linked to relative humidity and thus seems to be independent from the value of T-Tg, once Tg is exceeded.

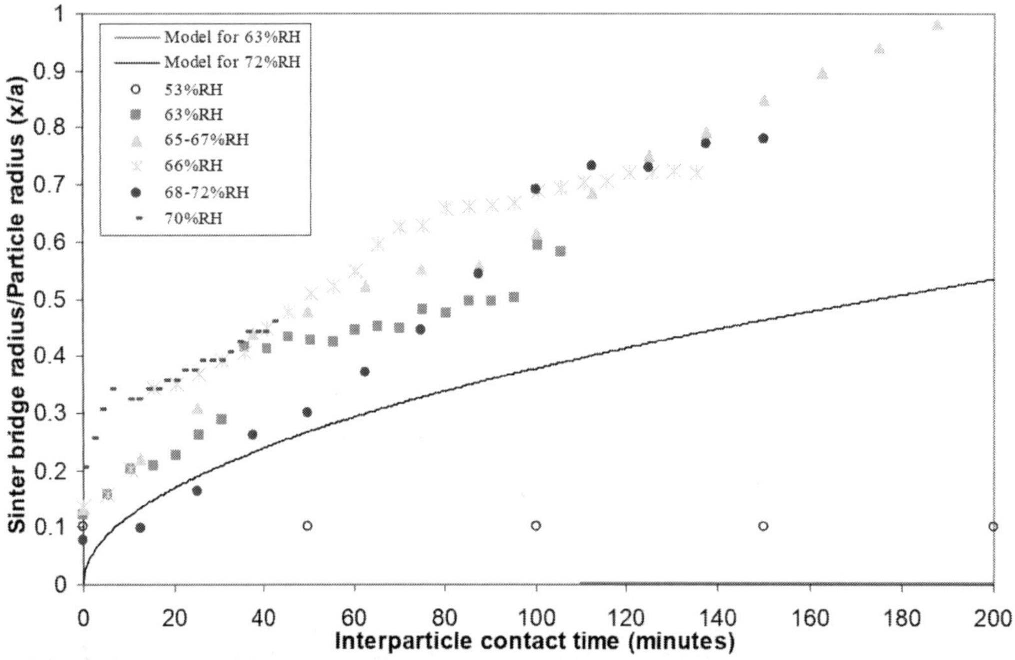

Figure 5: Measured and calculated ratio x/a (sinter bridge radius/particle radius) as a function of the contact time between the particles and normalised to the particle size. The dotted lines represent the

experimental data measured at room temperature and 6 different RH (from 53% to 72%). The curves represent the theoretical kinetics, calculated with Rumpf et al's model[1], for 63% and 72%RH.

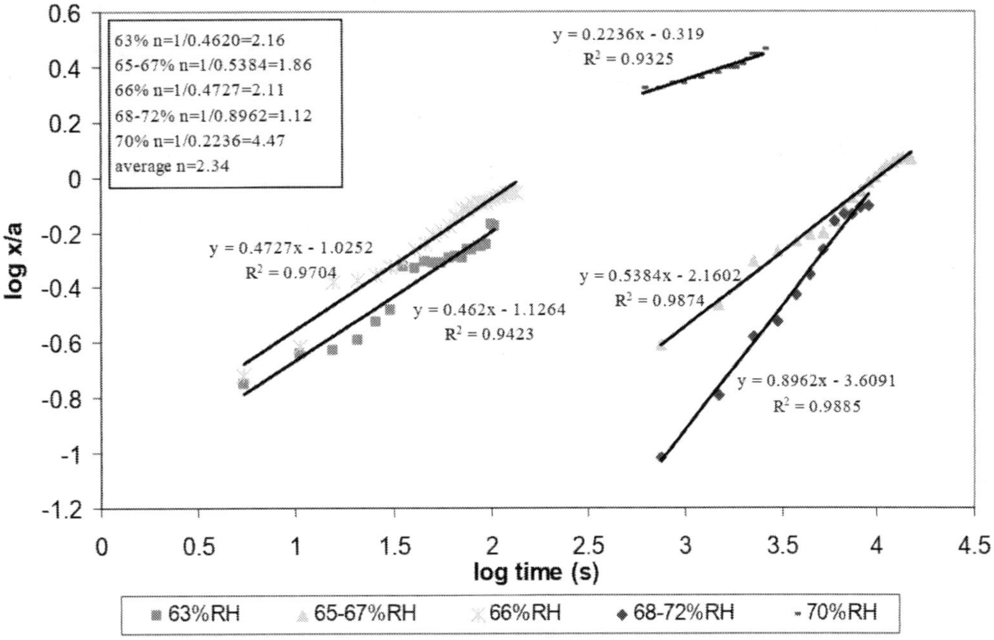

Figure 6: Logarithmic plots of the evolution of the ratio x/a as a function of time. The equations of the linear regression and the calculated values of n are shown in the inserts.

5. CONCLUSION

From the first series of experiments, we can conclude that Rumpf et al's model[1] gives a good qualitative representation of the sintering process of two amorphous particles. However, the use of this model for a quantitative description requires accurate determination of the viscosity of the material. Indeed, a slight uncertainty on the viscosity value may lead to big variations in the theoretical x/a value.

From the second set of experiments, we can conclude that Rumpf et al.'s model cannot be used to describe the sintering of couples of amorphous/crystalline particles in the case of maltodextrin DE21 associated to NaCl. Indeed, the kinetics does not depend on the T-Tg value, i.e. on viscosity of the maltodextrin. Since equation 5 takes into account the viscosity of the amorphous component to accommodate the process, it is not appropriate to use it here. Further experiments are on going to find possible explanations for the kinetics observed in the case of crystalline NaCl and maltodextrin DE21 particles. The viscosity of the sinter bridge may be the key parameter in this case.

ACKNOWLEDGMENTS

This work was financially supported by the EU 6th Framework programme through the Marie Curie Research Training Network Biopowders (MRTN-CT-2004-512247).

REFERENCES

1. Rumpf, H., Sommer, K., and Steier, K. Mechanismen der Haftkraftverstärkung bei der Partikelhaftung durch plastisches Verformen, Sintern und viskoelastisches Fließen. *Chem.-Ing.-Tech.* **48** (4), 300 (1976).
2. Debenedetti, P.G. and Stillinger, F.H. Supercooled liquids and the glass transition. *Nature* **410** (6825), 259 (2001).
3. Angell, C.A., Bressel, R.D., Green, J.L. et al. Liquid fragility and the glass transition in water and aqueous solutions. *Journal of Food Engineering* **22** (1-4), 115 (1994).
4. Williams, M.L., Landel, R.F., and Ferry, J.D. The temperature dependance of relaxation mechanisms in amorphous polymers and other glass-forming liquids. *Journal of the American Chemical Society* **77**, 3701 (1955).

5 Gordon, M. and Taylor, J.S. Ideal copolymers and the second-order transitions of synthetic rubbers. I. Non-crystalline copolymers. *Journal of Applied Chemistry* **2** (1952).
6 Johari, G.P., Hallbrucker, A., Mayer, E. The Glass-Liquid Transition of Hyperquenched Water. *Nature* **330**, 552 (1987).
7 Palzer, S. The effect of glass transition on the desired and undesired agglomeration of amorphous food powders. *Chemical Engineering Science* **60** (14), 3959 (2005).
8 Kuczynski, G.C. Self-diffusion in sintering of metallic particles. *Trans. Met. Soc. AIME* **185**, 169 (1949).
9 Frenkel, J. Viscous Flow of Crystalline Bodies under the Action of Surface Tension. *Journal of Physics* **9** (5), 385 (1945).
10 Peleg, M. On the use of the WLF model in polymers and foods. *Critical reviews in food science and nutrition* **32** (1), 59 (1992).

The effects of initial polydispersity of particle size on the growth kinetics and mechanism of wet granulation in a laboratory-scale high shear mixer by means of image processing and analysis

Alvaro Realpe and Carlos Velázquez. University of Puerto Rico, Mayaguez, Puerto Rico

Abstract

The effect of initial polydispersity of particle size (unimodal versus bimodal distribution) and binder characteristics on the growth kinetics and mechanism of wet granulation was studied. Wet granulation of pharmaceutical powders with initial bimodal particle size distribution (PSD) presented growth kinetics consisting of two stages: fast growth followed by slow growth. The fast stage is controlled by the amount of binder and high probability of coalescence due to the collisions of small and large particles. The second stage is characterized by slow agglomeration of powder mixtures with water content 13.6% v/w, and slow breakage of powder mixtures with water content of 9.9 and 11.7% v/w. The wet granulation of powders with initial unimodal PSD exhibited slow growth kinetics consisting of one stage, since similar particle sizes do not promote agglomeration. The experimental results were better described by a population balance equation using a coalescence kernel that favors growth rate by collision between small and large particles. In general, the probability of a successful collision increased with higher size difference between particles, smaller particle size, and higher binder content.

1 *Introduction*

Wet granulation in a high shear mixer is primarily a process for particle enlargement that is used in many industries such as pharmaceutical, mineral processing, agriculture, food, and detergent. The process of particle enlargement via a coalescence mechanism improves the flow properties of the particles, and decrease segregation of the ingredients, thus also improving the uniformity of the mixture. [1] A granule breakage mechanism could also happen during the wet granulation process, depending upon the viscosity and the amount of binder.

Ennis et al. [2] developed a *viscous Stokes' (St_v) number*, wherein factors such as particle size and density, and binder viscosity and amount are considered to quantify the probability of particle coalescence and, as a consequence, the granule growth rate. Ennis *et al.* [2, 3] considered the viscosity force of a liquid bridge, and modeled two colliding granules using the equation of motion (Eq. 1):

$$m_p \frac{dv}{dt} = \frac{3\pi \mu r_p^2}{2x} v \qquad 1$$

where m_p is the granule's mass; v is the relative velocity between the two particles; r_p is the granule's radius; μ is the viscosity of the binder liquid and $2x$ is the distance between colliding granules.

Recently, the importance of Ennis' theory has inspired researchers [4, 5, 6, 7] to further study the effects of particle size, binder viscosity and amount on the growth kinetics and mechanism of wet granulation. Badawy and Hussain [7] determined in experiments with anhydrous lactose that the growth rate increases with a decrease in initial particle size, in accordance with Ennis' theory, wherein smaller particles present low kinetic energy that is dissipated easily in the binder layer. Ennis et al. [2] and Adetayo et al. [6] observed a sequential two-stage mechanism during the wet granulation of fertilizers with a wide size distribution, while Mills et al. [7] failed to observe the sequential two-stage mechanism during the granulation of materials with a narrow size distribution.

Powders with bimodal PSD or wide size distribution have many small and large particles that favor the preferential coalescence between small and large particles, as suggested by Tardos et al. [8] with their layering mechanism or coalescence of coarse and fine particles (Figure 1). Hounslow et al. [9], and Boerefijn and Hounslow [10] also developed size-dependent agglomeration kernels that mathematically favored the coalescence of small and large particles. On the contrary, it is believed that powders with unimodal PSD or narrow size distribution have similar size particles which do not favor the preferential coalescence between small and large particles.

The main objective of the current work is to study the effects of initial polydispersity of particle size (unimodal versus bimodal), initial particle size, and the amount and viscosity of binder on growth kinetics and the mechanism of wet granulation. Particle size distribution (PSD) was determined by means of image processing and analysis. [11, 12] Near-infrared (NIR) spectroscopy was used, at the same time, to determine moisture content.

2 *Experimental method*

Material and equipment

Pharmaceutical powders such as lactose anhydrous supplied by Sheffield Products and lactose monohydrate manufactured by *FOREMOST FARMS USA #55-0072* and supplied by *Mutchler Inc.* (Cayey, PR), were used in the experiments of wet granulation. Figure 2 shows the three particle size distributions used in the wet granulation experiments, designated as PSD1 (fine size powders), PSD2 (medium-size powders) and PSD3 (coarse size powders).

The three levels of particle size were prepared by screening lactose monohydrate and anhydride using the following sieve trays Tyler's 270, 230, 200, 170, 140, 70, 50 and 40 mesh US Standard. The PSD of powders recollected in each sieve tray was measured (PSD1, PSD2 and PSD3) by means of image processing and analysis. [11, 12]

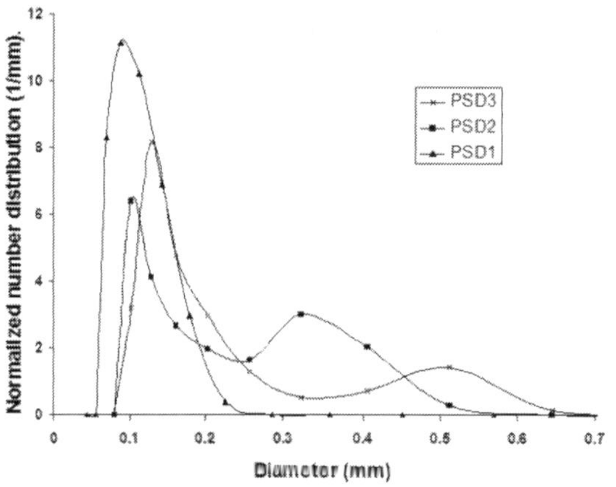

Figure 1 Initial particle size distributions for wet granulation

A viscometer was used to measure the viscosity of Povidone-water mixtures at room temperature. Three different amounts of solid Povidone (0%, 5% and 9%) were dissolved in water at room temperature to obtain homogeneous binder solutions with viscosities of 1, 2.49 and 4.85 cp, respectively. The Povidone was manufactured by ISP Technologies INC.

Figure 3 depicts a scheme of a laboratory-scale high shear mixer. A Ryobi^{TM} 12-inch drill press was used as an impeller motor, rotating vertically; its speed was fixed at 560 rpm. Two stainless steel blades were attached to the impeller axis. The liquid binder was added to the powder mixture at a rate of 161.7 ml/min by means of a peristaltic pump through a nozzle.

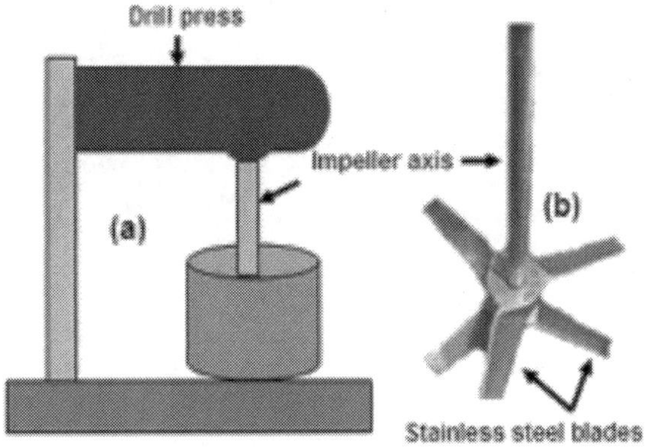

Figure 2 Laboratory-scale high shear mixer

3 *RESULTS AND DISCUSSION*

3.1 Effect of initial polydispersity of particle size (unimodal versus bimodal) on the mechanism and growth rate of wet granulation

The normalized number distribution (NND) was used to show the change in PSD over time, as indicated in Figure 4. The two-stage coalescence mechanism was observed in the wet granulation of pharmaceutical powders at 13.6% of binder with both initial unimodal and bimodal PSD. The bimodal PSD showed a fast growth followed by a slow one while the unimodal PSD showed an opposite mechanism, slow growth followed by a fast one as can be observed in Figure 5 through the evolution of mass median diameter of particles during wet granulation.

The initial bimodal PSD (PSD2 and PSD3) exhibits first a fast stage of granule growth. This stage is controlled by both, the binder amount added during the first 30 seconds and the high probability of collision between small and large particles that increases the growth rate. The opposite case is observed with initial mono-dispersed unimodal PSD (PSD1). The first stage of granule growth is slow, due to the high concentration of particles of similar size, which does not favor coalescence mechanism.

The strength of the liquid bridge between particles of similar sizes is weaker than between small and large particles due to the force exerted downwards by the weights of each particle. The higher strength of the liquid bridge between small and large particles, and lower kinetic energy of small particle, allows the particles to stay in contact long and, thus, the small particle will join the larger one by diffusion of small particle through binder layers.

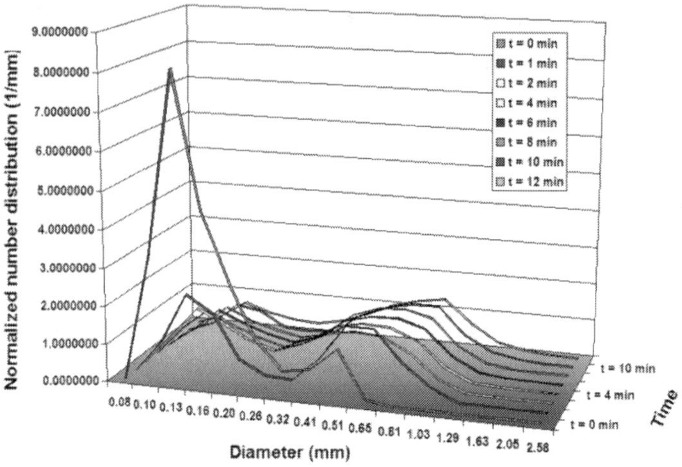

Figure 3 Effect of initial PSD (PSD3) shape on wet granulation at initial 13.6% v/w and 1 cp viscosity

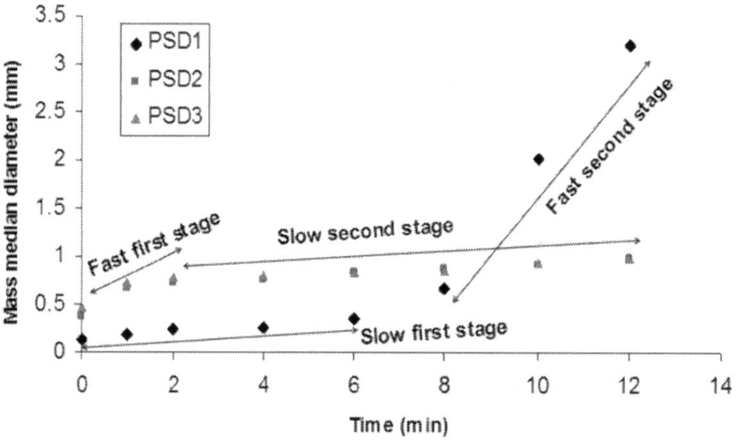

Figure 4 Mass median diameter evolution in wet granulation, at 13.6% v/w and 1cp viscosity

The gradient of each line in Figure 5 is shown in Table 1, and represents the granule growth rate, similar to that calculated by Mills et al. [5]. For wet granulation of powders with initial bimodal PSD, the growth rate increases with decrease in particle size, as indicated in Table 1. These results are consistent with Ennis' theory. [2]

Table 1 Effect of particle size on granule growth rate at 13.6% v/w and 1cp viscosity

Mass media Diameter (mm)	PSD	Stage	Growth Rate (mm/min)	Ratio: fast/low
0.46	PSD3: bimodal	First: fast	0.160	8.0
		Second: slow	0.020	
0.37	PSD2: bimodal	First: fast	0.175	6.7
		Second: slow	0.026	
0.13	PSD1: unimodal	First: slow	0.038	16.7
		Second: fast	0.634	

3.2 *Effect of initial particle size on granule growth rate in wet granulation*

Figure 5 shows how mass median diameter increased over time for all three size ranges of powders (PSD1, PSD2, and PSD3), at 13.6% v/w and 1cp viscosity. The figure depicts how particle growth rate increases with a decrease in initial particle size, as indicated by the mass median diameter in Table 1. Through Fig. 5 and Table 1, we can see that fine powders with initial unimodal PSD exhibit a faster growth rate in the second stage of wet granulation than the growth rate observed in the first stage of wet granulation with medium-sized and coarse powders. This phenomenon occurs due to a mono-dispersed unimodal PSD becoming a poly-dispersed unimodal PSD over the first eight minutes, favoring agglomeration of particles. The unimodal PSD grows larger in size than the initial bimodal PSD after eight minutes, since it reaches a volume ratio of 150, which is higher than the 112 reached by the initial bimodal PSD. The initial unimodal PSD consists of fine particles with high surface area favoring agglomeration of particles.

3.3 *Effect of binder content on granule growth and the mechanism of wet granulation*

Figure 6 depict the effect of binder amount on the growth of the particles. A change in the binder amount can alter the mechanism and growth rate in the second stage of wet granulation. For powders with initial bimodal PSD, the mechanism changes from a particle agglomeration at 13.6% v/w of binder, to a particle breakage mechanism at 9.9 and 11.7% v/w of binder. For powder mixtures with initial unimodal PSD, the variation in binder amount can also change the number of stages from two (slow and fast) at 13.6% v/w of initial binder, to one for 9.9 and 11.7% v/w of initial binder, as observed in Figure 7.

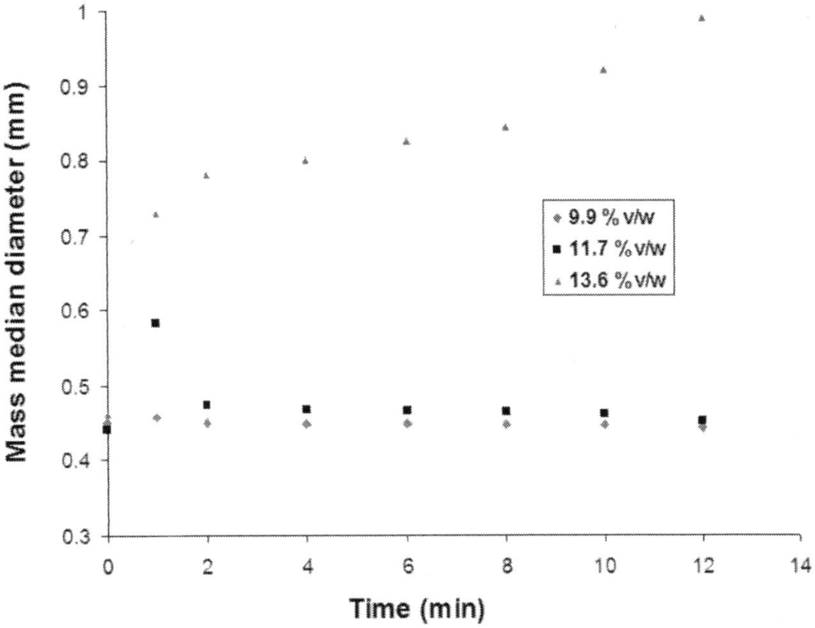

Figure 6 Effect of binder content (1 cp viscosity) on wet granulation of coarse size powders (PSD3) with bimodal PSD

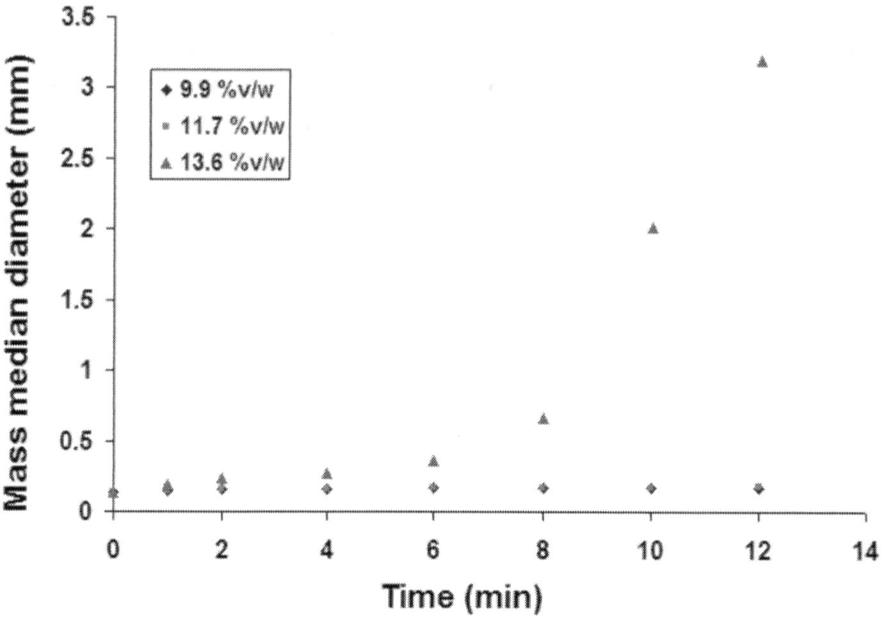

Figure 7 Effect of binder content (1 cp viscosity) on wet granulation of fine size powders (PSD1) with unimodal PSD: a) 9.9% v/w, b) 11.7% v/w and c) 13.6% v/w

3.4 Description of experimental data using the agglomeration kernels proposed in the literature

As indicated above, the particle coalescence that occurs during the wet granulation process is favored by the collision between small and large particles, as demonstrated by the higher granule growth rate for powders with bimodal PSD than for powders with unimodal PSD. Table 2 shows the three size-dependent kernels published in the literature used to predict our experimental results relative to the stage of slow granule growth. These experimental results were obtained for wet granulation of powders with both unimodal and bimodal PSD. The coalescence kernels were divided into three types: 1) kernels that favor growth rate by collision between small and large particles [10]; 2) kernels that favor growth rate by collision between particles of similar size (Golovin variation); and 3) kernels without preferential effect on growth rate by collision of particles of any size. [13]

Table 2 also shows the residuals obtained for each kernel by fitting the parameters with experimental data. The parameter estimation was performed using Parsival ®, a commercial software designed to solve a large class of integro-differential equations using the discretization technique. A detailed description is given by Wulkow et al. [14]. As anticipated, the Golovin variation and Smoluchoski kernels do not describe the granule agglomeration produced by collision between different-size particles, as observed in Figure 8, which reveals results only at 12 minutes. The Hounslow's kernel describes the experimental data obtained from wet granulation of fine and medium size powders, but it does not describe the first peak in the curve for wet granulation of coarse size powders at the endpoint of the experiment.

Table 2 Goodness of fit for kernels, relative to the experimental data

Kernel structure	*Reference*	*Average Residuals*		
		Unimodal PSD1	*Bimodal PSD2*	*Bimodal PSD3*
$(l+\varepsilon)^3$	Smoluchoski [13]	9.6083×10^{-2}	6.0243×10^{-1}	4.0120×10^{-1}
$(l\times\varepsilon)$	Golovin variation	8.7336×10^{-2}	6.0247×10^{-1}	4.0140×10^{-1}
$(l+\varepsilon)^2\times\sqrt{\left(\frac{1}{l^3}+\frac{1}{\varepsilon^3}\right)}$	Equipartition of kinetic energy. Boerefijn and Hounslow [10]	8.7284×10^{-2}	1.6876×10^{-1}	1.7822×10^{-1}

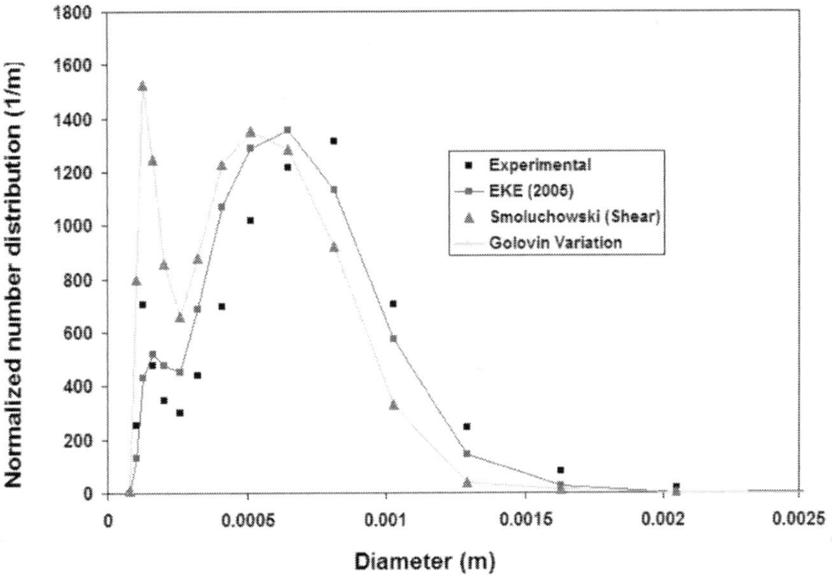

Figure 5 Comparison of simulations of coalescence kernels published in the literature versus the experimental data (bimodal PSD3)

CONCLUSIONS

Initial PSD shape, whether unimodal or bimodal, has a strong effect upon granule growth rate and the mechanism by which wet granulation occurs. A sequential mechanism consisting of two stages was observed during the wet granulation of powders with bimodal PSD, for three binder contents. The first stage of rapid granule growth is controlled by the binder amount and by the high probability of coalescence resulting because of collision of small and large particles. The second stage exhibits slow agglomeration when binder content is 13.6% v/w, and slow breakage when binder content is 9.9 or 11.7% v/w.

Wet granulation of pharmaceutical powders with initial unimodal PSD, at 9.9 and 11.7% v/w binder content, exhibited slow granule growth rate in two stages, stages which are less pronounced than the two stages for powders with bimodal PSD. In contrast, wet granulation of pharmaceutical powders with initial unimodal PSD at 13.6% v/w binder content exhibited two pronounced stages, due to the enlargement of particles that occurred during the first stage (the first six minutes) of wet granulation, which transforms the distribution shape from a mono-dispersed, unimodal PSD to a poly-dispersed, unimodal PSD. A volume ratio between small and large particles of approximately 150 was necessary to obtain fast granule growth. The experimental results were best predicted by a population balance equation using a coalescence kernel that favors growth rate by collision between small and large particles.

In general, the binder content, PSD shape, and interaction between particle size and binder content were factors with a strong effect upon granule growth rate and the mechanism of wet granulation; whereas binder viscosity and initial particle size affected the granule growth rate moderately.

REFERENCES

1. Cape C. 1980. Particle size enlargement. Handbook of Powder Technology, 1; Elsevier: Amsterdam.
2. Ennis B, Tardos G, Pfeffer R. 1991. A microlevel-based characterization of granulation phenomena. Powder Technology 65: 257-272.
3. Ennis B, Tardos G, Pfeffer R. 1990. The influence of viscosity on the strength of an axially strained pendular liquid bridge. Chemical Engineering Science 45: 3071-3088.
4. Mackaplow M, Rosen L, Michaels J. 2000. Effect of primary particle size on granule growth and endpoint determination in high-shear wet granulation. Powder Technology 108: 32-45.
5. Mill P, Seville J, Knight P, Adams M. 2000. The effect of binder viscosity on particle agglomeration in a low shear mixer/agglomerator. Powder Technology 113: 140-147.
6. Adetayo A, Litster J, Pratsinis S, Ennis B. 1995. Population balance modelling of drum granulation of materials with wide size distribution. Powder Technology 82: 37-49.
7. Badawy S, Hussain M. 2004. Effect of starting particle size on its agglomeration behavior in high shear wet granulation. AAPS PharmSciTech, 5 (3) Article 38.
8. Tardos G, Khan M, Mort P. 1997. Critical parameters and limiting conditions in binder granulation of fine powders. Powder Technology 94: 245-258.
9. Hounslow M, Pearson J, Instone T. 2001. Tracer studies of high-shear granulation: II. Population balance modeling. AIChe Journal 47 (9): 1984-1998.
10. Boerefijn R, Hounslow M. 2005. Studies of fluid bed granulation in an industrial R&D context. Chemical Engineering Science 60: 3879-3890.
11. Realpe A, Velázquez C. 2006. Pattern recognition for characterization of pharmaceutical powders. Powder Technology 169: 108-113.
12. Velázquez C, Realpe A, Obregón L. 2006. Image Analysis for Powder Concentration Estimation and Particle Shape and Size. Journal of Process Analytical Technology 3 (5): 20-24.
13. Smoluchowski M. 1917. Mathematical theory of the kinetics of the coagulation of colloidal solutions. Z. Phys. Chem. 92: 129-168.
14. Wulkow M, Gerstlauer A, Nieken U. 2001. Modeling and simulation of crystallization processes using Parsival. Chemical Engineering Science 56: 2575–2588.

Model Predictive Control of Continuous Drum Granulation of Limestone

Thomas Glaser[1,2], Constantijn F. W. Sanders[1], Fu. Y. Wang[3], Ian T. Cameron[3], James D. Lister[3,5], Jonathan M.-H. Poon[4], Rohit Ramachandran[4], Charles D. Immanuel[4], Francis J. Doyle III[1]
(1) University of California, Santa Barbara, CA, USA
(2) Universität Stuttgart, Stuttgart, Germany
(3) University of Queensland, St. Lucia, Australia
(4) Imperial College London, London, UK
(5) Purdue University, West Lafayette, IN, USA

Introduction

Applications of control to granulation processes in industry are rare. This paper details a methodology for a model predictive controller (MPC) that can be applied in a straightforward manner to an industrial granulation plant. Granulation is a complex process whereby fine particles agglomerate together due to adhesive forces. As Iveson et al. described in a review paper [3], the understanding of the fundamental processes that determine granulation behavior and product properties have increased in recent years. This knowledge can be used during process design as well as for choosing the right formulation and operating conditions. Furthermore it can be used to improve process control via closed-loop strategies. Wang et al. [1] showed that a simplified and discretisized one-dimensional population balance model (1D-PBM) is suitable for optimal control studies on a continuous drum granulator. In our work we will investigate the same continuous drum granulation process with the focus of deriving a model predictive controller for closed-loop control. Previously Sanders et al. [2] showed the benefits of MPC compared to a standard PID controller. An actual granulation pilot plant, which is located at the University of Queensland, Brisbane, Australia, was used for step response tests in order to analyze the dynamics of the granulation process. In particular the process of granulating agricultural limestone (Limestone, Aglime, Landmark, Australia) with water and 2.5% polyvinyl alcohol (PVA, Elvanol T66, Du Pont, USA) binder mix has been under investigation. The aim of this work is to design an MPC in order to control the process. With this aim in focus, three steps have been applied.

1. Step change experiments have been undertaken to gain input-output data sets which contain the plant dynamics.

2. A PBM developed by Wang et al. [1] has been fitted to the measurement data and controllability of the process was analyzed. Furthermore this yields a simulation test-bed to test the performance of the derived model predictive controller without the need to run labor and capital intensive experiments at the pilot plant.

3. System identification techniques have been used to derive a linear time invariant (LTI) model which was used as a prediction model for the model predictive controller.

Simulations show that satisfactory MPC tunings can be found for regulatory control. The flexibility of process constraint handling, the intuitive tuning of MPC, and the capability to handle multiple input and multiple output (MIMO) systems make MPC the first choice for closed-loop control of granulation.

Setup

The schematic of the pilot plant can be seen in Figure 1. The plant consists of a horizontal drum granulator (RD-101) with three inflow streams, the fresh solid (S2), binder (S3a & S3b) and recycle stream (S10). The outflow of the granulator is analyzed by an OptiSizer (VC-101), which determines the size

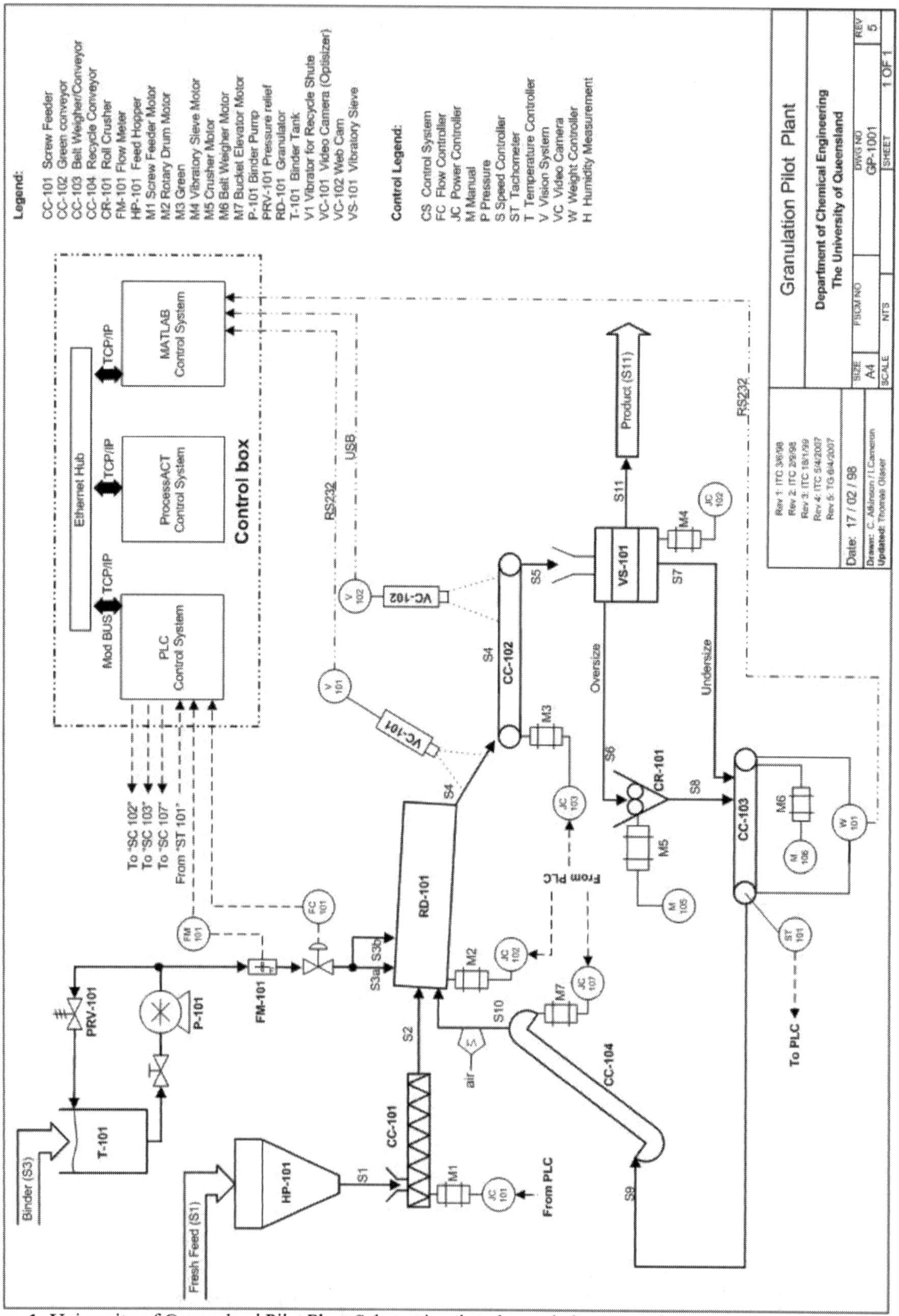

Figure 1: University of Queensland Pilot Plant Schematic; closed recycle loop granulation circuit with binder-/solid-supply, sieve, crusher and sensor-/actuator-interfacing.

distribution of the granules by image analysis, before they enter a sieve (VS-101), which separates the granules into over- (S6), under- (S7) and product-size (S11). The over-size gets crushed (CR-101) and forms together with the under-size (S7 + S8) the recycle stream (S9). If the plant is operated in closed-loop mode, this recycle stream is fed back into the granulator (S10) whereas it is discarded in open-loop operation. The recycle stream flow-rate and moisture content are measured (W-101) before it enters the recycle conveyor (CC-104). All process sensors and actuators are interfaced to a control box.
The structure of the 1D-PBM which has been used for the numerical simulations is depicted in Figure 10 and is based on the Wang model [1]. The model keeps track of the particle size distribution (PSD) as well as the average moisture content in the granules.

Data Analysis

The data presented here was gathered from a step test at the continuous granulation pilot plant. The nominal sampling time is 15 seconds. This sampling interval is too small to get a high enough number of particles for a representative PSD, as seen in Figure 2. This results in a strong variation between two data points during a complete run.

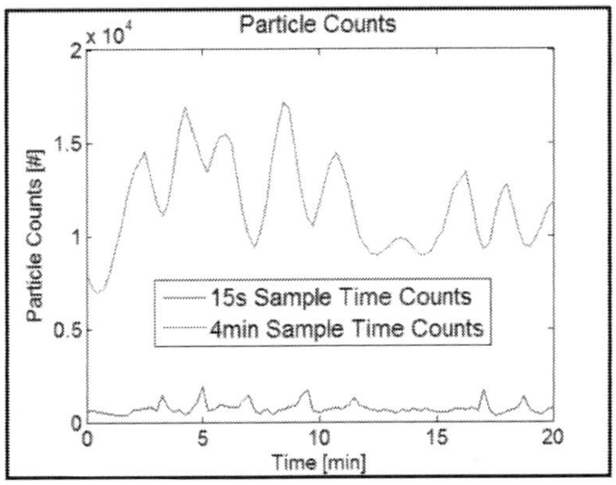

Figure 2: Particle counts per sampling

To improve the data quality, several filtering techniques have been applied. A moving average filter was used to change the sampling time from 15 seconds to 4 minutes. This results in much higher particle counts, as seen in Figure 2, and therefore in a smoother data set. However, there are still outliers in the data set. Subsequently a Hampel filter [6] was used. The Hampel filter is similar to the aforementioned moving average filter but much more effective when it comes to data cleaning applications. The Hampel filter was applied to each size class of the PSD along the time axis. The final result of this filtering can be seen in Figures 3 & 4, which shows the PSD response to a step in the solid feed-rate at $t = 0$.
In the following, this data is used to fit the Wang model parameters to the experimental data using unconstrained nonlinear optimization. Different optimization algorithms and cost functions have been used for the parameter fitting. The best result was achieved with the cost-function

$$J = \min_{a_0, a_1, a_2, a_3, n_1, n_2, n_3, k, g_m, \alpha} \sum_{i=1}^{2} \left| SS^i_{Mod} - SS^i_{Ex} \right|. \qquad (1)$$

Where SS^i_{Mod} is the steady state of the model and SS^i_{Ex} is the steady state of the experiment. $i = 1$ is before and $i = 2$ is after the step. The comparison of the simulated PSD with the PSD from the experiment can be seen in Figures 5 & 6.

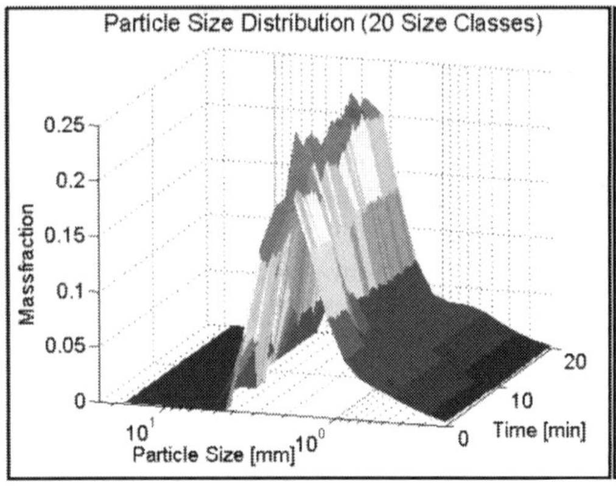

Figure 3: PSD shift after step change in fresh solid feed from $\dot{m}_{Solid}$ = 151 kg/h to $\dot{m}_{Solid}$ = 141 kg/h at t=0 and constant binder feed-rate at $\dot{m}_{Binder}$= 13 kg/h.

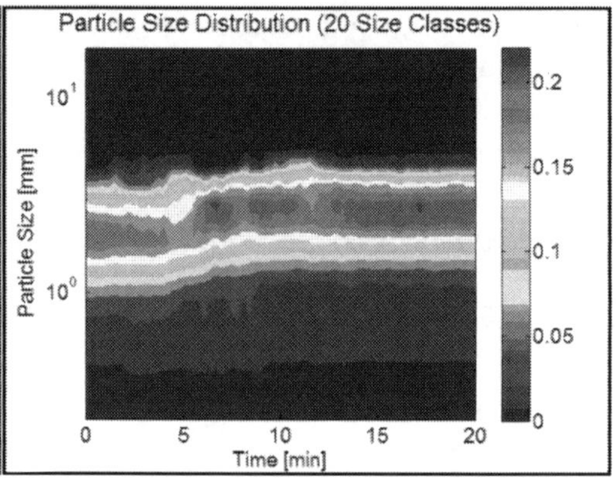

Figure 4: PSD shift contour plot after step change in fresh solid feed from $\dot{m}_{Solid}$ = 151 kg/h to $\dot{m}_{Solid}$ = 141 kg/h at t=0 and constant binder feed-rate at $\dot{m}_{Binder}$= 13 kg/h.

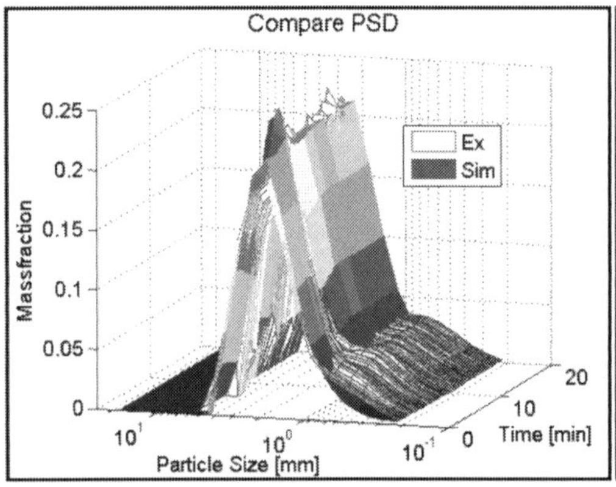

Figure 5: Comparison of simulated and measured PSD shift after step change in fresh solid feed from $\dot{m}_{Solid}$ = 151 kg/h to $\dot{m}_{Solid}$ = 141 kg/h at t=0 and constant binder feed-rate at $\dot{m}_{Binder}$= 13 kg/h.

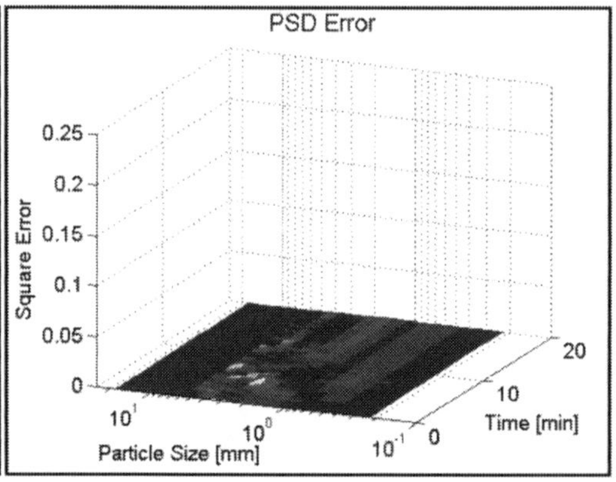

Figure 6: Error between simulated and measured PSD shift after step change in fresh solid feed from $\dot{m}_{Solid}$ = 151 kg/h to $\dot{m}_{Solid}$ = 141 kg/h at t=0 and constant binder feed-rate at $\dot{m}_{Binder}$= 13 kg/h.

Controllability

Before implementing the MPC design, controllability studies on the simulator were performed. The d_{50} and d_{84}/d_{16} are common values denoted to describe the shape of PSDs. The d_{50} describes the mean of the PSD and the d_{84}/d_{16} describes the distribution width. Both values can be derived from the cumulative mass of the PSD, as seen in Figure 7.

In Figure 8, an example of an input to output gain direction map is shown. It can be seen that there is no gain from the inputs to the d_{84}/d_{16} output. This means that the d_{84}/d_{16} is not controllable through these inputs and therefore not usable as a controlled variable (CV) since it is not affected by the manipulated variables (MVs). The same phenomenon can be seen for all the other available MVs. It should be mentioned that the available experimental data supports this simulation results. Consequently we dropped the d_{84}/d_{16} as a possible CV and only used the d_{50} in subsequent studies.

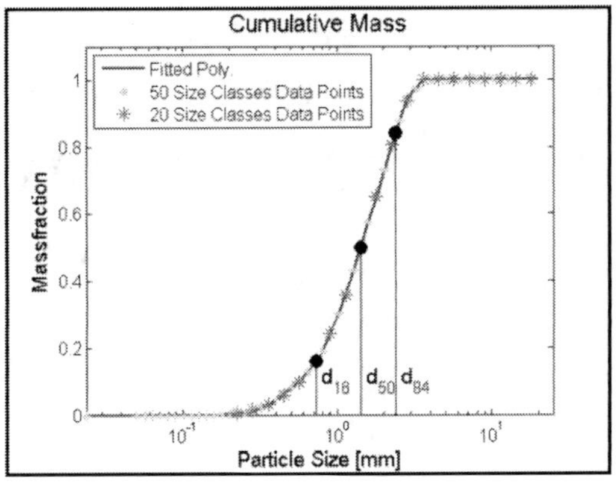

Figure 7: Deriving the mass based d_{16}, d_{50} & d_{84} from the cumulative mass. x% of the granule mass is $\leq$ the value given by d_x.

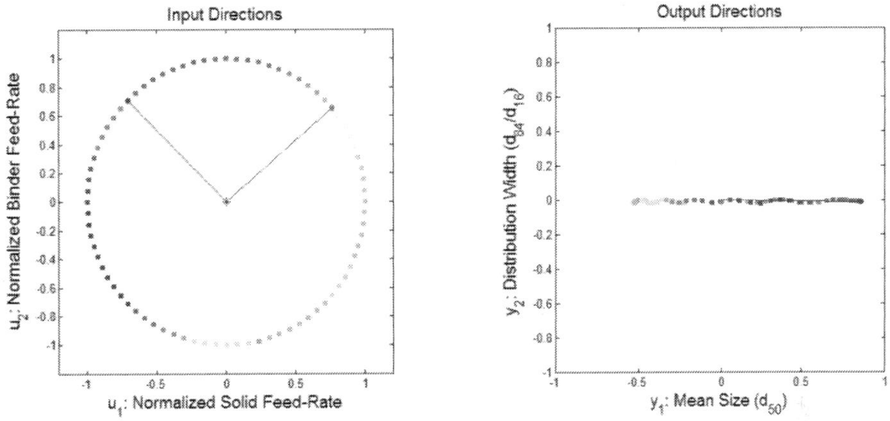

Figure 8: Gain-directions for different input-directions to the d_{50} and d_{84}/d_{16} output. The blue line marks the maximum gain and the red line the minimum gain. The input unit-circle flattens out to a horizontal line (loss of one dimension). This means that the d_{84}/d_{16} is not controllable by this inputs.

The MPC Model

The experimental step response of the d_{50} around a chosen steady state (SS) mirrors the shape of a first order transfer function (TF) with delay (t_d) (Figure 9) even though the process is nonlinear. Consequently a simple first order TF model ($PT_{t_d}T_1$) has been fitted to the experimental data.

$$G_{Plant} = e^{-t_d s}\frac{-k}{\tau s + 1} \qquad (2)$$

This has been done by deriving the three LTI model parameters: dc gain $k = 0.7692$, the time constant $\tau = 200$ and the delay $t_d = 213$ seconds from the step response data. The final model is given by

$$G_{Plant} = e^{-213s}\frac{-0.7692}{200s+1} \qquad (3)$$

The comparison between measurement data, LTI model and PBM-simulator can be seen in Figure 9.

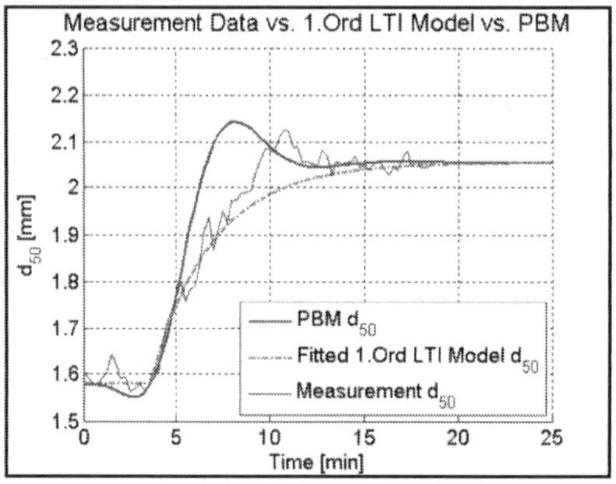

Figure 9: Comparison of plant-, simulator- and LTI model- d_{50} step response after step change in fresh solid feed from $\dot{m}_{Solid}$ = 151kg/h to $\dot{m}_{Solid}$ = 141 kg/h at t=0.

SISO MPC

For the first application of MPC, a simple single-input single-output (SISO) architecture was considered. Since the moisture to solid ratio highly [4, 5] influences the particle size distribution, the powder feed-rate has been chosen to be the MV of the single input system. The output particle size distribution is described with the mass based mean (d_{50}) particle size, resulting in a SISO system. The simulation setup which has been used for the validation of the different MPC tunings can be seen in Figure 10. It consists of the granulation plant simulator which is based on the Wang Model, and an additional MPC component. The recycle loop is not closed in this setup. The detailed structure of the MPC block shown in Figure 10 can be seen in Figure 11. The derived LTI model (3) is used as the prediction model. Figure 12 & 13 show a MPC tuning with a good compromise between speed and overshoot. The simulation results indicate that MPC is suitable for the process and applicable to the pilot plant.

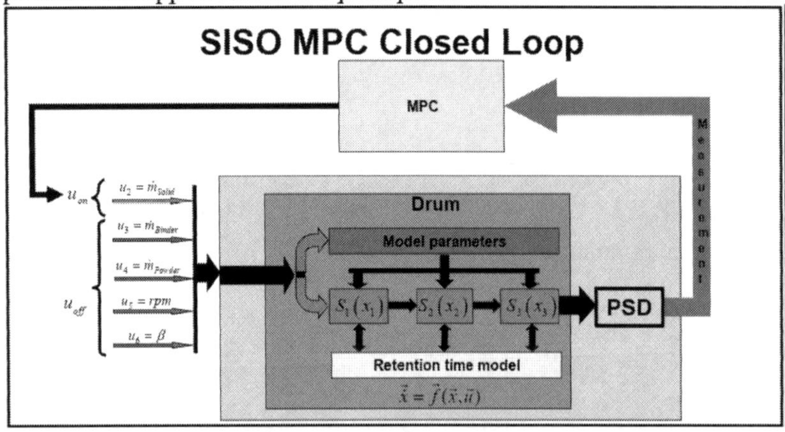

Figure 10: SISO MPC simulation setup; recycle loop is not closed, MV is solid feed-rate, CV is d_{50}.

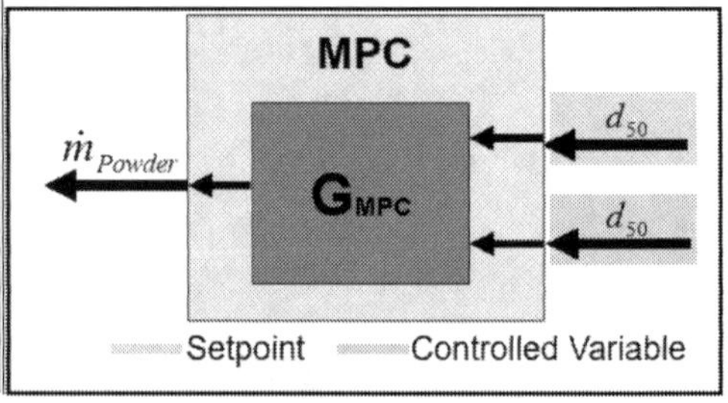

Figure 11: Detailed MPC structure for SISO setup.

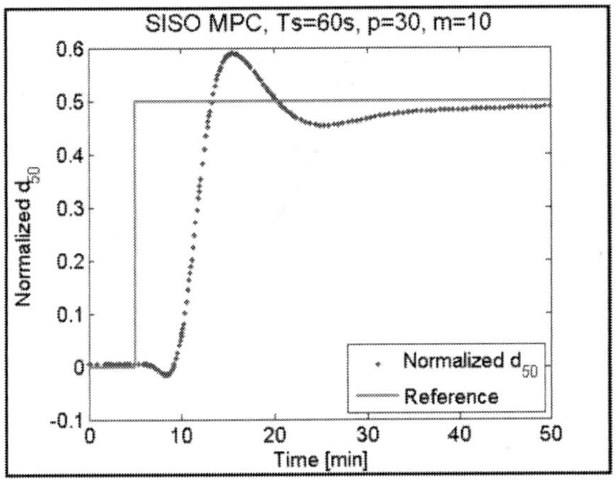

Figure 12: Normalized d_{50} step response after step change in d_{50} reference (red line). Sampling time Ts=60s, prediction horizon p=30 and control horizon m=10.

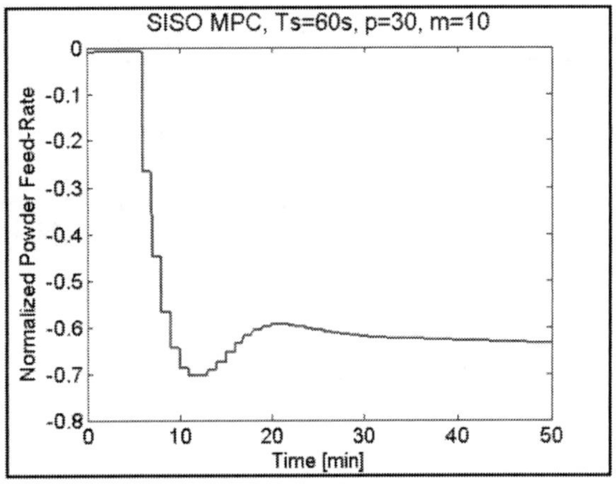

Figure 13: Binder feed-rate after step change in d_{50} reference. Sampling time Ts=60s, prediction horizon p=30 and control horizon m=10.

MIMO MPC

In order to leverage the full capacity of MPC, in particular the ability to handle multiple-input multiple-output (MIMO) systems, we next studied a more complex control structure. For this purpose we changed the MV from solid feed-rate to moisture to solid ratio. This is reasonable since it will enable us to operate the granulator closer to its optimal throughput by keeping the total solid feed-rate (fresh solid and recycle feed-rate) constant.

In this setup, shown in Figure 14, the recycle loop is closed in order to feedback the recycle stream into the granulator. This means that we have an additional input stream into the granulator which has a varying flow-rate and varying moisture content. To be able to handle MIMO systems our prediction model needs to be extended. Since the recycle stream depends only on the state of the process and cannot be directly influenced by the controller it will be modeled as a measured disturbance.

$$G_{MPC} = \begin{bmatrix} G_{Plant} \\ G_{Disturbance} \end{bmatrix} \quad (4)$$

The recycle stream influences the process in the same way as the fresh solid feed. The only difference is the moisture content of the two streams. This means that we can use the same model structure and parameters for the disturbance as we used for the plant. We only have to adjust the time delay in order to incorporate the additional transport delay resulting from the recycle conveyor. The recycle conveyor delay is constant and can be adjusted due to the speed of the belt. A one minute delay is reasonable and will be used here. The disturbance model therefore becomes,

$$G_{Disturbance} = e^{-273s} \frac{0.7692}{200s+1} \quad (5)$$

Furthermore, as shown in Figure 15, a gain scheduling unit has been added to the setup. The gain scheduling adapts the gain of the disturbance model. This leads to an improved controller performance since the gain changes proportional with the amount of the recycle flow. Finally we have a MIMO control setup based on a multiple-input single-output (MISO) MPC with gain scheduling.

A simulation result of this setup can be seen in Figures 16, 17, 18 & 19. Here the process has been disturbed by reducing the recycle flow by 50 % at $t = 50\text{min}$. The controller counteracts this disturbance and brings the

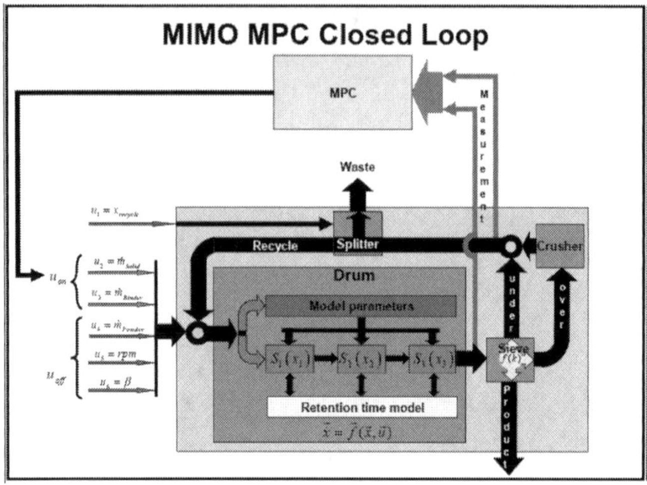

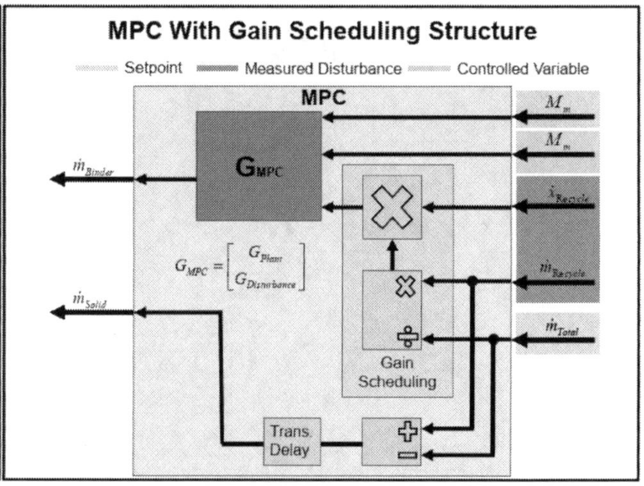

Figure 14: MIMO MPC simulation setup.

Figure 15: Detailed MIMO MPC structure with gain scheduling. $M_m = d_{50}$

process back to its set-point. By comparing these results with the previous results from the SISO setup, one might wonder about the slower dynamics. The slower dynamics are a result of the closed recycle loop that changes the process characteristics. The overall controller performance is still effective, see Figure 19, particularly if large disturbances are taken into account. As a comparison, the same simulation was run without controlling the binder spray-rate. All the other simulation parameters are the same as in the previous controlled case. The results can be seen in Figure 20 & 21. The process is unstable from the beginning and does not even need a disturbance to be moved away from its set-point, this results in almost no product output. In the real plant this setup leads to a condition of excess moisture content in the granulator. The result is a sticky paste in the granulator which forces the process to be shut down and cleaned.

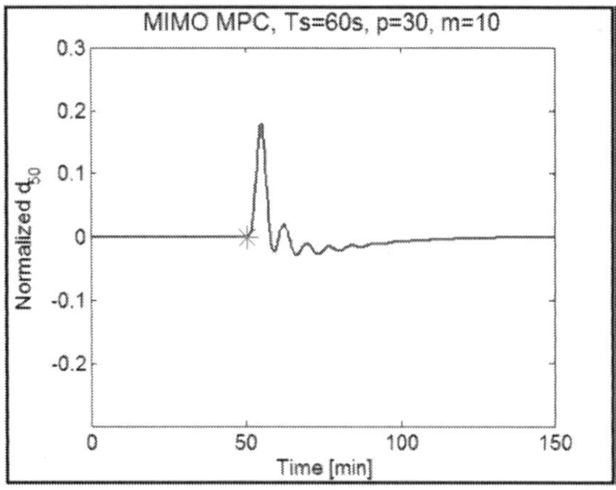

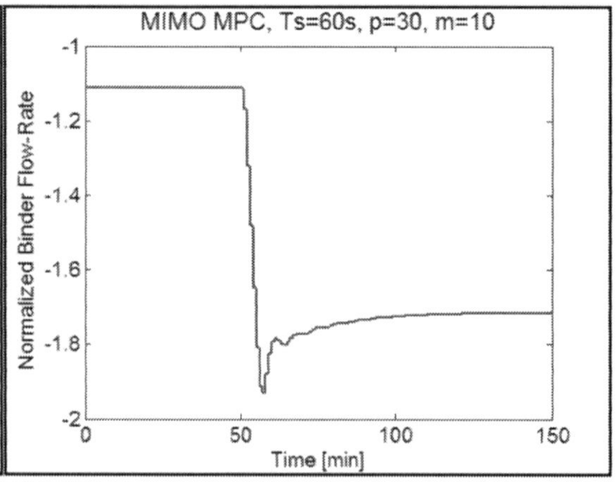

Figure 16: d_{50} after 50% Disturbance in recycle stream at t=50min. The red star marks the disturbance time stamp. Sampling time Ts=60s, prediction horizon p=30 and control horizon m=10.

Figure 17: Binder spray-rate after 50% disturbance in recycle stream at t=50min. Sampling time Ts=60s, prediction horizon p=30 and control horizon m=10.

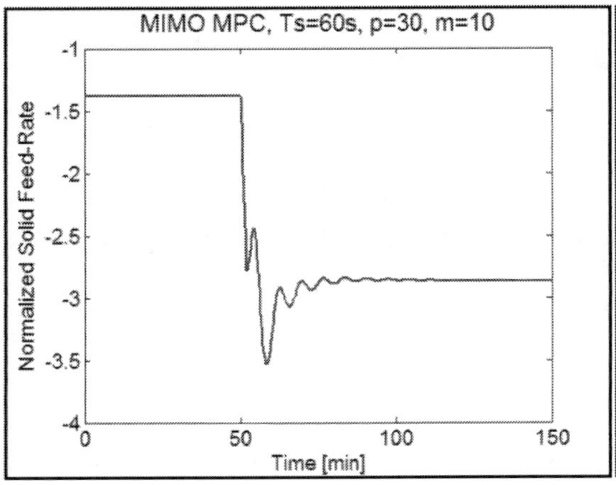

Figure 18: Solid feed-rate after 50% disturbance in recycle stream at t=50min. Sampling time Ts=60s, prediction horizon p=30 and control horizon m=10.

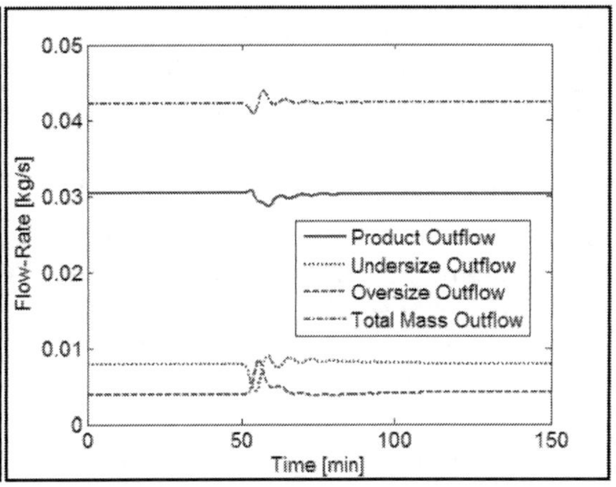

Figure 19: Granulator outflow, with MPC, separated in total, over-, under- and product-size. The product outflow hardly changes although there has been a disturbance in the recycle feed-rate at t=50min.

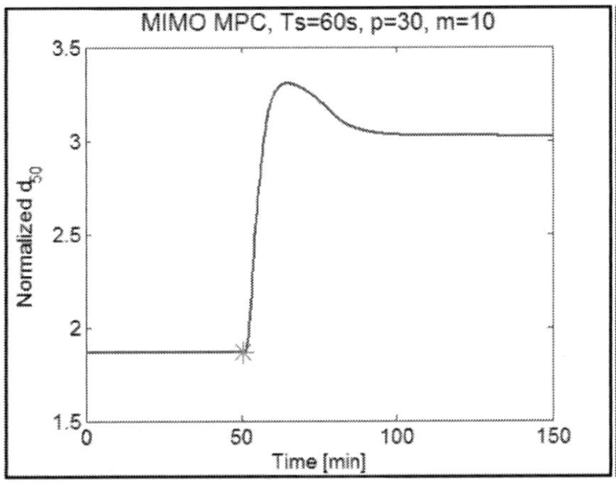

Figure 20: d_{50} Simulation without MPC controller. The process drifts away from its set-point and enters a region where the real plant cannot be operated. The red star marks the disturbance time stamp.

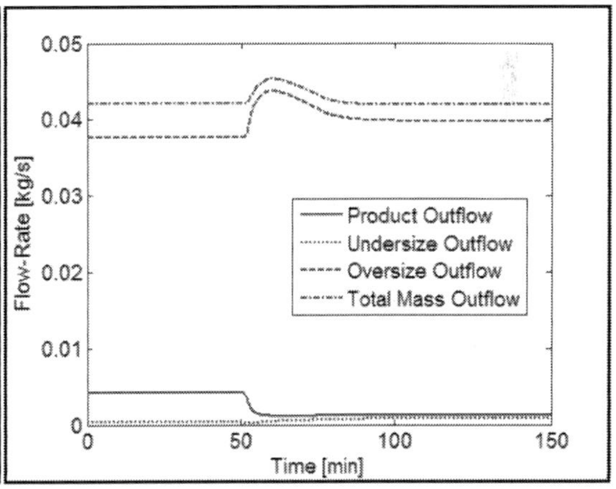

Figure 21: Granulator outflow, without MPC, separated in total, over-, under- and product-size. Almost no product output compared to the controlled case shown in Figure 19.

Conclusion

We have demonstrated that SISO and MIMO MPC is a suitable control strategy for granulation processes. Furthermore, we have found that MPC can improve the robustness of the process and, as a result, increase the efficiency. Instability and expensive shut down times of granulation plants due to incorrect plant operation can be avoided. In addition, the product to waste ratio can be increased by permanently operating the plant at its optimal set-point.

Acknowledgments

Special thanks to David Page, University of Queensland, St. Lucia, Australia, for insightful conversations pertaining to the experimental setup of the pilot plant. The authors also acknowledge financial support by the International Fine Particles Research Institute (IFPRI), from the Australian Research Council (Discovery Grant DP0345777) and from the Engineering and Physical Sciences Research Council (Grant EP/C511301/1).

References

1. F. Y. Wang, X. Y. Ge, N. Balliu, I. T. Cameron (2006), "Optimal control and operation of drum granulation processes," *Chemical Engineering Science 61*, pp. 257-267.

2. C. F. W. Sanders, M. J. Hounslow, and F. J. Doyle III (2006), "Model Predictive Control of Wet Granulation Using an Experimentally Validated Population Balance Model," in *AIChE Annual Meeting*, San Francisco.

3. S. M. Iveson, J. D. Lister, K. P. Hapgood, B. J. Ennis (2001), " Nucleation, growth and breakage phenomena in agitated wet granulation processes: a review," *Powder Technology 117,* pp. 3-39.

4. A. A. Adetayo, J. D. Lister, S. E. Pratsinis, B. J. Ennis (1995), "Population balance modeling of drum granulation of materials with wide size distribution," *Powder Technology 82,* pp. 37-49.

5. J. Zhang, J. D. Lister, F. Y. Wang, I. T. Cameron (2000), "Evaluation of control strategies for fertilizer granulation circuits using dynamic simulation," *Powder Technology 108,* pp.122-129.

6. R. K. Pearson (2002), "Outliers in Process Modeling and Identification," *IEEE Transactions on Control Systems Technology, Vol. 10, No. 1,* January, pp. 55-63.

277c Passive Acoustic Monitoring Of High-Shear Wet Granulation

Albert W. Alexander, AstraZeneca, 1800 Concord Pike, FFX FRC L138C, Wilmington, DE 19803

The control of high-shear wet granulation processes is one of the most difficult challenges in pharmaceutical processing. Granulation involves added a binder solution (often just water) to a powder mixture while undergoing high shear, and leads to many competing processes and constantly changing micro- and macro-environments within this highly dynamic system. Many different approaches have been attempted to model and control granulation, including power consumption/torque monitoring, NIR spectroscopy, and various attempts at physically monitoring granule growth (i.e. high-speed video microscopy, FBRM). Each of these methods has specific advantages and drawbacks but none has been shown to be clearly superior across variations both in equipment types and formulation variables.

Another method that has been used with some success is passive acoustic monitoring. One advantage to this approach in comparison to many of those mentioned previously is that the sensor attaches to the outside of the granulator, which eliminates problems that may arise from probe fouling. The system works by collecting sound data as the granulation proceeds, the theory being that as the granulator contents change over time (take up water/binder solution, grow in size, agglomerate, etc.) that the sound characteristics (both in terms of the intensity/loudness and the frequency distribution, or pitch) will alter in ways that reflect changes in the material within the granulation bowl. Sound capture covers frequencies that range from human hearing well into the ultrasonic region. Ongoing work has shown good reproducibility in terms of the evolving sound profile for a placebo mixture run in granulators ranging in size from 1 liter up to 10 liters and comparable responses that indicate good scale-up possibilities. Furthermore, granulations have been stopped and the contents analyzed to correlate changes both in size distribution and granule characteristics (porosity, tensile strength, etc.) to changes in the acoustic profile. The use of acoustics gives another means to investigate the change ongoing within high-shear granulation with an aim towards eventually developing more robust means of characterization and control.

277d Experimental Study Of Wurster Granulation For Pharmaceutically-Relevant Excipient Blends

Christopher Mancinelli[1], Pavol Rajniak[1], Frantisek Stepanek[2], Leon Farber[1], Brian Hill[1], and Rey Chern[1]. (1) Merck & Co, Inc., WP78-110, PO Box 4, West Point, PA 19486, (2) Department of Chemical Engineering, Imperial College London, South Kensington Campus, London, SW7 2AZ, United Kingdom

The effect of the physicochemical properties of an aqueous hydroxypropylcellulose (HPC) binder solution and different pharmaceutical excipients on the agglomeration kinetics and granule properties was investigated for combinations of common pharmaceutical excipients. Pure component and two-component blends of microcrystalline cellulose, lactose monohydrate, dicalcium phosphate and mannitol were granulated in a pilot plant Wurster granulator using a 15% HPC solution. Experimental conditions and solution concentration were chosen based on knowledge gained through a previous investigation of this agglomeration process [1]. The current work extends the previous study by encompassing a wider range of pharmaceutically relevant excipient combinations and with additional granule characterization beyond the scope of previous work. Granule growth kinetics were found to vary widely for the different blends. The properties of granules produced in pure component experiments correlated well with and were predictive of granule properties of the blends. The mechanical properties of compressed tablets, as measured in compaction simulation experiments, were inversely related to the granule growth rate. Attrition testing was also employed to determine the strength of granule blends. Physically-based criteria [2] were evaluated, which use the morphological properties of these excipients (size and surface roughness) together with physical properties of the binder solution for prediction of the coating vs. agglomeration regime at given flow conditions (collision velocity). This criteria based simulation was effective at the relative growth rates of blends.

[1] Rajniak, P., Mancinelli, C., Chern, R.T., Stepanek, F., Farber, L., Hill, B.T., Experimental study of wet granulation in fluidized bed: Impact of the binder properties on the granule morphology, International Journal of Pharmaceutics (2006), 334 (2007) 92-102.

[2] Ennis, B.J., Tardos, G.I., Pfeffer, R., 1991. A microlevel-based characterization of granulation phenomena. Powder Technol. 65, 257-272.

277e Suspension and Dispersion of Particles in Gels

Micaela Caramellino and Piero M. Armenante. Otto H. York Dept. of Chemical Engineering, New Jersey Institute of Technology, 323 King Blvd., Newark, NJ 07102

Solid–liquid multiphase mixing is of fundamental importance for the pharmaceutical industry since the achievement of the appropriate solid dispersion level influences both the synthesis of the active pharmaceutical ingredients and the manufacturing process of final dosage forms. The achievement of a spatially and temporally uniform dispersion of the active agent is a frequent goal of many pharmaceutical operations involving dispersed solids in liquids.

In this work the manufacturing process required to fabricate a film strip incorporating particles of an active material has been analyzed. This involves, at first, the preparation of a dispersion of the drug particles in an aqueous medium and, secondly, the homogenization of the latter with a gelling agent to obtain a suspension of the active.In addition, the film strip manufacturing process necessitates the drug particles to have an appropriate size compatible with the film thickness. The particle size distribution needs also to be narrow. Therefore a key aspect that has been addressed is the particle de-agglomeration process. The operating conditions that generate the necessary particle size distribution have been studied.

Vessels of different volumes were considered: 20, 11 and 5 L. All the tanks were baffled, cylindrical and flat bottomed. The same geometrical proportions have been used for the different tanks examined: the baffle width-to-tank diameter ratio of 0.1 and the ratio of liquid height, H, to tank diameter, T, of 1 (H=T). The system was provided with a centrally mounted pitched-blade turbine (PBT) used to suspend the solids, and a high speed homogenizer that breaks up the agglomerates formed by primary particles.

A water-solid system was considered at first in order to characterize its behavior for different agitation speeds and solid concentrations. Solids of the same size and density as the typical active ingredients to be eventually used in the film strips were used as the dispersed phase. For the first experiments, surrogate silica particles with primary particle size of 16 nm and initial agglomerate size of ~30 microns were selected.

The effect of high-shear agitation on particle de-agglomeration and the hydrodynamic forces generated in the solid-liquid system have been characterized in terms of: final agglomerates size, time to reach the asymptotic value of the mean agglomerate size, and quality of the particle size distribution.

The results show that the disruptive forces generated by the homogenizer are strong enough to break up the agglomerates and to reduce their size to the submicron range. Moreover, the ratio between the total tank volume and the volume effectively affected by the homogenizer's action influences the time needed by the system to reach an asymptotic value for the mean agglomerates diameter.

307a Opportunities, Challenges, And Approaches For Modeling In The Pharmaceutical Industry

Prashant B. Kokitkar, Engineering Technology Center, Eli Lilly and Company, Lilly Corporate Center, Indianapolis, IN 46285

Pharmaceutical industry has traditionally lagged behind the chemical industry in terms of the use of modeling in process development, control, and optimization. With increased use of process analytical technology (PAT) and increased emphasis on process engineering and optimization, the role of modeling has become more prominent in pharmaceutical process development. The data obtained with on-line or at-line PAT instruments often requires a logical model to interpret it and acquire knowledge from it. This talk will present some of the challenges and opportunities present in pharmaceutical modeling and will provide examples of modeling approaches available.

MODELLING OF BREAKAGE AND DISSOLUTION OF CRYSTALS WITH VARYING PARTICLE SIZE DISTRIBUTION WITHIN THE QUALITY BY DESIGN FRAMEWORK

P. Rajniak, K. Gallagher, R. Chern, C. Ikeda, J. Moser
Merck & Co., Inc., West Point, PA

Abstract

An important aspect of the Quality by Design (QbD) framework for pharmaceutical product development and manufacture is to develop a full understanding of how material attributes and process parameters affect critical quality attributes, which in turn determine product performance. Theory and modeling based on first principles approach can be a very effective tool to accomplish this task.

In this contribution, we attempt to demonstrate the use of mechanism-based mathematical modeling in developing a fundamental understanding of both the effect of the crystal size variability of an active pharmaceutical ingredient (API) and the impact of different encapsulation process conditions on dissolution rate of a direct encapsulated drug product.

During development of a drug product capsules, dissolution rate was shown to correlate with the particle size distribution (PSD) of the API crystals. A dissolution model was developed for predicting API dissolution profiles in agitated vessels from a known PSD of API. The model assumes that dissolution can be internal or external mass transfer limited, particles are spherical (equivalent spheres), and dissolution occurs under sink or non-sink conditions. The model was employed for *a priori* prediction of dissolution of API and validated against dissolution profiles of different batches.

Since some particle size reduction (breakage) was observed during encapsulation, a population balance (PB) model [1] of the crystal breakage was developed to predict the extent of particle size reduction. The PB model assumes a size dependent breakage kernel and a uniform breakage distribution function.

A procedure was developed for merging the breakage model with the dissolution model. A simple "trial and error" procedure was subsequently used to determine the breakage parameters by fitting the combined model to experimental dissolution profiles of capsule plugs of different densities.

A quantitative correlation between the plug density and the breakage rate constant was developed to describe the higher extent of crystal breakage at more aggressive (higher tamping) encapsulation conditions. The correlation, together with the combined model was used for prediction of dissolution of capsules prepared at various encapsulation tamping conditions from API of arbitrary PSDs.

As both the dissolution model and the breakage model are based on first principles, it was possible to use the models for direct evaluation of the model parameters by fitting to existing experimental data without need of any additional statistically designed experiments (DOE). It saved a lot of additional experimental work. Another advantage of models based on first principles is that they can be used also for prediction of the breakage and dissolution for interpolated and (to some extent) also extrapolated PSD and encapsulating conditions.

Theoretical Background

Mathematical model for dissolution of drug substance and powders with arbitrary PSD

The model simulates dissolution of drug substance based on the „shrinking core mechanism" without disintegration of dissolving particles. Dissolution of drug substance with arbitrary particle size distribution (PSD) can be modelled using extension of the monodispersed model [2] for arbitrary number of classes of the PSD:

General model for dissolution of particles

$$\frac{dC_{b,i}}{dt} = \frac{6k}{\rho_p d_{0,i}} \left(C_{all,i} - C_{b,i}\right)^{2/3} C_{all,i}^{1/3} \left(C_{sol} - \sum_{i=1}^{M} C_{b,i} \right) \qquad i = 1,2,\ldots,M \qquad (1)$$

Mathematical model for breakage of drug substance and powders with arbitrary PSD

The population balance model simulates particle size reduction of powders with arbitrary PSD. The population balance model is extension of classical model [2] and assumes a size dependent breakage kernel and a practical breakage distribution function.

Population balance model for size reduction

- mass balances for different size classes $\frac{dx_i}{dt} = -r_i + \sum_{j=i+1}^{M} b_{ij} r_j \quad i = 1,\ldots,MN$ (2)

- definition of breakage rate $r_i = S_i x_i = S(D_i)^{\gamma} x_i \quad i = 1,\ldots, M$ (3)

- breakage distribution function $b_{12} = 1 \quad b_{ij} = \frac{1}{\eta^{j-i-1}} - \frac{1}{\eta^{j-i}} \quad i = 1,\ldots, j-1$ (4)

$j = 3,\ldots, M$

Validation of mathematical models

Experimental particle size distributions (PSDs) of coarse (un-milled) Lot #1 and fine (wet-milled) Lot #2 of the same API are shown in **Figure 1** together with computed PSD of broken coarse drug substance and computed PSD of the 70:30 blend of broken coarse : fine drug substance.

A priori (without adjustable parameters) prediction of the dissolution profiles for both, coarse and fine drug substance is compared with experimental profiles in **Figure 2.** Theoretical prediction of dissolution for the 70:30 blend of broken coarse : fine drug substance (PSD is shown in Figure 1) is also shown in **Figure 2.**

Figure 1 (left) Experimental PSDs of different lots of API crystals and computed PSDs after breakage and blending. **Figure 2** (right). A priori simulations of dissolution of different lots of API crystals using External Diffusion Model

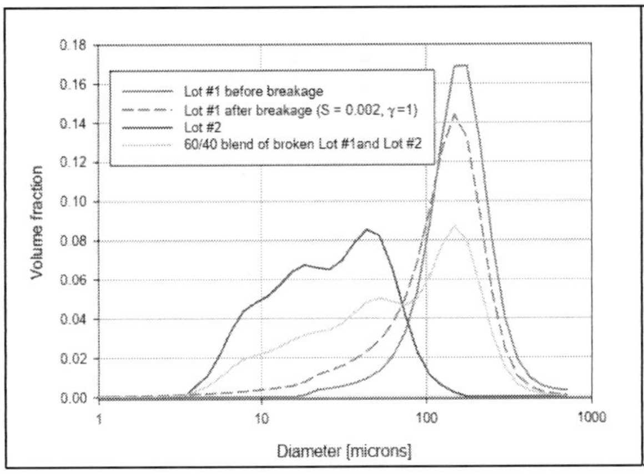

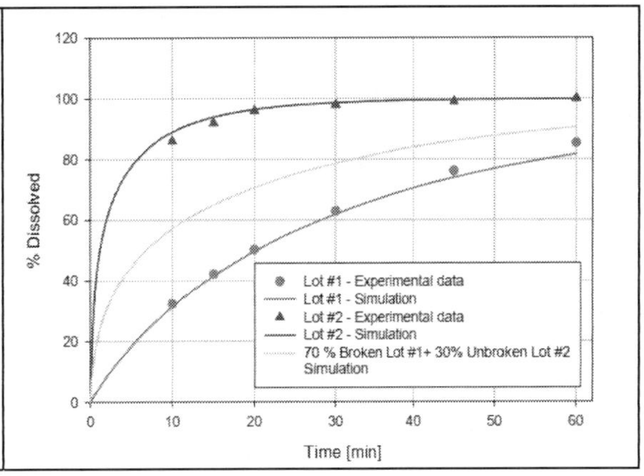

Prediction of capsule dissolution and optimization of model parameters

A procedure was developed for simulation of breakage of coarse API crystals, blending of the broken drug substance with the fine (wet-milled) drug substance and dissolution of the crystal blend from final drug product capsules. Note that it was shown that the capsule shell and excipients in the final formulation have negligible effect on drug product dissolution. The procedure combines the breakage model and the dissolution model. A simple "trial and error" procedure was employed for optimization of the breakage parameters by fitting the combined model predictions to experimental dissolution profiles of capsule plugs of different densities. The procedure is following:

1. Assuming the breakage constant S and breakage exponent γ
2. Simulation of breakage of the coarse drug substance using the breakage model
3. Simulation of blending of (70%) broken coarse and (30%) fine drug substance
4. Simulation of dissolution of the blend (capsule content)
5. Calculation of sum of squares of differences between computed and experimental dissolution profile
6. Repeated calculations for new values of the breakage constant S and breakage exponent γ

Comparison of experimental and simulated dissolution profiles for capsules with different density of the capsule plugs is shown in **Figure 3.** The optimum value of the breakage exponent was γ =1 and the optimum values of the breakage constant for different plug densities are illustrated in **Figure 4**. A correlation between the plug density and the breakage rate constant reflects higher extent of the drug substance particle breakage at more aggressive encapsulation condition. However, using the final formulation blend of 70% coarse with 30% fine API, the effect of breakage on dissolution was relatively small across a wide range of encapsulation tamping conditions.

Figure 3 (left). Prediction of dissolution of 70:30 capsules using combined Breakage-Blending-Dissolution Model with exponent in breakage model γ = 1

Figure 4 (right). Optimized values of the Breakage rate constant at different plug densities and breakage exponent γ = 1

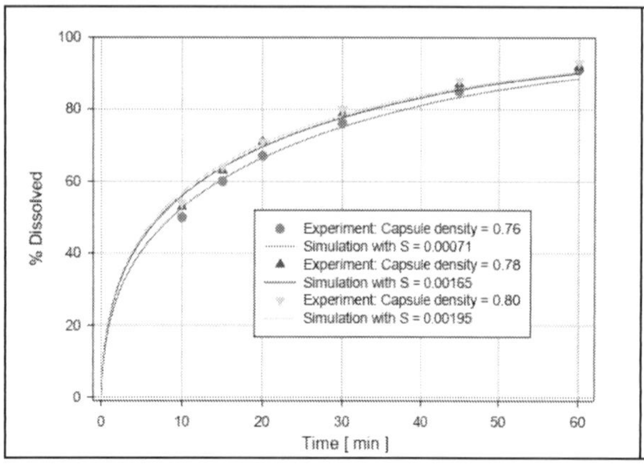

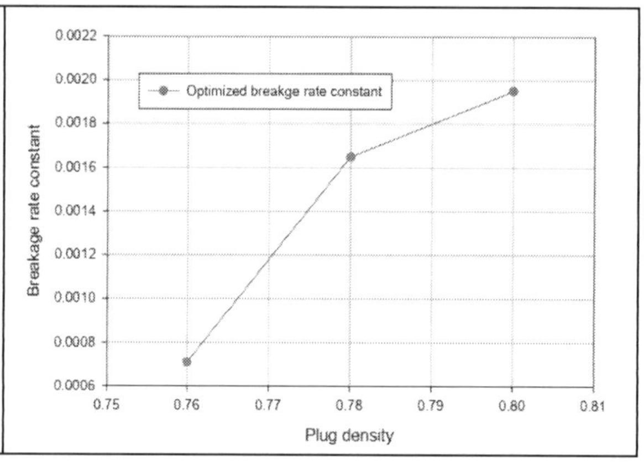

Conclusions

1. Combined Breakage-Blending-Dissolution mathematical model is a useful tool for analysis and prediction of the breakage and dissolution of the API with arbitrary PSDs.
2. The correlation between the capsule plug density and the breakage constant together with the combined model can be used for prediction of dissolution of capsules prepared at various encapsulation conditions from different blends of coarse and fine drug substance with arbitrary PSDs.

Notation

M = number of fractions (size classes) of PSD

$d_{0,i}$ = initial diameter of spherical (equivalent sphere) API crystals of the i-th class

$C_{b,i}$ = transient concentration of the i-th fraction[kg/m^3 of solution]

t = time [s]

ρ_p = density of API [kg/m^3 of API]

$C_{all,i}$ = concentration of the i-th fraction when its initial API are completely dissolved in the solution with volume V_{sol} [m^3 of solution]

$$C_{all,i} = \frac{x_{0,i} m_{sample}}{V_{sol}}$$

C_{sol} = solubility of drug substance [kg/m^3 of solution]

$k = k_{dif}$ = diffusion rate constant [m^3 of solution/(m^2 of drug substance .s)]

x_i = volume fraction of the i-th sieve cut (fraction)

r_i = breakage rate of the i-th fraction

S_i = breakage rate kernel (size dependent)

D_i = diameter characterizing the i-th sieve cut

S = absolute breaking rate parameter;

γ = exponent of the power law kernel

M = number of sieve cuts

η =diameter screen ratio

b_{ij} = breakage distribution (volume fraction of the broken material going from the j-th fraction to the i-th fraction, i.e. $\sum_{i=1}^{j-1} b_{ij} = 1$

References

[1] Shan G. et al.: Chem. Eng. J. 88 (2002) 53-58.
[2] Reid K.: Chem. Eng. Sci. 20 (1965) 953-958.

307c Application of Modeling in a Qbd Context

James Wertman, GSK, Upper Merion, PA 19406

The role of mechanistic modeling in pharma API process development will be discussed and illustrated by examples from our Quality by Design initiative. When a process is understood a mechanistic level, and this understanding is captured in a DynoChem model, it is possible to save significant development time in determining viable operating ranges. Application to a chemical reaction step will be used for illustration.

307d Coater-Dryer Models

Alexander J. Marchut, Wei Chen, Steven H. Chan, Shih-Ying Chang, Olav Lyngberg, Donald Kientzler, and San Kiang. Bristol Myers Squibb, New Brunswick, NJ 08903

Multi-scale modeling was used to assist with the design and scale up of a process to spray coat tablets with an Active Pharmaceutical Ingredient (API). Three models were implemented- a Discrete Element Model (DEM), a Population Balance Model (PBM), and a Thermodynamic Model. The DEM and the Thermodynamic Model are first principles models; the DEM is used to track the motion of individual tablets to study their mixing and the Thermodynamic Model is used to predict their temperature and moisture. The PBM model is a more coarse-grained model that predicts content uniformity by tracking the coating distribution. The models link operating conditions to the coating quality and uniformity and can be used for virtual experiments to aid in actual experimental condition selection and to map out the design space. The advantages of multi-scale modeling to pharmaceutical process development are demonstrated by the timescale range and mechanistic depth provided by combining the models.

POTENTIAL OF S88, BATCHML AND OTHER STANDARDS IN STREAMLINING PROCESS MODELING

Alistair Gillanders, DynoChem Inc., Wilmington, DE

Abstract

Existing standards developed for the manufacturing world have significant potential to streamline the information supply chain in pharma process development, including the application of both shortcut and detailed process modeling. Getting access to data for modeling can sometimes be more difficult than the modeling itself. This presentation outlines the potential and discusses the use of ISA S88 and BatchML in particular to integrate information from different systems.

Understanding the Modeling Workflow

When we build models of any type there are common elements to the workflow, only by understanding this workflow can we identify ways of streamlining it. The steps in this high level workflow get called different things within different communities and across different modeling technologies; but the common aspects are generally very widely applicable.

Figure 1 outlines one view of the steps in the typical workflow; and the text below offers examples and explanation of the different elements.

A statement of objective

We build models for a business purpose, not just to build models. Examples might include:

- e.g. predict or fit the physical properties of a novel material
- e.g. understand a reaction mechanism
- e.g. estimate a process mass balance
- e.g. predict the behavior of an operation or equipment item for different operating conditions
- e.g. scale-up or scale-down an operation whilst achieving equivalent performance
- e.g. assess the safety of a new material/reaction

Selection of a modeling technology

By mathematical approach

- e.g. high level heuristics / correlations (non-integrating)
- e.g. differential equation solvers (integrating)
- e.g. statistical models
- e.g. various others such as neural networks, stochastic models and genetic algorithms etc.

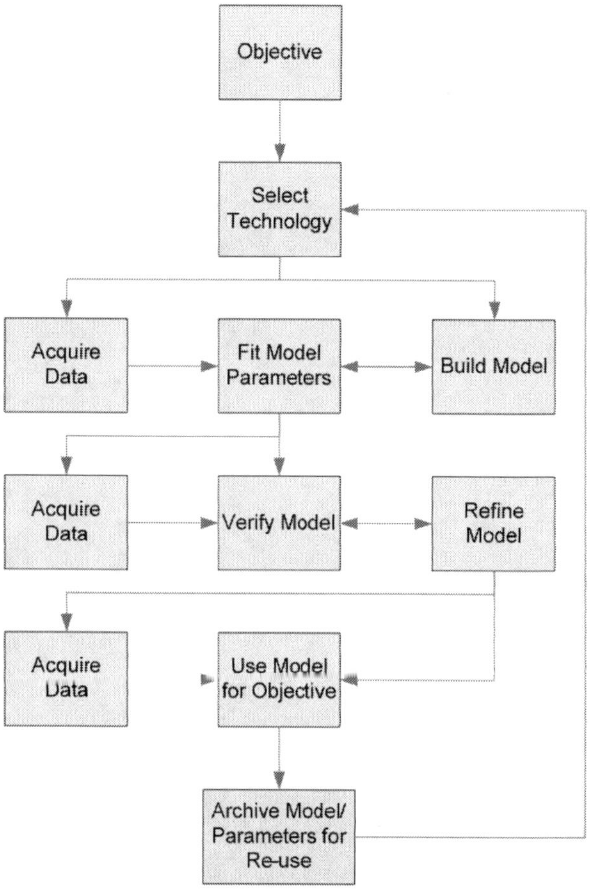

Figure 1: A generic modeling workflow

By scope

- e.g. for specific material(s), reaction(s) or transformation(s)
- e.g. for a specific unit operation (generally equipment type specific)
- e.g. for a specific unit procedure (reaction, filtration, drying etc)
- e.g. for a specific overall recipe
- e.g. for a specific synthetic route or manufacturing route

By modeling focus

- e.g. physical property prediction / correlation
- e.g. steady-state process simulation
- e.g. dynamic process simulation

Acquiring data

This turns out to be a very common element; which really should not be a surprise. And it is axiomatic that, whatever modeling technology you use, the experienced practitioner will acknowledge that "getting adequate data generally takes much longer than actually building, verifying and/or using the model". For example one industry study [11] states:

"Drug development professionals spend, on average, five hours per week looking for data. Some respondents spend eight hours or more on data retrieval each week."

This is one of the main areas that can have a major impact in streamlining any modeling workflow, regardless of the modeling technology to be used.

- e.g. data on equipment, materials, lab experiments and/or plant batches that can be used to initially create a model and find its key characteristics.
- e.g. data that allows an existing model and fitted parameters to be verified against different conditions than originally used to build the model.
- e.g. data to allow a verified model to predict process behavior in a scenario where the results are unknown but desirable; maybe for scaling-up to production plant
- e.g. data to explore process performance against a range of variables for defining a design space, examining a surface response, optimizing operating parameters against an objective function and so on; maybe for a QbD based CMC, or a plant troubleshooting / continuous improvement project.

Of particular value would be to eliminate, as far as possible, data entry and especially data transcription. Data entry remains one of the greatest causes of errors in data dependant systems, examples continue to be seen in critical systems [2].

Building and fitting models/parameters

Any model will generally involve fitting characteristics (model parameters) against some experimental results to identify values that work for the specific experiment and that can be used to gain confidence in the hypothesis the model is based on.

A few models have become so pervasive that fitted values for common targets are published or available in databases for others to use (e.g. see table 1 below). However, often you must hypothesize your own model, and estimate/measure data for your own materials, processes and equipment to obtain suitable characteristics. This is generally true even for the widely used models when targeting novel materials or equipment; a situation which is a norm rather than an exception in the pharmaceutical industry.

Additionally specific batch or experimental results are often valuable to focus models towards a particular operating procedure. For example imposing temperature profiles, addition/flow profiles and volume profiles can all help make a model more closely equivalent to something run in the lab, pilot plant or plant. This is frequently done to facilitate the easier fitting of complex characteristics by reducing the number of variables that must be handled.

Table 1: Examples of models and characteristics often published and/or in databases

Model coefficients for T and/or P dependant physical properties *for common materials.*	e.g. Antoine coefficients for vapor pressure. e.g. Polynomial coefficients for heat capacity.
Binary interaction parameters for different thermodynamic models *for common materials.*	e.g. Wilson, UNIFAC and NRTL parameters.
Sieder-Tate model heat transfer characteristics for *common heat transfer devices.*	e.g. Process side coefficients for different impeller configurations. e.g. Service side coefficients for different jacket / coil configurations.
Reaction kinetics for reactions involving *common reagents / substrates.*	e.g. Power law and Langmuir-Hinshelwood kinetic parameters for published reactions

Verifying and refining models/parameters

This generally has the same requirements as for fitting model characteristics but deliberately modifies some aspects of the setup or procedure to verify that the model can predict system behavior for the new conditions and is not constrained to simple interpolation. For example data from a different scale, a different operating temperature, different compositions and so on.

In this situation the new batch/experimental data is used to compare against the model to give an indication of the confidence level for predicting that type of change in conditions. Once a model is fitted and verified it is offering predictability with an understood level of confidence and within understood constraints.

Note that some modeling approaches are more sensitive to extrapolation than others; in particular pure statistical approaches tend to be higher risk than mechanistic models where the mathematics used has built-in understanding of the relevant underlying scientific principles.

Using models

Whatever level of verification is done, from none to extensive, you now have a model and associated parameters that can be used to study new scenarios to answer questions that it would be difficult to answer any other way. This generally requires acquiring yet more data to properly specify the new scenarios you want to study.

This is sometimes straightforward but can often be just as complex as previous work; for example when scaling-up an exothermic reaction system you obviously need to understand the heat transfer performance of the target reactor. This can be challenging, for example, when the reactor does not yet exist and is a frequent cause for secondary modeling projects to provide equipment and material characteristics that can subsequently be used by other models.

Key Standards Review

There are two main manufacturing standards that can help simplify the acquisition and sharing of data for building process models. If we can improve productivity related to "getting adequate data" for our models we will certainly have a significant impact on streamlining our overall modeling efforts. These standards are (see http://www.isa.org/):

ANSI/ISA-S88 Batch Control: Parts 1-4
ANSI/ISA-S95 Enterprise – Control System Integration: Parts 1-3

These are multi-part international standards (via IEC and ISO); plus the WBF (http://www.wbf.org) maintains XML schemas for both S88 (BatchML [6]) and S95 (B2MML [7]). Both standards continue to grow with additional parts currently in development. It is not within the scope of this paper to describe these standards fully, however a brief overview will help you understand their potential for helping to streamline process modeling.

ANSI/ISA-S88 Overview

ANSI/ISA-S88 Batch Control is a well established standard for the definition and control of batch processes. It has also been very successfully applied in both continuous and discrete manufacturing where procedural elements are required [1]. Within the pharmaceutical industry it is widely adopted within DCS's (distributed control systems) and MES's (manufacturing execution systems) in manufacturing.

Arguably the key benefit of S88 is the separation of recipe definitions from procedures for implementing them on specific equipment. It achieves this by addressing four core concepts:

- **Describing Equipment**: the *physical model* is a hierarchical description of equipment in an organization; it conceptually goes up to the Enterprise level but mainly focuses on equipment contained within a batch plant (a process cell in S88 vocabulary).
- **Describing Recipes**: the *process model* defines the recipes and parameters required to make a product. A master recipe is for a specific process cell. A control recipe is a copy of a master recipe with parameters defined for a specific batch.
- **Controlling Equipment**: the *procedural model* describes how procedural elements, that a recipe might use, are implemented for specific equipment. Note that both recipes and procedures actually use a common procedural representation of a process. The critical distinction is that a recipe procedure element describes what is desired and the equipment procedural element understands how to achieve that for a specific piece of equipment.
- **Mapping**: describes how these three concepts are combined and managed. This ranges from equipment independent recipes mapping to master recipes (specific to a plant); from master recipes to control recipes (specific to a plant and batch); from control recipes onto equipment procedural elements that know how to implement recipe elements for the designated items of equipment. Figure 2 shows this (from figure 7 of ANSI/ISA–88.01–1995).

Finally, also important for streamlining process modeling, is the more recent S88 Part 4: Batch Production Records (ANSI/ISA-88.00.04-2006) which addresses how to capture a wide variety of data that represents actual results associated with a specific batch; it is reasonable to consider a batch in this context to represent any experiment in a development environment.

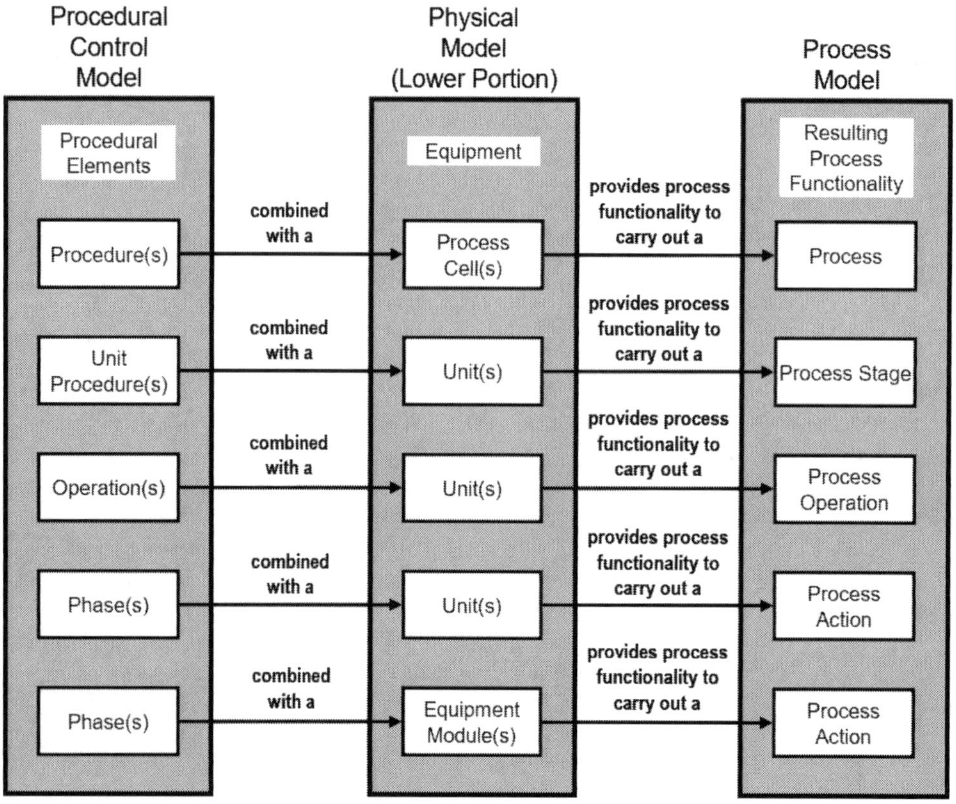

Figure 2: ANSI/ISA-S88 procedural control/equipment mapping to achieve process functionality

ANSI/ISA-S95 Overview

Figure 3 is used within ANSI/ISA-S95 to help explain the role of the different parts of a manufacturing enterprise and, in particular, where the standard is focused. For example S95 Parts 1 and 2 are focused on the interface between Level 3 and Level 4 systems whilst Part 3 is focused on the activities within Level 3.

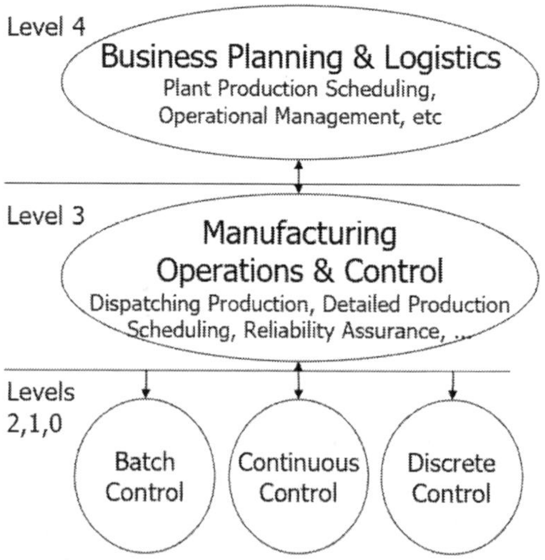

Figure 3: ANSI/ISA-S95 Functional Hierarchy

In this picture S88 is, in theory, at levels 0-2 and is only for batch control. That said S88 Part 4 is really at level 4 and S88 Part 5, in development, is focused on discrete manufacture. Plus there are several implementations of S88 for continuous processes [1].

Essentially S95 defines the activities, object models and terminology for communication between business level systems and plant level systems. This encompasses a very wide range of functionality often delivered by disparate systems that need to interoperate seamlessly.

With respect to streamlining process modeling there are a number of key points of value that S95 based systems offer. Of particular note are:

- The extension of the physical model from S88 to a broader context with much clearer definition of the levels above the work center (a process cell in S88 vocabulary is a batch work center in S95 vocabulary).
- The mechanisms for product definition management (recipe management for batch processes) provide centralized management of master and control recipes across work centers and hence the procedures that define how a batch/lot of product is made.
- The mechanisms for capturing actual results from a plant level activity and passing these up to business systems e.g. to production/experimental records systems like S88 Part 4.

Using the Standards

So how can using these standards help streamline process modeling? Most important is an approach to unifying results data management; this is an essential foundation to any viable single interface for accessing results from development, manufacturing and analytical sources.

To effectively combine information from such a wide variety of sources there are really only three obstacles, but they are big ones: a common approach to providing context for data to support identification, searching and retrieval; a common way of communicating the data between systems; and a common vocabulary for interpreting the data. The standards offer a path forward for two of the three; the third, a common vocabulary, is a gap that we need to address as an industry although the standards do offer a framework for a solution there too [10].

Unifying Data from Lab to Plant

Manufacturing, development and analytical all talking the same language…is this really possible? [3] Because of the uptake of these standards in the manufacturing arena, largely driven by the control community, there is already fairly strong support within manufacturing.

However development tends to use a different organization for their data that has grown out of other work processes (e.g. the lab notebook). Whilst this meets the needs of development it can limit the ease of access and transferability of data downstream in the business process.

Fortunately there is no need to abandon these development focused approaches; after all they are frequently systems of record for regulatory purposes and represent a significant investment financially and in resources. Instead we could start with an independent system to publish batch and experimental results for use by modeling (knowledge creation) tools, and for sharing the results from those models.

Providing Context

A common approach to contextualizing content for identity, searching and retrieval basically means a common way of providing reference information for any set of results. These contexts can take many forms and it is important to agree which to use and what they mean for each domain so they can be applied consistently. Some examples are outlined in table 2.

Table 2: Example contexts that could be based on the S88 and S95 data structures.

Context	Standard	Comments
Activity ID	S88 control recipes have a unique ID that is usually a batch or lot ID.	An experiment ID is conceptually identical to a batch ID.
Synthesis Route and Steps	Being rooted in manufacturing the standards do not recognize this per se (see process route). But this is a gap that is relatively easy to resolve.	A synthesis step is subtly different to a process step. For example multiple synthesis steps may telescope into one process step.
Process Route and Steps	An S95 product definition is a sequence of product segments. Each product segment may be implemented on a work center by a product definition (recipe version).	The outcome of development is a definition of the process route and steps so, although often not tracked until late development today, it should be possible to use this much earlier.
Location	The physical model in S95 provides a comprehensive way of identifying locations from the enterprise down to individual work centers.	Extending work center types to include a laboratory would probably be needed [5].
Equipment ID or Tag	The S88 physical model captures this very well and is extendable to cover the unique needs of development.	In a laboratory not all equipment will be tracked by a unique ID. It is normal for ad-hoc configuration of apparatus to be used, but a specification for the equipment could still be captured.
Recipe elements	S88 Recipes capture the procedures and parameters for making a product.	Recipe elements are the basic building blocks of a recipe and generally go down to the unit operation level in development.
Material	S95 has some simple material specification capability that may be adequate here; S88 has nothing.	This is potentially related to synthesis and process routes. However a safety test on a reagent does not depend on where the reagent is used.

A Common Way of Communicating

In both development and manufacturing there is an ongoing trend towards automating raw data collection. This provides a rich resource to mine for information for modeling and other purposes. However the full potential can only be achieved if that information is accessible to a broader range of systems than the "information silo" of the originating data acquisition system.

So providing context to find data is just a start, to publish that data two other issues must be addressed: how to interface between acquisition systems (those connected to devices) and those concentrating data for wider publication; and how to package complex data for that transfer. Using XML provides solutions for the former (e.g. Web Services) which build on the BatchML and B2MML schemas from the WBF [6,7]. And S88 Part 4 offers a comprehensive definition of how to package information about any type of production (or experimental) activity; including all the contextual information already discussed as well as detailed content such as datasets containing time-series process trends and even the complex arrays typical of PAT results.

Filling the Gaps – A Vocabulary Lesson?

Although there are clearly a number of issues that need to be resolved, probably the major remaining obstacle is agreeing the kernel of an open common vocabulary. This is needed on several levels, from process data trend names, through units of measure to common recipe element and parameter names; e.g. for unit operations in particular.

For example a temperature profile from the same reaction executed in a pilot plant and a reaction calorimeter may be referencing what should be comparable data with very different definitions.

- Plant: Tag = B15E-TI207-3, Alias = ReactorTemperature, Units = DegC
- Lab: Tag = unknown, Alias = Tr, Units = °C

In fact in one area we can still fall back on an existing standard, UN/CEFACT provides a comprehensive vocabulary for units of measure that was created specifically to facilitate the type of system interoperability we need [8]. However, no open guidelines exist for a standard vocabulary for process data elements or recipe elements and parameters. Some companies have created internal libraries [9, 10] to address this but a standard open approach would benefit the whole industry and should readily address the majority of elements; extensibility could add support for any local exceptions that are needed within proprietary implementations.

References

1. Brandl, Dennis, "Design Patterns for Flexible Manufacturing", *ISA*, 2007.
2. Letter to the Secretary of State entitled "State Department Needs to Resolve Data Reliability Problems that Led to Inaccurate Reporting to Congress on Foreign Arms Sales", *Government Accountability Office*, Report GAO-05-156R, Jan 28, 2005 (see http://www.gao.gov/).
3. McKenzie, Paul et al, BMS, "Can Pharmaceutical Process Development Become High Tech?", *AIChE Journal*, December 2006, Vol. 52, No. 12, pp 3990-3994.
4. Basu P. "Pharmaceutical Process Development is Different" *Chem Eng Progress*, September 1998, pp75.
5. Brandl, Dennis, Personal communication and comment to SP95, August 2007.
6. BatchML XML Schemas and Documentation V02, *The World Batch Forum*, 2003, http://wbf.org/associations/2718/files/BatchML-V02.zip.
7. B2MML v04 XML Schemas and Documentation, *The World Batch Forum*, 2007, http://wbf.org/associations/2718/files/b2mml/B2MML-V0400.zip.
8. Information Content Management Group, "Codes for Units of Measure Used in International Trade", *UN/CEFACT*, Recommendation No. 20, Revision 4, 16 March 2006.
9. Pfizer Inc, US Patent Application US 20070050070A1, *US Patent Office*, 1 March 2007.
10. Shanley, Agnes "Changing Pharma's DNA: An Interview with Paul McKenzie of BMS", *Pharmaceutical Manufacturing*, September 2007.
11. Morris, Ken et al, "Making the Most of Drug Development Data", *Pharmaceutical Manufacturing*, November 2005.

307f Batch Plus Modeling As A Tool To Drive R&d-Based Process Expansion

Samrat Mukherjee, Shekhar Viswanath, and Jeffrey Breting. Process R&D, Abbott Laboratories, Dept R452, Bldg R8-3, 1401 Sheridan Rd., North Chicago, IL 60064

Batch Plus, an AspenTech utility, was used by the R&D organization to investigate the feasibility of available options – process modifications and/or plant expansion - for a commercial manufacturing process anticipating a throughput increase. The Batch Plus model aided the quick identification of the time- and size-limiting equipment in the existing process. Thereafter, new process scenarios were churned out for the existing process by proposing additional equipment such as reactors, mixed tanks and filtration units, such that the incremental capital cost per incremental throughput is minimal. Having the Batch Plus model built for the existing process also facilitated the simulation of process scenarios involving new process modifications, which have been demonstrated only in lab scale, to apriori evaluate their core impact on the throughput. The variable and additional capital cost components that are influenced by the cycle time, batch size and additional equipment, were taken into account for assessment of each new process scenario. The outcome of this exercise was a set of recommendations for future R&D activities and plant-expansion efforts, towards achieving the goal of higher throughput at the lowest cost.

Application of agent technique in bioprocess modelling for plant-wide process improvement

Ying Gao[a], Katie Kipling[b], Jarka Glassey[b], Mark Willis[b],
Gary Montague[b], Yuhong Zhou[a], Nigel Titchener-Hooker[a]
a. Biochemical Engineering Department, University College London
b. School of Chemical Engineering and Advanced Materials, Newcastle University

Abstract

A systematic framework is developed in this work to analyse bioprocesses based on a whole process understanding considering the interactions between process operations. This yields a capacity to predict process performance in the case of process variations, and to improve process efficiency within the process design space.

An agent-based approach is proposed to provide such an environment for the necessary integration of process models, to handle process interactions, to simulate the overall process behaviour and to lead to the fast evaluation of process improvement options. A multi-agent system comprising of a process knowledge base, process models and a group of functional agents will be introduced in this work. Agents are designed to link the unit operation models, to represent the unit operations, and to simulate the interaction between units.

The adoption of an agent-based approach to bioprocess modelling provides a flexible infrastructure for the integration of process operations, supports timely assembling of process models and the updating of process descriptions. Agents have the ability to detect and act on the most up-to-date bioprocess information. In this system, agent components run on top of the process models and datasets, they cooperate with each other in performing their tasks for the description of the whole process behaviour, predict critical process performance, and control of process parameters in a timely manner to maintain the product quality and to ensure an efficient manufacturing process.

An extra-cellular protein production process is used as an example to demonstrate the implementation of the agent-based approach, and in particular the application of agents in detecting of deviation in upstream unit operation and predicting the influences on the subsequent downstream processing whilst providing guidance in terms of possible corrective actions designed to improve process efficiency and ensure product quality.

Key words: Bioprocess modelling, agent-based system, bioprocess interaction, process improvement.

1. Introduction

In the biopharmaceutical industry, timely development and transfer of cost effective products to the market is of great importance. The industry also requires to move quickly from small-scale data to manufacturing-scale confidence and to improve manufacturing processes so as to achieve optimal process performance. At the heart of this is the need to develop predictive tools such as unit operation models for the description of the whole manufacturing process and the subsequent evaluation of process improvement options.

Developing bioprocess models is time-consuming and there are several factors that should be considered in order to achieve more efficient process descriptions and improvements. Bioprocesses comprise multiple unit operations, and with strong interactions between them. Integration of the unit operations is important for any plant-wide process description and the evaluation of process improvement options. Generally unit operation models describe individual unit operations. When it is necessary to integrate the individual unit operation models together to form a whole bioprocess description, extensive efforts are required in the modification of each unit operation model. This will also be the case when attempting to reuse the unit operation models for describing different process sequences. In addition, an iterative cycle of model improvement is normally involved to enable the unit operation models to be further developed as updates in the knowledge base are realized. This will make the integration of whole process operations more complicated. A flexible environment is therefore required for the integration of unit models into a whole bioprocess description.

In order to improve plant-wide bioprocess efficiency, the strong interactions present between units operation (Davies, *et. al.* 2000; Meirels, *et. al.* 2003; Zhou and Titchener-Hooker, 2003) must be adequately modelled as these directly affect overall productivity. Consideration of the whole process is also required in order to identify the optimal process conditions. Multiple objective optimization approaches are suited to determine the overall plant optimum due to the inherent need for trade-offs between the multiple objectives of operation in the bioprocess manufacturing process (Halsall-Whitney, *et. al.* 2003). Tools which can integrate the unit operation models with a process optimisation strategy will help the industry to identify and evaluate process improvement opportunities.

Ttraditional bioprocess modelling considers the process units separately, making it difficult to investigate interactions between unit operations, and also limits the capacity to achieve iterative model development. What is required is a system that allows whole process descriptions to be evolved using the available system information and to be updated as new information arises. Agent-based techniques are ideally suited to such a problem. By contrast to traditional software programs, agents can help to solve such problems by collaborating and communicating with other software agents and other resources in a network (Bradshaw, *et. al.* 1997). The communication ability allows agents to exchange information with each other and to facilitate the interoperation between application programs. Agents can also detect and act on the most up-to-date bioprocess information, and cooperate with each other to solve the time-critical problems that may occur in a manufacturing process.

In this paper, an agent-based approach is proposed to provide a flexible environment for the necessary integration of bioprocess models, to handle process interactions and hence to lead to the fast evolution of bioprocess descriptions and plant-wide manufacturing improvements. This paper is organized as follows. In section 2, we introduce the basic concepts of a multi-agent system and discuss the potential advantages of applying an agent-based approach to bioprocess description. Section 3 presents the framework of a multi-agent system (MAS) for a whole bioprocess description while the application of a multi-agent system is illustrated in section 4 through an example scenario.

2. Agent-based solutions

An agent is a computer system that is situated in some environment, and that is capable of autonomous action in this environment in order to meet its design objectives. (Wooldridge and Jennings, 1995). Agents have the potential to assist in a wide range of activities relevant to engineering problems. They can integrate disparate operating environments, handle the complexity of solutions through decomposition, and model the interrelationships between components. Agents have proven to be a useful technique in many industrial and business sectors, including telecommunication, air traffic control, traffic and transportation management (Wahle and Schreckenberg, 2001), supply chain management (Julka, *et. al.* 2002; Garcia-Flores and Wang, 2002), and medical care (Jennings, *et. al.* 1996). In these situations, agents are used in the modelling, monitoring and management of time-critical information in their particular field and provide decision support for the management system.

A multi-agent system (MAS) consists of a number of agents that work together to find answers to problems that are beyond the individual capabilities or knowledge of any single entity. The agents interact with one another, typically by exchanging messages through the computer network infrastructure. In order to interact successfully, these agents need to have the ability to cooperate, coordinate and negotiate with each other to achieve common goals. These agents share information, knowledge and tasks among themselves. With the capability of communication and coordination between agents to achieve common goals, multi-agent systems are likely to have an increasingly important role to play in engineering domains in that they significantly enhance the ability to model, design and build complex software systems necessary to deal with dynamic, real-life problems of which bioprocess manufacture is a good example.

Multi-agent systems provide an excellent environment for modelling whole bioprocesses characterised by multiple unit operations, and with strong interactions between steps. The modular approach of system development enables the user to define a multi-agent architecture providing the ability to integrate unit operation models together. The communication ability between agents is well-suited to tackling unit interactions, and the ability to respond to time-critical information allows agents to detect changes in process operations and to represent the dynamic bioprocess operation behaviour. Applying a multi-agent system architecture will also make the system flexible, extendable and reusable, which can support the integration of unit operation models already available, as well as the addition of others when they become available. Traditional bioprocess modelling approaches are not so adaptable.

Using the agent technique, the complex task of whole bioprocess simulation can be divided and assigned to several sub-agents, which will reduce the burden on any single agent. This can also ease the optimization problem when the whole process sequence needs to be considered, and multiple objective functions have to be investigated simultaneously. Furthermore, the multi-agent system is also suited to bioprocess modelling when multiple levels of information and unit operation models exist in the system knowledge base, and these are frequently updated as manufacturing knowledge accumulates.

3. Framework for agent-based plant-wide bioprocess description

In this section, we describe the proposed agent-based framework which aims to provide a flexible infrastructure to support the integration of unit operation models, and to simulate the interaction between unit operations, which together may lead to a timely description of the whole bioprocess behaviour and the fast evaluation of process improvement options. The system architecture and function of the agents are introduced here.

3.1 Multi-agent system architecture

Figure 1 illustrates the multi-agent system architecture proposed in this work. The system comprises a process knowledge base, unit operation models, a group of functional agents and a user interface. The process knowledge base itself contains data and information from the different unit operations along the process sequence. The data sets include experimental data which is obtained from small-scale experiments such as USD mimics, and historical manufacturing data. Information on the unit operation conditions and equipment details are also provided in the knowledge base. Data and information can be

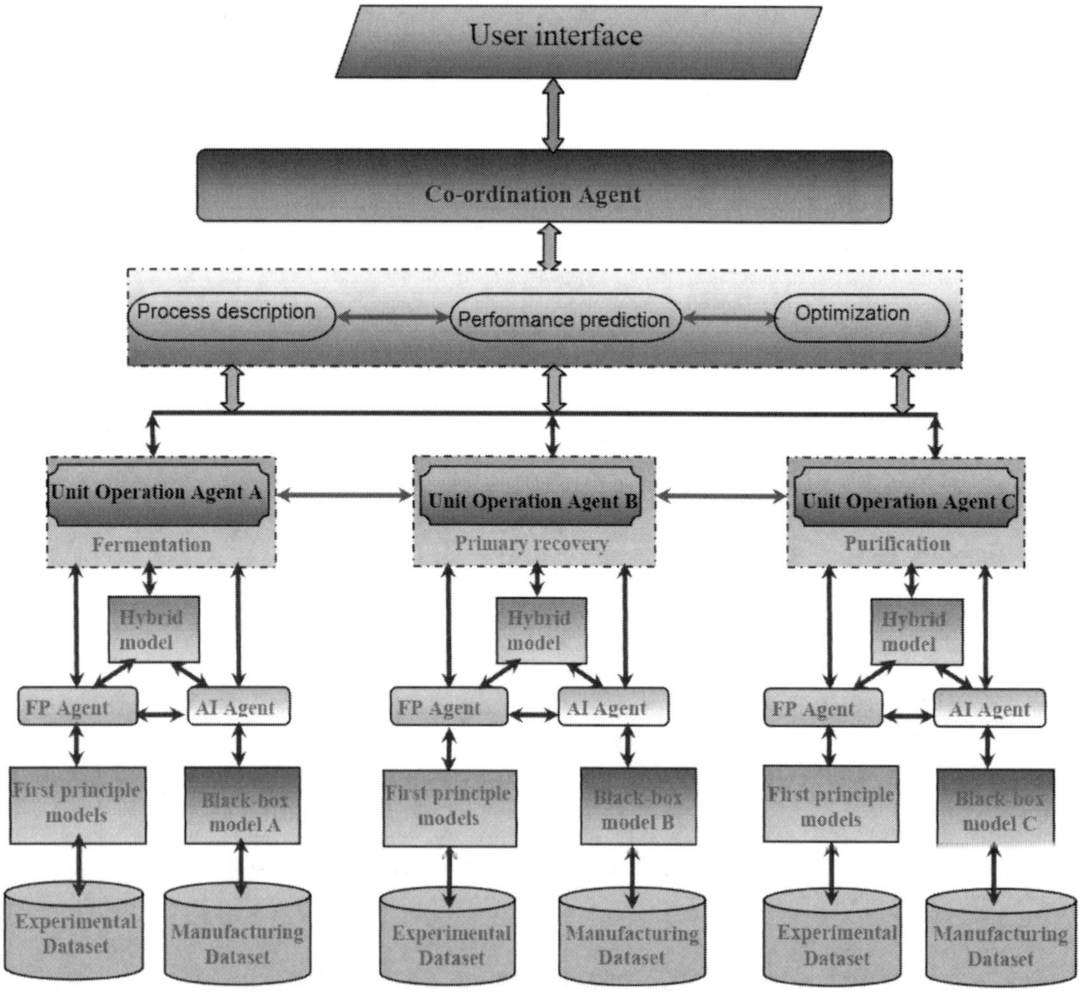

Figure 1 Multi-agent system framework

classified and organised at different levels in the knowledge base for convenient utilization.

The unit operation models can be developed based on a mechanistic understanding and may use parameter values derived from small scale experiments or obtained directly from the data-based methods in which case they will essentially be black-box by nature. Experimental data combined with the first principle models will mimic steady-state unit operation behaviour, and provide for the fast generation of unit operation descriptions. Data-driven models can be developed by applying data mining techniques to the manufacturing data and will result in a dynamic model which can describe and predict manufacturing process behaviour. For each unit, different levels of models may be constructed according to the complexity of the factors considered and the level of sophistication of the final model sought.

Agents are organised in a hierarchical architecture within the framework. The lower level agents will be able to access process knowledge base and datasets, and to represent the unit operation behaviour utilising the unit operation model. The higher level agents will be able to integrate unit operations, then communicate and co-operate to simulate the interaction between units and thus construct the whole process description. These agents can also evolve together to provide an overall process description that can be subsequently used to optimise the process in order to achieve best overall plant performance. The function of individual agents involved in this system is explained further in the next section.

The hierarchical architecture of the agent-based system has the benefit of making the agents work most efficiently. Rather than exposing each agent to all of the detailed process information, the hierarchical structure can help to avoid the problems of information overload, and also makes it convenient for bringing unit operation models and different levels of information into the agent system. Updating and adding new information and unit operation models to such a system can be achieved easily without changing the system structure, and the agents will be able to detect and act on the most up-to-date process information and unit operation models. This facilitates the timely assembling of process models and the updating of process descriptions, especially when the process information is updated or new process models become available to the knowledge base, i.e. as process data accumulates with the running of more manufacturing campaigns.

3.2 Function of agents

The proposed agent-based system comprises a coordination agent, unit operation agents and a group of first principle (FP) agents and artificial intelligent (AI) agents. The functionalities of each agent are now described briefly.

- FP agent

The FP agents access both the experimental dataset and the first principle models developed in combination with experimental data. As there are different levels of models in the unit operation model pool, the FP agent can choose the specific level of model according to the particular purpose and requirements of the process description.

- AI agent

In this work AI agents are designed to perform the tasks of process simulation and process performance prediction using black-box models developed based on the manufacturing data. The AI agent may also be used in updating the models and process description based on the latest data available in the manufacturing dataset.

- Unit operation agent

A group of unit operation agents are used to represent the individual operation units in the bioprocess, such as fermentation, centrifugation and chromatography. The unit operation agents can work as a team and evolve together in the evaluation of how process conditions (i.e. the fermentation harvest time, centrifugation flowrate and the loading conditions of chromatography) impact on the whole process performance. Such unit operation agents can also act as a bridge to link first principle and data-driven models with the cooperation of FP and AI agents, to integrate the best features of these two different types of models and to derive a hybrid model for a more sophisticated process description.

- Co-ordination agent

The co-ordination agent, located at the top of the multi-agent system, can decompose a given task into a number of subtasks and dispatch the subtasks to relevant agents for execution, in order to achieve its goals. A whole process description, process performance prediction and the investigation of the interactions between unit operations can be achieved through the collaboration of sub-level agents. Process optimization can be carried out to determine optimum plant operating policies through the co-ordination agent by integrating the models for different units and embedding the resulting description within a multi-objective optimisation algorithm. This takes into account the interactions between units and provides an overall indication of performance likely to be achieved and the operating policy required to do so.

4. Application of the agent-based system framework

In this section, we use an example scenario to demonstrates key aspects of the tool and in particular the application of agents in detecting of deviation in process operation and providing guidance for corrective actions. An extra-cellular protein production process is used as an example process in this study. Figure 2 shows part of the flowsheet for the production and isolation of an extra-cellular recombinant protein from a *S. cerevisiae* strain. In this process, the micro-organism is cultured in a fed-batch fermenter. Cells are removed using a high-speed disc-stack centrifuge, subsequent processing of the protein supernatant will include multiple chromatographic purification and ultra-filtration steps for product concentration.

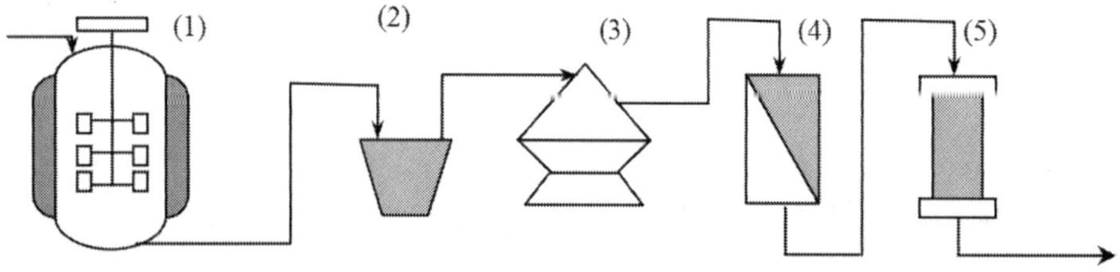

Figure2 Flowsheet of an extra-cellular product recovery process. (1) Fermenter, (2) Harvest tank, (3) Disc-stack centrifuge, (4) Filter, (5) IE chromatography

The proposed agent-based framework was implemented using JADE (Bellifemine, 2001) as the agent development platform. JADE is a software framework fully implemented in the Java language and it simplifies the implementation of multi-agent systems through a middle-ware that complies with the FIPA specification. FIPA ACL (Labrou, *et. al.,* 1999) is used as the communication language in JADE.

Unit operation models for fermentation, centrifugation and chromatography operations were developed using Matlab and C/C++. Fermentation, Centrifugation and Chromatography agents were used to represent each unit operation. A Co-ordination agent was created to coordinate the unit operation agents, and a User agent was configured to communicate between the user and the agent system, to receive user commands and to present results to the user.

In this scenario, the agent is used to advise on actions relating to the generation and processing of a manufacturing batch. The agent monitors current performance data for any deviations from historic data sets. An alert will be triggered if a deviation of over 10% is detected. The alert is in the form of dialogue between the unit operation agents and is facilitated by the Co-ordination agent. The latter provides useful advice on possible control actions which could be pursued in order to correct for the deviation.

The main kind of deviation that may be expected in the fermentation would be whether

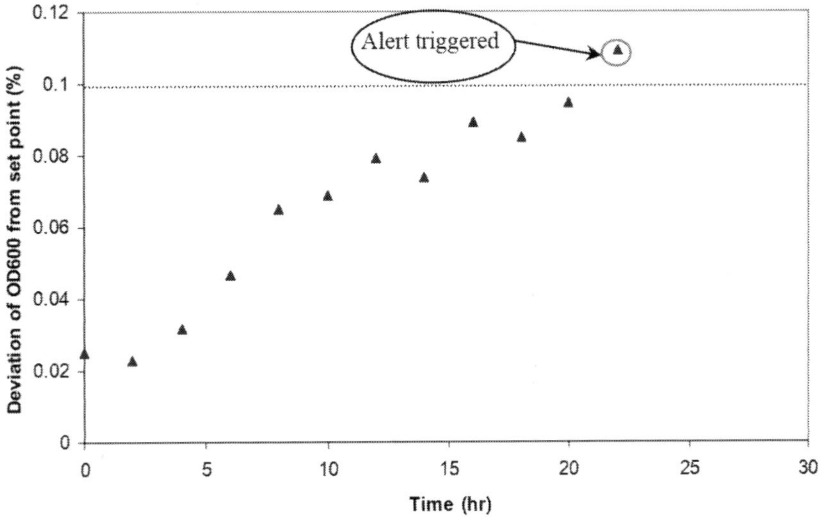

Figure 3 Time course of fermentation biomass (detected by OD600 measurement) showing a batch where the growth deviation from historic data and eventually differs by more than 10%, an alarm is triggered.

the culture growth rate changed relative to historical trends. This affects subsequent processing of the batch. Hence if the final biomass concentration deviates from the target level this may necessitate a change of the condition of centrifugal solid removal. In manufacturing, set points will exist for each stage of the process and it is the function of the agents to cooperate in the detection of a deviation and in proposing appropriate control actions to bring the problem back within the defined process envelop.

A 1000L yeast fermentation is studied in this example, the fermentation is expected to harvest at 30hr, and the target biomass concentration is 120g/L. As the fermentation progresses, data such as the on-line OD600 which can be used to refer the level of biomass product is evaluated using the multi-agent system. This task is performed through the collaboration of the User agent, Co-ordination agent, Fermentation agent and Centrifugation agent. The Fermentation agent compares the current operation data with historic manufacturing data. If the deviation is within the normal operation boundary (less than 10%), the agent will report that the operation situation is fine. If more than a 10% derivation is detected, the Fermentation agent will trigger an alert. At this point the agent uses the available fermentation models to predict the trend of cell growth and to compute the biomass concentration which is anticipated at normal harvest time. This output information will be sent to the Centrifugation agent in order to predict the centrifugation performance at the normal operation flowrate of 300L/hr. If the predicted level of centrifuge clarification cannot reach the target (in this case set at 95%), then an alteration in the centrifuge operation flowrate is suggested in order to cope with the change of material properties arising from deviating fermentation unit. The Centrifugation agent will send feed-back information to the Fermentation and Co-ordination agent about selecting a suitable centrifuge operation flowrate at the anticipated solids concentration. The Co-ordination agent will then propose possible actions in order to cope with the deviation in the fermentation operation.

Such a scenario is shown in Figure 3. The activities of the agents are described in the following:

a. The Fermentation agent checks the fermentation condition at every two-hour intervals, and reports the result to the Co-ordination agent. Fermentation agent reports that the operation is fine till 20hr of fermentation.

b. At 22 hours of fermentation, the Fermentation agent triggered an alert, as a greater than 10% deviation of OD600 is detected from the set point.

c. The Fermentation agent compares the current biomass concentration to the target required at the end of the fermentation, and predicts the trend of the biomass concentration.

d. The fermentation agent predicts that the biomass concentration will reach the target after 26 hr of fermentation. The Fermentation agent sends this information to the Co-ordination agent to suggest earlier harvesting.

e. The Fermentation agent also predicts the biomass concentration 30hr into the fermentation which is the normal time for cell harvest. Based on the current cell growth rate, 15% higher biomass concentration is predicted at 30hr. This information is send to the Centrifuge agent.

f. The Centrifuge agent simulates the operation performance based on this prediction, and a 91% level of clarification is predicted if the centrifuge is operated at 300L/hr.

g. The Centrifuge agent suggests that altering the operating flowrate to 230L/hr in order to maintain the clarification level at over 95%. The suggested centrifuge operation flowrate to cope with the higher biomass concentration expected at the end point of the fermentation is sent to both the Co-ordination and the Fermentation agent.

h. The Co-ordination agent combines the information from the Fermentation and Centrifugation agents, and returns the two options to the user so that they can act on the deviations detected: one option is to harvest at 26hr and keep the centrifuge operation at 300L/hr, this will maintain the normal fermentation and centrifugation performance, but will be result in harvesting being 4 hours earlier in the schedule; the other option is to harvest at the normal harvest time (30hr) but to lower the centrifuge flowrate to 230L/hr in order to compensate for the higher level of biomass. Higher product yield will be expected in the second option, but the centrifuge will need to be operated one hour longer. A typical dialogue between the Co-ordination and the User agents is shown in Figure 4.

```
( INFORM
    :sender     Co-ordination Agent
    :receiver   User Agent
    :content    "( Fermentation time(hr): 22
                  Solid concentration (OD600): 26.71
                  Reaction: Deviation detected, OD is high!
                  Operation suggestion:
                  A. Harvest at 26hr, biomass concentration will reach the fermentation target;
                     keep centrifuge operation at 300L/hr;
                     yield will be as expected, but operation schedule is now 4 hour earlier.
                  B. Harvest at 30hr, biomass concentration will be 15% higher than expected;
                     centrifuge operation flowrate should be altered to 230 L/hr,
                     to maintain the clarification over 95% in centrifuge operation;
                     a higher product yield is expected, but centrifuge time will be 1 hour longer. )"
    :language   FIPA SL
    :ontology   MAS-ontology
)
```

Figure 4 Message between Co-ordination and User agents in the event of a deviation being detected. The agent proposes two possible choices of action to follow in this event.

6. Conclusion

A systematic framework for the generation of an agent-based whole bioprocess description is proposed in this paper. The agent approach overcomes some of the limitations of traditional forms of bioprocess modelling, especially the difficulty of capturing the interactions between unit operations, and the lack of flexibility in updating and extending existing programs. A multi-agent system is proposed to simulate overall process behaviour and the interaction between process operations. In such a system, agents are utilized to bridge between process models and a knowledge base, to access datasets and process models, to represent the unit operations, and to simulate the interaction between units. The process models are deployed in a series of hierarchies which make it convenient for them to be brought into the agent system. Agent components run on top of process models and datasets, they communicate with each other to exchange process information, cooperate in performing the tasks of whole process description, manufacturing optimization and assist in suggesting appropriate decisions and actions.

The proposed modelling framework is particularly suitable for describing bioprocess manufacturing which is characterised by multiple unit operations and with interactions between them. The agent-based framework provides a flexible environment for the necessary integration of process models for the description of whole process behaviour and the inclusion of the interactions between unit operations. The systematic framework is also suitable for describing bioprocess manufacturing which has different levels of information and multiple objectives, and therefore allows for rapid development of a process description based on the most up-to-date information. Moreover, the modular approach of system development makes the system reusable and extendable.

The implementation of the multi-agent system is illustrated using a sample scenario. Future work will be carried out to include rule-based decision making strategies in the agent system in order to provide guidance in adjusting operation conditions to cope with deviations which might be occurred in the manufacturing process. Further development of the multi-agent system will be undertaken to combine information from both USD experiments and manufacturing processes to achieve more comprehensive process descriptions.

References:

Bellifemine, F., Poggi, A., Rimassa, G. (2001) JADE - A FIPA2000 compliant agent development environment, *In Proc. Agents Fifth International Conference on Autonomous Agents (Agents 2001)*, 216-217

Bradshaw, J. M., Dutfield, S., Benoit, P., Woolley, J. D. (1997) KAoS: Toward an industrial-strength open agent architecture, *Software Agent*, MIT Press. 375 – 418

Davies, J. L., Baganz, F., Ison, A. P., Lye, G. J. (2000) Studies on the interaction of fermentation and microfiltration operations: Erythromycin recovery from Saccharopolyspora erythraea fermentation broths, *Biotechnol. Bioeng.*, 69(4), 429- 439

Garcia-Flores, R., Wang, X. Z. (2002) A multi-agent system for chemical supply chain simulation and management support, *OR Spectrum*, 24(3), 343-370

Halsall-Whitney, H., Taylor, D., Thibault, J. (2003) Multicriteria optimization of gluconic acid production using net flow, *Bioproc. Biosys. Eng.*, 25(5), 299-309

Julka, N., Srinivasan, R., Karimi, I. (2002) Agent-based supply chain management—1: Framework, *Computers Chem. Engng.*, 26(12), 1755-1796

Labrou, Y., Finin, T., Peng Y. (1999) Agent communication language: The current landscape, *IEEE Intelligent Systems*, 14(2), 45-52.

Meirels, M., Lavoute, E., Bacchin, P. (2003) Filtration of a bacterial fermentation broth: harvest conditions effects on cake hydraulic resistance, *Bioproc. Biosys. Eng.*, 25(5), 309-314

Wahle, J., Schreckenberg, M. (2001) A multi agent system for on-line simulations based on real-world data, *Proceedings of the Haiwai International Conference on System Science (HICSS) ,(IEEE Computer Society, 2001)*, 3, 3037

Wooldridge, M., Jennings, N. R. (1995) Intelligent agents: Theory and practice, *The Knowledge Engineering Review*, 10 (2), 115-152

Zhou, Y. H., Titchener-Hooker, N. J. (2003) The application of a Pareto optimization method in the design of an integrated bioprocesses, *Bioproc. Biosys. Eng.*, 25(6), 349-355

An Integrated Quality-by-Design (QbD) Approach for Pharmaceutical Powder Blending Unit Operation: Process Monitoring, Endpoint Determination, and Concentration Quantification

Huiquan Wu, Mobin Tawakkul, and Mansoor A. Khan[a]

Division of Product Quality Research (DPQR, HFD-940), OTR/OPS/CDER/FDA, White Oak LS64, 10903 New Hampshire Ave, Silver Spring, MD 20993

[a] Corresponding author: Dr. Mansoor A. Khan. Email: Mansoor.Khan@fda.hhs.gov; Tel: 301-796-0016.

Abstract: The objective of this study was to develop an integrated QbD approach using process monitoring to determine the blending end-point and constituent concentrations of final blending mixture. A mixture design was created to include 26 powder formulations consisting of ibuprofen as the model drug and three excipient components (HPMC, MCC, and Eudragit L100-55). The mixer was stopped at various time points to enable NIR scan of the powder mixture and sampling for UV assay. Moving block standard deviation was applied to the NIR spectra for trend analysis and blending end-point determination. To quantify the concentrations of well-blended powder mixture, two types of multivariate calibration models (PLS and PCR) were constructed and validated. The first model was constructed to correlate the UV spectral data with the targeted concentrations of a binary mixture of ibuprofen and Eudragit L 100-55. This model was then used to predict the concentrations of ibuprofen and eudragit L100-55 in the final well-blended powders. The second model was constructed to correlate the NIR spectra of well-blended 4-component powder mixture and the concentrations of blended powders. The present study showed that the time to reach blending end-point ranges from 5 to 20 minutes. The multivariate PLS and PCR models based on UV spectra has R^2 value of 0.99 and could predict the concentrations of both ibuprofen and Eudragit L100-55 well. The multivariate PLS and PCR models based on NIR spectra could predict the concentrations of both ibuprofen and excipients well for the final well-blended 4-component mixtures. The discrepancy between the prediction results from the two multivariate calibration models was discussed based on the fact that NIR and UV have different scale of scrutiny.

Key words: Quality-by-Design (QbD), design of experiments, multivariate statistical data analysis, powder blending, process monitoring, blending end-point determination, concentration quantification, moving block standard deviation.

INTRODUCTION

A satisfactory mixing of pharmaceutical powders is important as it is directly related to the content uniformity of drug in the final solid dosage form. In powder mixing, it is necessary to know the composition of a mixture in various places throughout its volume in order to describe the distribution of components adequately. However, the theory of solids mixing has not advanced much beyond the most elementary of concepts and, consequently, is far behind that which has been developed for fluids. This lag can be attributed primarily to an incomplete understanding of the ways in which particulate variables influence such systems and to the complexity of the problem itself[1]. Conventionally, the blending end-point is determined by wet assays (such as HPLC) of the blending samples, which is labor intensive and time consuming. When there are large differences between formulation components' properties, the mixing process could be prolonged substantially to achieve blending homogeneity. For powder blending process of such formulation system, innovative monitoring techniques may provide advantages such as fast data acquisition, reliable results generation, and representative sampling, etc.

A number of online techniques have been developed for monitoring powder blending process over the last few decades, such as LIF (Light Induced Fluoresecence)[2], NIR (Near-Infrared Reflectance Spectroscopy)[3-5], polychromatic light[6], and effusivity[7]. All of these previous works collectively advanced our understanding of powder blending process from different aspects. However, none of them have used an integrated approach of combining statistical experimental design, multivariate data analysis, on-line or at-line process monitoring with UV spectroscopy confirmation for both qualitative and quantitative understanding of 4 component powder formulation blending process. From process design and process control perspective[8], knowledge gained from such process understanding should include but are not limited to process kinetics, blending homogeneity, and simultaneous quantification of both API and excipients, etc, as those are essential aspects for controlling the blending process and ensuring the quality of blending materials.

Recent regulatory documents and guidance, such as FDA's Process Analytical Technology (PAT) Guidance[9], FDA's cGMPs for the 21st Century[10], ICH Q8 and Q9[11-12], highlighted the critical importance of using integrated approach for in-depth process understanding. Powder blending as one key unit operation for the solid dosage form manufacturing, dictates at least two critical quality attributes: blending uniformity and content uniformity because uniformity is necessary to produce a tablet having a reproducible dissolution profile, uniform taste and color. These two quality attributes are highly related to product safety especially for narrow therapeutic index (NTI) drugs and high potency drugs. On the other hand, knowledge of exact concentrations of both API and excipients in the final dosage form is critical for ensuring product performance, product quality, and patient safety. However, current industrial and regulatory practices have been heavily focused on the API for drug product, the determination of excipients' concentrations in the final dosage form have been largely relied on the initial physical dispersion of pharmaceutical components prior to mix operation. Few alternative methods have been developed in this regard[13].

From a process perspective, key questions for blending operation may include the following:

(1) How do the formulation component compositions impact the time course of blending operation and the blending end-point?

(2) How to validate or confirm the PAT blending process monitoring results via other fast and convenient spectroscopic methods?

(3) How to link the scale of scrutiny and the homogeneity of both API and excipients?

To examine the above technical challenges at the laboratory scale, an extreme vertices design was created to include 26 formulations which consist of a four-component formulation system for a blending process study in this work. The focus was on blending process monitoring, end-point determination, and concentration quantifications. Moving block standard deviation method was applied to extract the information of blending process kinetics. Multi-variate calibration models were developed to predict concentrations of all components in the final blended powder mixtures.

EXPERIMENTAL

Materials and Methods

The following pharmaceutical materials were used as-received for this study, without further processing or purification prior to the powder mixing: USP 70 grade Ibuprofen (Albemarle Corp., LA. Lot No. 062342); Hydroxypropyl Methylcellulose (HPMC), Methocel E15 Pemium LV (Dow Chemical, Midland, Michigan. Lot No. TH03012402); USP/NF Microcrystalline Cellulose (MCC) (JRS Pharma LP, Cedar Rapids, Iowa. Lot No. E5D6B17); Eudragit L 100-55 (Methacrylic Acid-Ethyl Acrylate Copolymer (1:1), Methacrylic Acid Copolymer Type C NF) (Degussa, Germary. Lot No. B040804021). Class 1B HPLC grade Methanol (Fisher Scientific, USA. Lot No. 051796) was used for dissolving the powder samples prior to the UV analysis.

Experimental Design

The extreme vertices design was employed with the following constraints applied to the weight fractions (W) of corresponding formulation ingredients: 0.25 $\leq W_{ibuprofen} \leq$ 0.75; HPMC: 0.01$\leq W_{HPMC} \leq$0.03;MCC: 0.19 $\leq W_{MCC}\leq$ 0.57; 0.05$\leq W_{Eudragit\ L100\text{-}55} \leq$ 0.15.The weight fractions for the 26 formulations were computed by JMP 5.1 software (SAS Institute, Cary, NC).

Powder Blending Experiments

Weighted formulation components (a total mass of 5 grams) were placed into a 20 ml scintillation vial for geometric mixing for 5 seconds. The vials were then placed inside a basket of a Tubula mixer (Willy A. Bachofen AG, Maschinenfabrik, Basel Switzerland). The mixer was stopped at predefined time points to allow NIR scanning of the powder mixture. After NIR scanning, total of 20 mg powder was sampled from the surface of the powder bed. The sample was placed in a 20 ml scintillation vial which was kept in desiccator at constant relative humidity of 10% and room temperature (21 ^{0}C) for UV assay later on. Density of the powder formulation was measured by Electrolab Tap Density Tester (USP) model ETD-1020 (GlobePharma, New Brunswick, NJ).

NIR Spectroscopy

A Luminar^{TM} acoustic-optic tunable-filter (AOTF) based NIR spectrometer (Brimrose Corporation of America, Baltimore, MD), equipped with a transflectance probe, and was used for scanning the powder mixture. Several measures were evaluated to ensure a representative and consistent sampling from the probe measurement perspective.

UV Spectrometer

The UV absorbance spectra were collected using the Agilent UV-VIS 8453 Spectrophotometer (Santa Clara, CA) attached to a sipper system connecting the peristaltic pump and tubing.

Data Analysis Methods

For the averaged NIR spectra at each pre-defined blending time points, the standard deviations of three NIR spectra acquired at three consecutive pre-defined blending time points were computed along the NIR wavelength region [1100, 2500 nm]. Then these standard deviations were plotted vs. wavelength to observe the process trend, such as whether a global or local blending end-point is approaching. On the other hand, the multivariate models were constructed to correlate the UV spectral data with the powder concentrations, and the NIR spectra with the powder concentrations. All multivariate data analysis was performed by using Unscrambler 9.2 software (Camo Technologies, Woodbridge, NJ).

RESULTS

Powder Blending Process Monitoring

The quality of mixtures must ultimately be judged upon the basis of some measure of the random distribution of their components. Such an evaluation depends on the selection of a quantitative method of expressing the quality of randomness or "goodness of mixing". Previous study[14] showed that NIR spectra is information rich and carries both chemical and physical information of the pharmaceutical materials studied. In this study, a moving standard deviation method was employed to calculate the standard deviation of the NIR absorbance spectra acquired at three consecutive pre-defined time points over the entire wavelength region (1100 nm to 2300 nm), as shown in Fig.1 for formulation #3. Therefore, two neighboring plots will always have two consecutive time points overlapped. When two such plots converge over the entire wave length region, mathematically the blending process has approached its end-point. An ideal situation would be that two neighboring plots become flat lines and parallel to the wavelength axis, as shown in Fig.1 for "358" and "5810" plots. However, deviation from such an ideal situation may be easily observed.

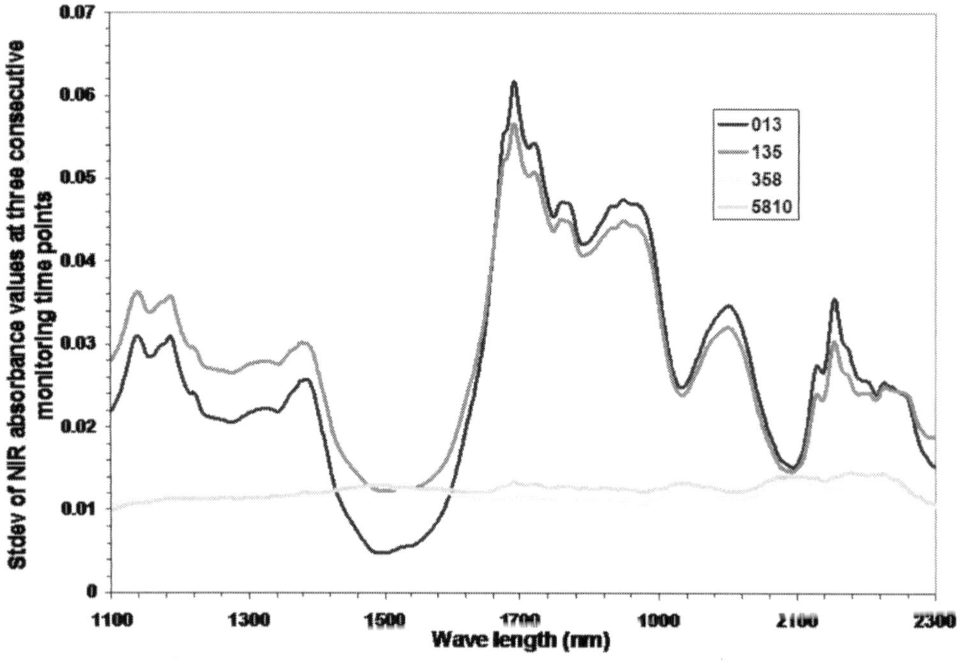

Fig.1 Evolution of stdev of averaged NIR spectra at three consecutive monitoring time points during blending. Legend stands for three consecutive time points of monitoring.

Furthermore, a plot of the stdev of the moving standard deviation vs. blending time can be made for each formulation, as shown in Fig.2 for formulation #3. The utility of this plot is discussed below.

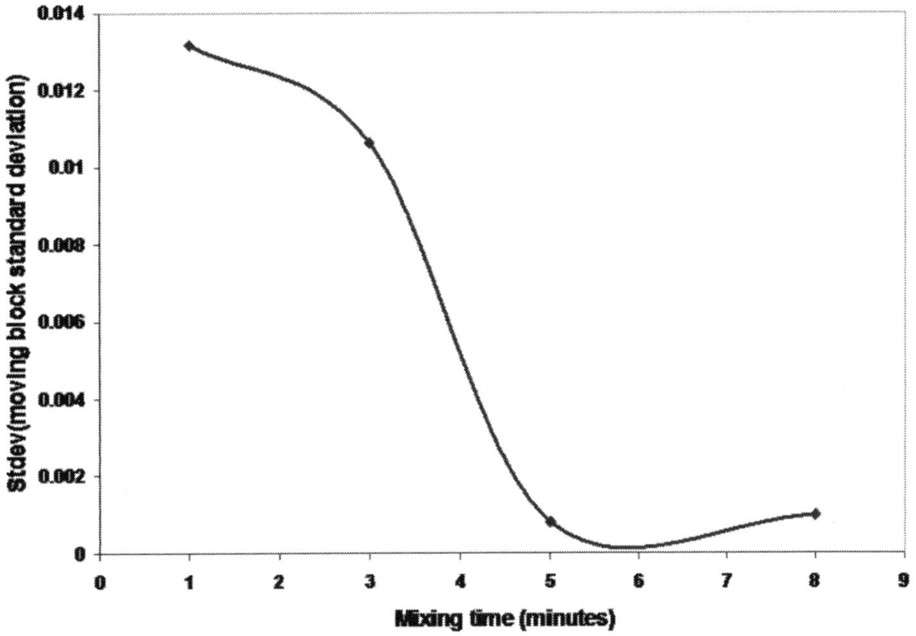

Fig.2 Integrated spectral standard deviation shown decreasing as a function of time.

Powder Blending Process End-point Determination

Powder blending process end-point can be determined by various measures. Mathematically, it could be determined by the inflection point on the plot of stdev of the moving block standard deviation vs. blending time, as illustrated in Fig.2. Or, it could be determined by comparing component concentrations at various time points through integrating UV measurement and multivariate calibration method, as discussed in the following section.

Concentration Quantification of Final Blended Mixture

In this work, the concentrations of final blended mixture were predicted by using two different multivariate calibration models established. One was the calibration model established to correlate the NIR absorbance data of blended powder with the concentrations of components which are known from the formulation component weights, as shown in Fig.3. The other one was the calibration model established to correlate the UV absorbance data with the concentrations of powder components weighted. It was found for most samples, the prediction error using PCR model for ibuprofen concentration will be smaller than 6%, although higher prediction errors were obtained for a few samples.

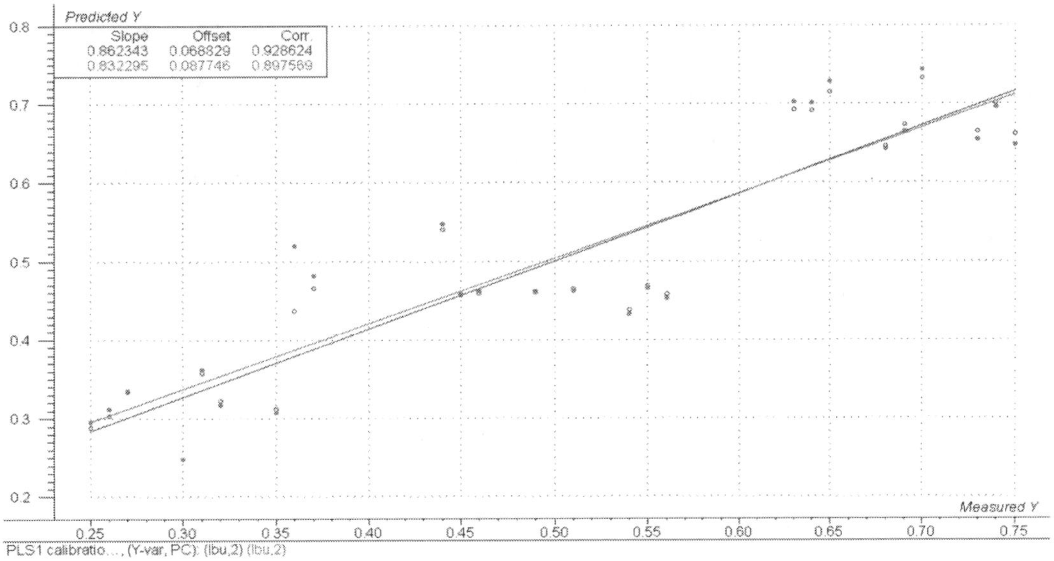

Fig.3 NIR calibration model for predicting the ibuprofen's concentration

DISCUSSION

According to the complete-random-mixture (CRM) model[15], relative variance of measured API content is inversely proportional to its true value. Consequently, measurement uncertainties of low-dose pharmaceutical mixture measurements are larger than their high dose counterparts. An enlarged sampled volume is significant for reducing measurement uncertainty and enhancing the precision for determination of low-dose API contents. Theoretically, the same principles apply to excipients if measurement of excipients in the powder formulation is feasible and of interest. In this work, HPMC and Eudragit are minor components in terms of their concentrations. Therefore, the measurement uncertainty would be expected to be higher for HPMC and Eudragit. The illuminated spot size (d) for the NIR probe is approximately 4 mm. Penetration depth (H) that depends on density of powder will be approximately

0.5-2 mm. Let D_b and D_{tap} be the bulk density and tap density of the powder formulation, respectively, the region of effective mass (M) sampled by NIR probe for each formulation can be estimated as follows, if we take the minimum penetration depth H_{min} as 0.5 mm:

$$M_{\min} = \pi(\frac{d}{2})^2 H_{\min} D_b \quad (1)$$

$$M_{\max} = \pi(\frac{d}{2})^2 H_{\min} D_{tap} \quad (2)$$

Based on the density measurement results, the estimated effective sampling mass for NIR probe was: [2.34g, 2.84g] for M_{min}, and [3.36g, 3.93g] for M_{max}.

On the other hand, the effective mass used for UV spectrometer measurement was 20 mg only. Therefore, on average, the effective mass sampled (or the scale of scrutiny) by NIR probe is about 119~196 times larger than that used for UV analysis. This suggests that the measurement uncertainty for the NIR probe should be much lower than that of UV spectrometer. To certain extent, NIR probe may measure the macro-mixing while UV spectrometer may measure the micro-mixing of the powder. As proposed by Danckwerts[16-17], the concepts of macro-mixing and micro-mixing may be applicable here to explain some of the discrepancy between the prediction results from the two multivariate calibration models, given that NIR and UV have different scale of scrutiny, in addition to the different measurement errors by the two techniques. Ongoing research in this regard will be reported in a later publication.

CONCLUSIONS

An integrated multivariate calibration approach was developed for powder blending process monitoring, blending end-point determination, and concentration quantification for a 4-component pharmaceutical powder blending system. The time to reach blending end-point is related to a number of factors such as the composition, the scale of scrutiny, and the homogeneity for both API and excipients in the formulation. The present study shows that the blending time end-point ranged from 5 to 20 minutes. The multivariate PLS and PCR model based on UV spectra has R^2 value of 0.99 and could predict the concentrations of both ibuprofen and Eudragit L100-55 well for binary powder mixture. The multivariate PLS and PCR models based on NIR spectra could predict the concentrations of both ibuprofen and excipients well for the final well-blended 4-component mixtures. The different scales of scrutiny estimated from the amount of samples used for UV and NIR measurement may be partially responsible for the discrepancy between the prediction results from the two multivariate calibration models.

ACKNOWLEDGMENTS

Mr. Maury White is acknowledged for technical assistance on some of the experimental work of this study. Dr. Robbe C. Lyon at DPQR/OTR/OPS/CDER/FDA is acknowledged for technical discussion.

Disclaimer: *The views and opinions expressed in this paper are only of the authors, and do not necessarily reflect the views or policies of the FDA.*

REFERENCES

1. Lachman L, Lieberman, HA, Kanig, JL, 1986. The Theory and Practice of Industrial Pharmacy, 3rd ed. Lea & Febiger, Philadelphia, PA.
2. Ashton MD, Schofield C, and Valentin FHH, 1966. The use of a light probe for assessing the homogeneity of powder mixtures. Chemical Engineering Science 21:843-849
3. Hailey PA, Doherty P, Tapsell P, Oliver T, and Aldridge PK, 1996. Automated system for the on-line monitoring of powder blending process using near-infrared spectroscopy. Part I. System development and control. J Pharm Biomed Anal 14:551-559
4. Sekulic SS, Wakeman J, Doherty P, Hailey PA, 1998. Automated system for the on-line monitoring of powder blending process using near-infrared spectroscopy. Part II. Qualitative approaches to blend evaluation. J Pharm Biomed Anal 17: 1285-1309
5. El-Hagrasy AS, Morris HR, D'amico F, Lodder RA, Drenen III JK, 2001. Near-infrared spectroscopy and imaging for the monitoring of powder blend homogeneity. J Pharm Sci 90(9): 1298-1307
6. Beretzky Á, Joó K, Erős I, and Pintye-Hódi K, 2005. Examination of homogeneity with X-ray beams. Int J Pharm 291:155-159
7. Léonard G, Bertrand F, Chaouki J, and Gosselin PM, 2007. An experimental investigation of effusivity as an indicator of powder blend uniformity. Powder Technology xx
8. Wu, Huiquan, Hussain, Ajaz S., and Khan, Mansoor A. (2007).Process Control Perspective for Process Analytical Technology: Integration of Chemical Engineering Practice into Semiconductor and Pharmaceutical Industries. *Chem. Eng. Comm.*, 194, 760-779.
9. FDA (2004). Guidance for Industry, PAT — A Framework for Innovative Pharmaceutical Development, Manufacturing, and Quality Assurance. Available at: http://www/fda.gov/cder/guidance/6419fnl.pdf.
10. FDA (2004). cGMP final report. Pharmaceutical cGMPS for the 21st Century—A Risk-Based Approach, Final Report. http://www.fda.gov/cder/gmp/gmp2004/CGMP%20report%20final04.pdf.
11. FDA/ICH (2006). Guidance for Industry. Q8 Pharmaceutical Development. Available at: http://www.fda.gov/cder/guidance/6746fnl.pdf.
12. FDA/ICH (2006). Guidance for Industry. Q9 Quality Risk Management. Available at: http://www.fda.gov/cder/guidance/7153fnl.pdf.
13. Wu, H., Heilweil, E., Hussain, A., Khan, M., 2007. Process Analytical Technology(PAT): Quantification Approaches in Terahertz Spectroscopy for Pharmaceutical Application. J. Pharm. Sci., DOI 10.1002/jps.21004

14. Wu H, Lyon RC, Hussain AS, *et al.*. 2003. Application of Principal Component Analysis in Assessing Pharmaceutical Formulation Design: Exploring the Casual Links between the Tablet Processing Conditions and Drug Dissolution Rate. In: Proceedings of Symposium on Innovations in Pharmaceutical and Biotechnology Development and Manufacturing, 2003 AIChE Annual Meeting [TE001], pg. 635-642. San Francisco, Nov. 17-21, 2003.
15. Pan T, Barber D, Coffin-Beach D, Sun Z, Sevick-Muraca EM, 2004. Measurement of low –dose active pharmaceutical ingredient in a pharmaceutical blend using frequency-domain photo migration. J Pharm Sci 93(3): 635-645
16. P.V. Danckwerts, 1953. Continuous flow systems: distribution of residence times. Chem. Eng. Sci. 2: 1-11.
17. P.V. Danckwerts, 1953. The dedinition and measurement of some characteristics of mixtures. Appl. Sci. Res. A3 : 279-296.

318b Quality By Design Using An Integrated Active Pharmaceutical Ingredient-Drug Product Approach To Development

Mary T. am Ende, Oral Products Center of Emphasis, Pfizer, Eastern Point Road, MS 8156-01, Groton, CT 06340, Vincent E. McCurdy, RFT Program Office, Pfizer, Eastern Point Road, Groton, CT 06340, Thomas P. Garcia, Regulatory CMC, Pfizer, Eastern Point Road, Groton, CT 06340, Frank Busch, Process R&D, Pfizer, Eastern Point Road, Groton, CT 06340, and Mark R. Berry, Non-Clinical Statistics, Pfizer, Eastern Point Road, Groton, CT 06340.

The design and development of a solid oral drug product was pursued using a holistic, quality by design (QbD) approach that integrated the active pharmaceutical ingredient (API) and drug product processes. The basic principles of this approach include i) conducting a team-based risk assessment process; ii) planning experiments to mitigate those risks; iii) prioritizing the experimental plans to intentionally focus development efforts in a staged manner; and iv) conducting the experiments with development of functional relationships between process parameters and quality attributes. This QbD approach is exemplified through a case study that transcends API synthesis through to the final drug product.

A critical quality attribute of the drug product was determined to be tablet content uniformity. The particle size of the API, which was a critical quality attribute for the drug substance, was identified as being a critical process parameter for the drug product. API particle size is controlled through crystallization temperature, and the normal operating range is 15-25oC. The final particle size specification was determined to be D[4,3] not more than 35 microns with a D(v,0.9) not more than 100 microns. A second critical process parameter identified for the drug product is the point at which the tablet press is shut off at the conclusion of the batch. The granulation possessed excellent flow properties, which resulted in the maintenance of weight uniformity until approximately 300 grams of blend remained. At this point, a mechanically induced segregation of the blend occurred, due to the lack of head pressure on the blend being fed to the tablet dies. As a result, the press shut-off implemented to eliminate the risk of high potency tablets was set at a level approximately 1.5 kg of blend remaining. Acceptable control of the process was accomplished through the stratified sampling of in-process tablet cores throughout the compression run. Control of these process parameters, as demonstrated by the results of testing of stratified samples of tablet cores, has resulted in the production of tablets that consistently exceed quality standards for content uniformity.

318c The Use Of Routine Process Capability In The Determination Of Proven Acceptable Ranges (Pars) Critical Process Parameters (Cpps) And Mapping Of The Design Space For An Api Process

Kevin D. Seibert, Chermical Product Research and Development, Eli Lilly & Co., 1400 W. Raymond St., Indianapolis, IN 46221, Shanthi Sethuraman, MS&T Statistics, Eli Lilly & Co., 1400 W. Raymond St., Indianapolis, IN 46221, and Chad Wolfe, CPR&D, Eli Lilly & Co., 1400 W. Raymond St., Indianapolis, IN 46221.

Consistent with the Quality by Design initiative, routine process capability can be used to establish the criticality of process parameters for an API process in relation to its impact on the CQAs. In addition, routine process capability can also be used to assist in the mapping of the design space and control space for an API process prior to the installation and validation. An example depicting the consistent assignment of criticality to process parameters, as well as a DOE driven mapping of the design space and control space will be presented.

318d Mathematical Modeling Of The Impact Of Raw Material Variabilities On Granule Particle Size Distribution During Fluidized Bed Granulation

Pavol Rajniak, Ecevit Bilgili, and Rey Chern. Merck & Co, Inc., WP78-110, PO Box 4, West Point, PA 19486

The desired state for pharmaceutical manufacturing is envisioned as a maximally efficient, agile, flexible sector that produces high-quality drug products without extensive regulatory oversight [1]. The initiative, called Quality by Design (QbD), is regarded as a framework along with Quality Risk Management and Quality Systems initiatives to attain that state. As described in [1,2], determination of a design space for raw material attributes and process parameters is at the crux of Quality by Design framework, which can be accomplished by using a combination of tools: first-principles (theory and modeling) approach, statistically designed experiments (DOEs), and scale-up correlations. In practice, a heavy reliance on DOEs in pharmaceutical solid-dosage form development and manufacturing is apparent historically (see e.g. [3,4]), while first-principles approach has attracted much less attention although it could be much more effective if successfully developed.

An important aspect of QbD framework is to develop a full understanding of how material attributes and process parameters affect critical product attributes, which in turn determine product performance. In this paper, we attempt to explore and demonstrate the use of mechanism-based mathematical modeling (first principles approach) in developing a fundamental understanding of the effects of raw material variabilities on granule quality. Specifically, fluid bed granulation of two active pharmaceutical ingredients (APIs) is considered. Granule size distribution is one of the most important attributes of the granulation that impacts tablet quality attributes. The granule size distribution is a function of both raw material attributes and process parameters. Here, we focus on the effects of the former.

Temporal growth profile of the granule size distribution under a given set of fluidized bed conditions was fit successfully using a population balance model. The model has a remarkably simple mathematical structure: one fitting parameter (agglomeration rate constant) in a size-dependent agglomeration kernel. Then, either the initial particle size of one of the major API or the process parameters were varied, and the impact on the granulation size was simulated using the same fitting parameter. Although the model has not been advanced to the level for quantitative validation, the simulations agree with experimental observation that the higher the mean size of the API is, the larger the resultant granule mean size will be. More importantly, the simulations suggest that the impact of variability in API particle size is less than that of variation of the process parameters in pilot batches, where the API size was not varied. With further refinement, the model could be validated via a minimal number of experiments, and is expected to be capable of defining the boundaries of a design space for API particle size. As the model is based on first principles, it is possible to extrapolate the validated model outside the experimentally studied domain. It is our hope that this paper will stimulate more interest in the use of theory/modeling within the QbD framework for developing and manufacturing pharmaceutical solid dosage forms.

References 1. M. Nasr, Quality by Design (QbD)– A Modern System Approach to Pharmaceutical Development and Manufacturing – FDA Perspective, FDA Quality Initiatives Workshop, North Bethesda, MD, Feb. 2007. 2. M. Nasr, FDA's Pharmaceutical Quality Initiatives, IFPAC-2007, Baltimore, MD, Jan. 2007. 3. M. Yelvigi, Design Space and Control Space Application Strategy, IFPAC-2007, Baltimore, MD, Jan. 2007. 4. P. Dickinson, Science and Risk-Based Product Development: An Example from the FDA Pilot Program, IFPAC-2007, Baltimore, MD, Jan. 2007.

318e Synchronized Approach to Developing Optimal Control and Design Spaces

Jerry D. Mitchell, Bernard McGarvey, Kumar Abhinava, Kevin Seibert, Shanthi Sethuraman, and Kristi Griffiths. Engineering Tech Center, Eli Lilly & Company, Indianapolis, IN 46285

During the development and production lifecycle of any pharmaceutical product, the manufacturing process generally goes through several iterations designed to improve quality, consistency, throughput, cycle time, and cost. A key enabler for facilitating these types of improvements is the synchronization of well-understood Design and Control Spaces. When the Control Space and the Design Space are being established in conjunction with one another, there will be fewer incompatibilities between the process and equipment and thus fewer problems during process installation, validation, production operations, process optimization, and future site-to-site transfers.

This presentation proposes a methodology based on capturing performance and design data from existing Manufacturing unit operations and using these results to help establish the Design and Control Spaces for new products.

Correlation Between Excipient Characteristics And Product Properties Through Numerical Simulations Of Tablet Compaction

A. Koynov, S. Zheng and A. M. Cuitiño

Introduction

Compaction is a powder densification procedure of paramount importance to numerous engineering applications. It is the basic operation used to form ceramic parts, pulvimetallurgy components, and pharmaceutical tablets (Blumenthal et al. 1997; Lewis and Simpkin, 1994). Compaction is also important in soil stabilization, avalanche and slide control, and the storage of mining and agricultural products Ewsuck (1997). Powder densification by compaction occurs as a natural process in a number of geological phenomena Skinner and Porter (1997); Allen and Allen (1990). Besides the vast technological interest of powder densification, the study of compaction is a source of valuable insights into the basic physics of the 'granular state' Jaeger and Nagel (1992); Edwards and Oakeshott (1989).

Compaction has been the focus of extensive research efforts. Powder densification is a complex phenomenon, however, and many aspects of the problem remain poorly understood. As a consequence, compaction procedures are currently designed with little rational guidance, and implemented at a high cost in time and expenditure. The results are often less than satisfactory. In the pharmaceutical industry, for instance, the tableting of each new product is designed on ad-hoc bases, essentially by trial and error, a state of things which delays the marketing of the product, and causes a large number of defective tablets during production Lordi (1997).

Since the experimental study of the powder consolidation processes is severely impeded by hardware requirements necessitated by the pressures involved, numerical modeling remains as the main recourse for the study of the system. Recent advances in computer hardware and software have made ever more detailed simulations of granular system possible. Motivation for this type of simulations is provided by the desire to root models on a sound physical basis. There is, therefore, an increasing need for robust and versatile computational techniques for incorporating a clear physical description into numerical simulations.

A *discrete* model (also called *particle* model) would appear to be ideally suited to describe the deformation of granular materials. Unfortunately, no thermal agitation drives these systems to probe their phase space; as a consequence, the presence of very many degrees of freedom cannot turn to our advantage, as it does in thermal physics Jaeger and Nagel (1992). There remains the possibility of computational simulations based on the discrete model. Two different approaches have been advanced, and implemented in a number of flavors. In the first approach, the dynamic equations of a set of interactive particles are integrated in time using an explicit algorithm Cundall and Strack (1979). Dissipative effects are incorporated in the model to account for the macroscopic nature of the particles, but otherwise the method is identical to molecular dynamics. One problem with this approach stems from the short time steps required by the explicit integration. The problem is circumvented by speeding up the simulations (as is usually done in molecular dynamics). In many cases of interest this procedure distorts the behavior of the granular material.

In compaction simulations, for example, the model predicts features with the appearance of a transformation front; in actuality, the observed features are sound waves propagating through the powder Kong and Lannutti (2000). The sound speeds are unrelated to the boundary conditions, of course, whereas the speed of the transformation front is completely determined by the boundary conditions. In the second approach the equations of static equilibrium are imposed on the system of particles at each time step Bagi (1993).

With this approach, ways have been devised Bojt´ar and Bagi (1993) to deal with snap-through buckling and other important mechanisms of particle rearrangement Kuhn et al. (1991); Guyon et al. (1990). The equations must be solved using an implicit method, however; this is an expensive computational process, which precludes the performance of large-scale simulations.

Although these two approaches have afforded many interesting insights into the micromechanics of granular materials, they require an inordinate number of degrees of freedom, of the same order of magnitude as the number of particles. As a result, their application has been limited to small systems, and they remain virtually unused in real applications. For engineering purposes other, much cruder approaches to the computational simulation of granular materials are employed. In the design of compaction procedures, computational simulations are customarily based on conventional finite elements with a plasticity material model Ewsuck (1997); Aydin et al. (1996). These continuum material models, however, are inherently incapable of accounting for the inhomogeneous nature of particle deformation.

In this work, we present the results of a mixed discrete/continuum or *granular quasicontinuum* (GQC) approach, which retains most of the advantages of discrete approaches, while circumventing some of their drawbacks. Quasi-continuum formulations have been recently used in the context of processes involving *atomistic scale* such as crystal defects, plastic deformation and fracture Tadmor et al. (1996); Shenoy et al. (1999); Knap and Ortiz (2000). More specifically, they have been utilized to study the interaction among cracks, grain boundaries and dislocations in metallic crystals Miller et al. (1998a,b); Shenoy et al. (1998), to simulate the formation and annihilation of Lomer-Cottrell in fcc single crystals Rodney and Phillips (1999), and more recently to analyze dislocation nucleation and metallic phase transformation during indentation Shenoy et al. (2000); Smith et al. (2000).

For the purposes of brevity, we have omitted the formulation of the method, which will be available in a shortly forthcoming publication. What follows is a sample of the results obtained for the uni-axial compaction of a powder bed composed of two different types of particles

Particle Pouring

Powder pouring sets the initial conditions for compaction, in the form of a pre-compaction structure, and is therefore bound to greatly influence the characteristics of the final compact. Many factors influence the structure which obtains during pouring; among these factors granule shape and the ratio between granule and die size play a more or less important role. However, the dominant factor appears to be intergranule cohesion, especially for very fine powders. Therefore, we have concentrate on pouring of cohesive particles. The numerical simulations are based on the so-called ballistic deposition method. This method has been successfully applied to a great variety of aggregation problems wherein the result is ruled by short-range attractive forces (or the lack thereof). Under these conditions the magnitude of these forces is irrelevant to the process; the key factor is whether they are present or not. As predicted by the ballistic deposition method, cohesive aggregates remain locked far away from equilibrium (in the sense of a global minimum) due to the limited capacity for accommodation (or diffusion) inherent to the process, thereby leading to open structures.

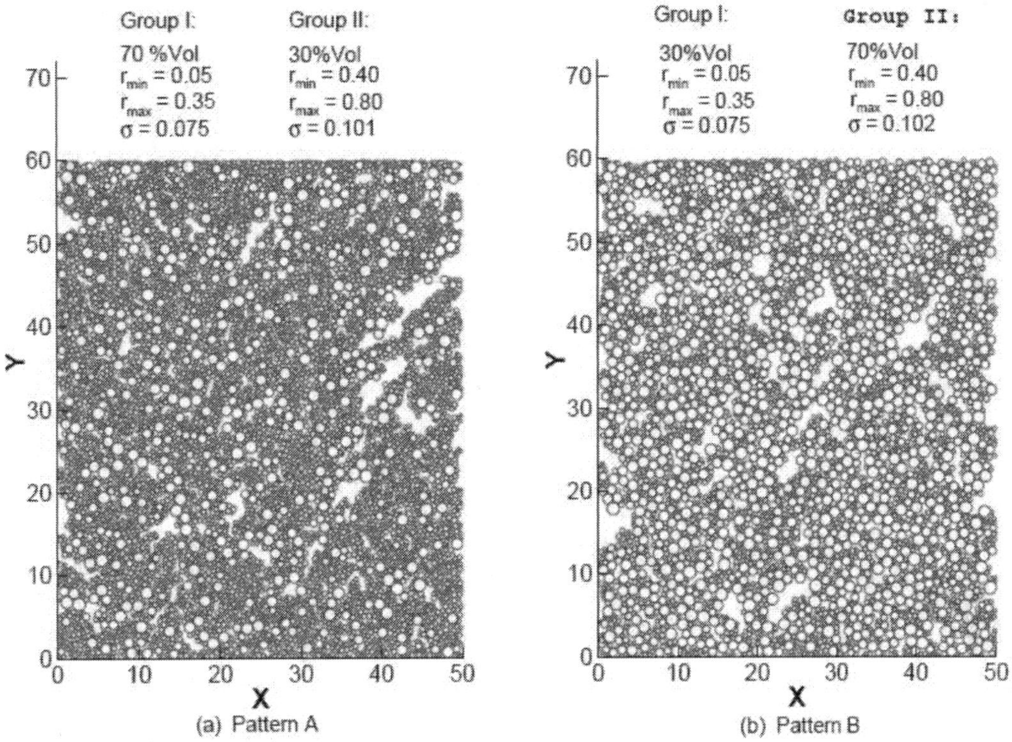

Figure 1: The initial powder configuration for (a) pattern A with 70% in volume of fine particles and (b) pattern B with 30% of fine particles

In order to ascertain the influence of particle size on the structure and the compaction behavior of the powder, simulations of pouring into a rectangular frictionless pit (width W= 50 mm and height H = 60 mm) of two groups of particles are conducted. One group (labeled group I) of fine particles and a second group (labeled group II) of coarse ones. On average particles in the group II are three times larger than the ones in group I. Using these particle sets, two different mixtures of particles are generated as shown in Fig. 1. The first pattern, referred as A, is composed of 70%_ in volume of the fine particles and 30% of coarse ones. The second pattern (labeled B) has the opposite composition.

Both configurations exhibit a distinctive feature characterized by an open structure, which is due to the existence of the cohesive forces. These patterns are also affected by particle size distribution and degree of mixture. In both cases, three distinct regions are observed: *i)* high density region where particles show a nearly closest packing structure, *ii)* low-density region characterized by a disordered array of particles and *iii)* regions with relatively large pores. A closer inspection of the differences between the two configurations reveals that pattern A exhibits larger high-density regions with elongated pores, while pattern B shows larger low-density regions with rounder pores. The reasons for such differences may be found on the details of the deposition process.

In a powder mixture dominated by fine particles (pattern A), an irregularity or defect originated in at a large particle may be either be quickly filled by large number of available fines resulting in a dense local packing, or the upper lips of the defect may grow continuously by aggregation of fines. The shortage of larger particles avoids a premature closure of the defect inducing a vertical growth of the upper lips, resulting in a configuration dominated by elongated pores aligned with pouring direction as observed in Fig. 2a. In contrast, in a mixture dominated by coarse particles (pattern B), a defect is rapidly closed (instead of filled) by a large particle, resulting a local disordered structure and rounder pores as shown in Fig. 2b.
These differences are also translated into the average density of the powder bed as well as in the density distribution as shown in Fig. 2 for both cases. Local variations of the relative density range from in the 0.60 to 0.85 with an average value of 0.715 for A and 0.724 for B. Although the average density difference is relatively small (~1.25%), the resulting mechanical behavior is markedly different as we shall see in Fig. 2, indicating that relative density is not enough to characterize the mechanical properties of a powder bed.

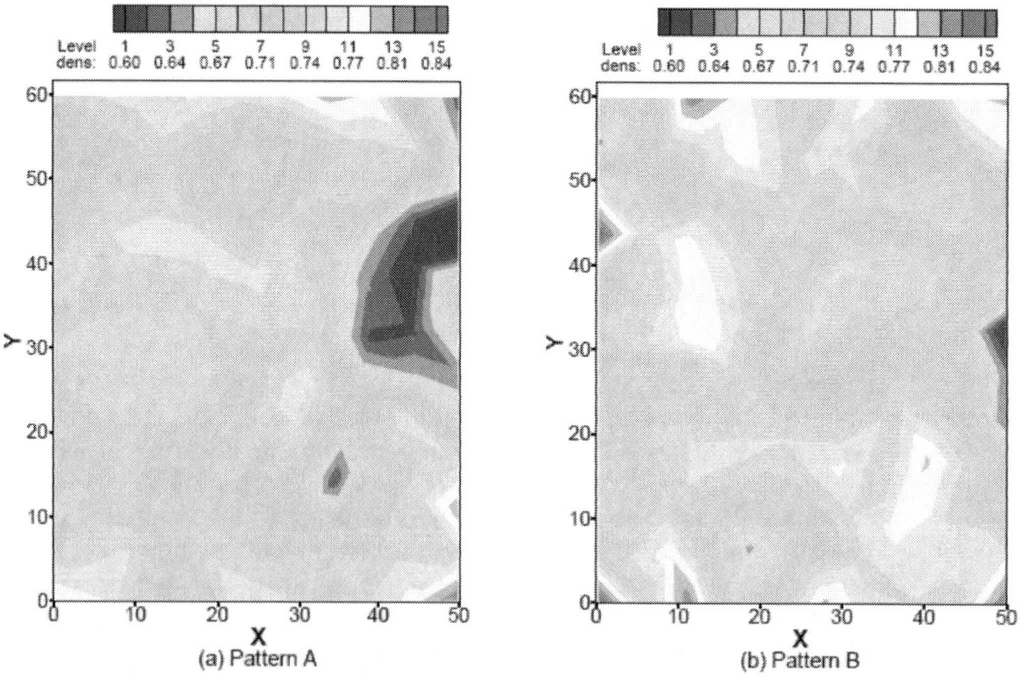

Figure 2: The initial density distribution contours for (a) pattern A and (b) pattern B.

Uni-axial Compaction Simulations

Once an initial configuration of particles is established, either by numerical simulations – the present case– or by extracting data directly for experiment, a interpolant mesh needs to be designed. The mesh utilized in this case is composed by 6-node (3 corner and 3 mid-side nodes) triangular elements with quadratic interpolation. Fig. 3 shows the mesh used in these simulation, which embraces all the particles in the pit. The nodal points are not required to coincide with the particle locations, in fact the same mesh is used to both pattern A and B.

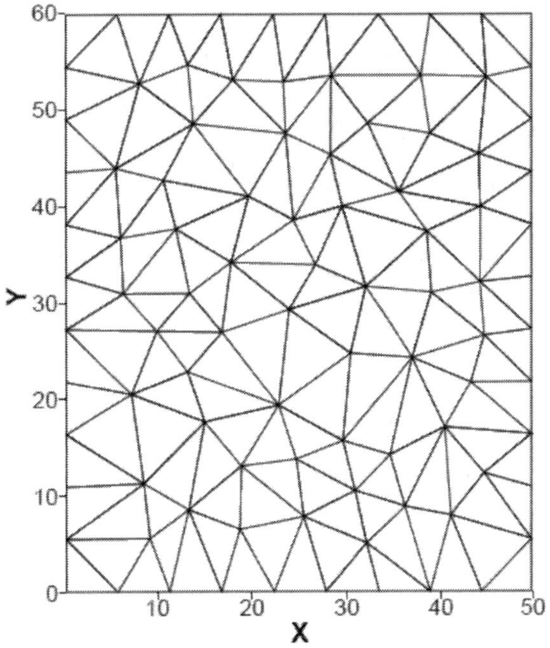

Figure 4: Typical finite element mesh used in the calculations

After the mesh is created, particles are assigned to the corresponding elements. A particle belongs to a given element if its centroid is contained in the elemental volume. This rule avoids the problems of *lost* particles or repetition. Also, it is clear that a good mesh design should include all particles in the system, otherwise some particle interactions will be missed.

While multiple models for inter-particle interaction are considered, the two cases shown here include a Leonard Jones type potential, as well as a complex interaction, in which the cohesive forces are modeled using Leonard Jones, while the repulsive ones are based on a Hertzian potential (details on the interaction potentials will be available in a coming publication).

The Figures below show the particle motion and ensuing density distributions during the compaction of both configuration A and B.

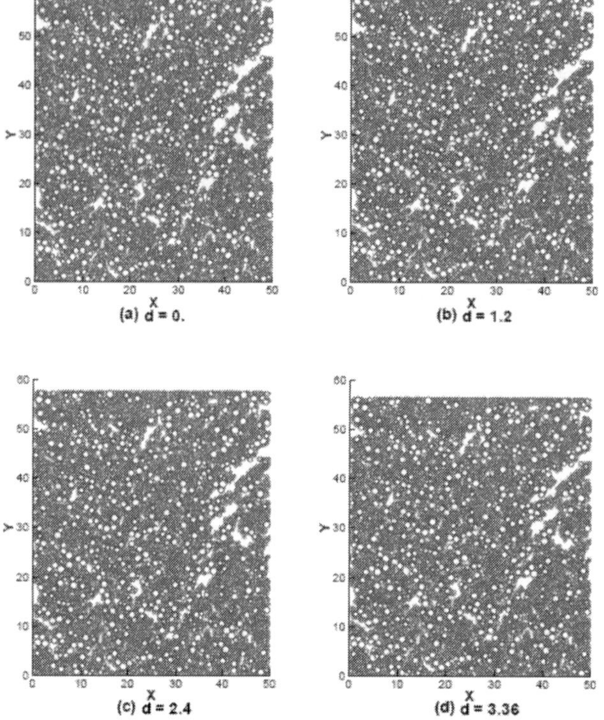

Figure 5: Predicted particle motion for mixture A at the displacement of (a) 0.0, (b) 1.2, (c) 2.4 and (d) 3.35.

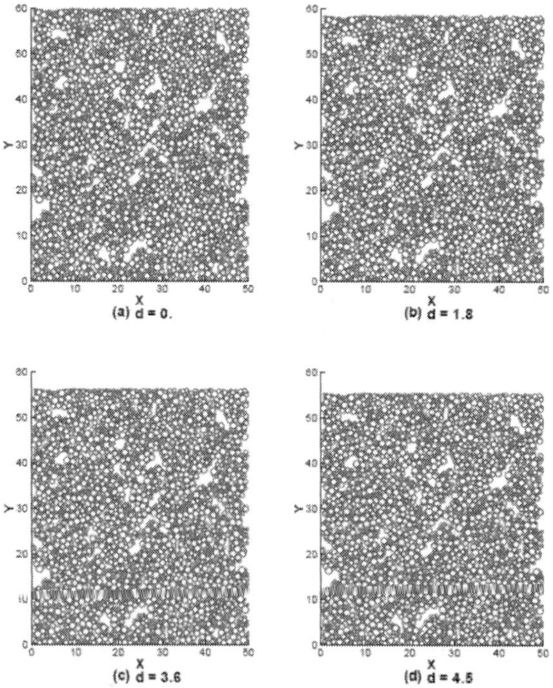

Figure 6: Predicted particle motion for mixture B at the displacement of (a) 0.0, (b) 1.8, (c) 3.6 and (d) 4.44

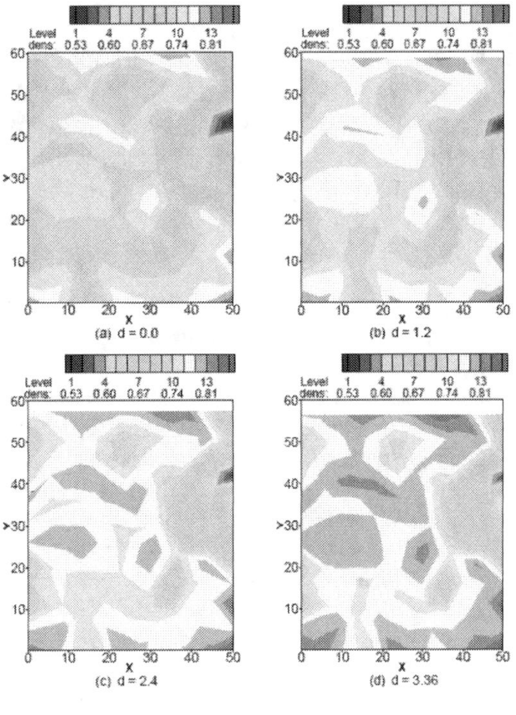

Figure 7: Evolution of density distribution of mixture A at the displacement of (a) 0.0, (b) 1.2, (c) 2.4 and (d) 3.35.

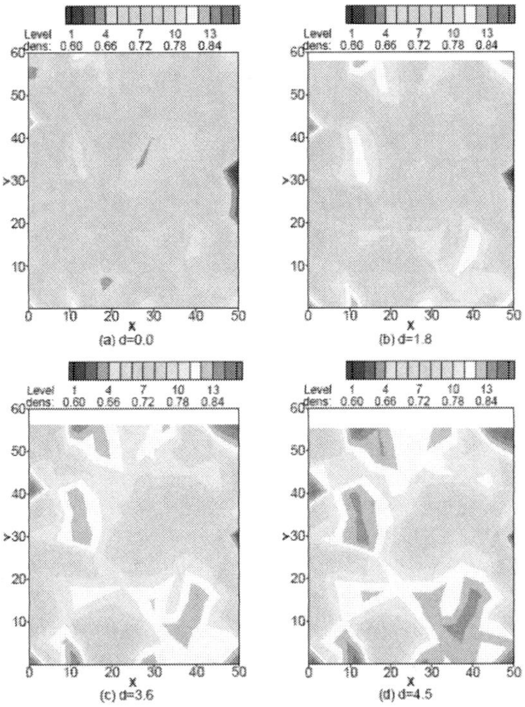

Figure 8: Evolution of density distribution of mixture B at the displacement of (a) 0.0, (b) 1.8, (c) 3.6 and (d) 4.44

Conclusions

We introduce a formulation for simulating the mechanical behavior of typical granular beds composed by particles of variable sizes and different materials. The numerical techniques combines the framework of the finite element method with interparticle constitutive laws. This quasicontinuum approach effectively bridges multiple scales from particle size to the powder bed dimensions.

References

Allen, P. A. and Allen, J. R. (1990). *Basin Analysis. Principles and Applications* . Blackwell Scientific Publications, Oxford, U.K., Oxford.

Aydin, I., Briscoe, B. J., and Sanliturk, K. Y. (1996). The internal form of compacted ceramic components: a comparison of a finite element model with experiment. *POWDER TECHNOLOGY*, 89(239):239–254.

Bagi, K. (1993). A quasi-static numerical model for micro-level analysis. *MECHANICS OF MATERIALS*, 16(101):101–110.

Baˇzant, Z. P. and Desmorat, R. (1994). Size effect in fiber or bar pullout with softening slip. *JOURNAL OF ENGINEERING MECHANICS*, 120(9):1945–1962.

Blumenthal, W. R., Sheinberg, H., and Bingert, S. A. (1997). Compaction issues in powder metallurgy. *MATERIAL RESEARCH SOCIETY BULLETIN*, 22(12):22–33.

Bojt´ar, I. and Bagi, K. (1993). Numerical analysis of loose and bonded granular materials. *MECHANICS OF MATERIALS*, 16:111–118.

Cuiti˜no, A. M., Alvarez, M. C., Roddy, M. J., and Lordi, N. A. (2000). Experimental characterization of the pore structure during compaction of granular viscous solids.

Cundall, P. A. and Strack, O. D. L. (1979). *The distinct element method as a tool for research in granular materials, Part II.* NSF Report on grant ENG76–20711, Dept. of Civil and Mineral Engineering, University of Minnesota.

Edwards, S. F. and Oakeshott, R. B. S. (1989). Theory of powders. *Physica A*, 157:1080–1090.

Espinosa, H. D., Lu, H.-C., and Xu, Y. (1998). A novel technique for penetrator velocity measurement and damage identification in ballistic penetration experiment. *JOURNAL OF COMPOSITE MATERIALS*, 32:722–.

Ewsuck, K. G. (1997). Compaction science and technology. *MATERIAL RESEARCH SOCIETY BULLETIN*, 22(12):14–16.

Gioia, G. and Cuiti˜no, A. M. (2000). Two-phase particle rearrangement in cohesive powder densification.

Guyon, E., Roux, S., Hansen, A., Bideau, D., Troadec, J., and Crapo, H. (1990). Nonlocal and nonlinear problems in the mechanics of disordered-systems - application to granular media and rigidity problems. *REPORTS ON PROGRESS IN PHYSICS*, 53(4):373–419.

Hutchinson, J. and Evans, A. (2000). Mechanics of materials: Top-down approaches to fracture. *ACTA MATERIALIA*, 48(1):125–135.

Jaeger, H. and Nagel, S. (1992). Physics of the granular state. *SCIENCE*, 255(5051):1523–1531.

Jullien, R. and Meakin, P. (1989). Simple-models for the restructuring of 3- dimensional ballistic aggregates. *JOURNAL OF COLLOID AND INTERFACE SCIENCE*, 127(1):265–272.

Knap, J. and Ortiz, M. (2000). Quasicontiuum of crystal defects.

Kong, C. and Lannutti, J. (2000). Localized densification during the compaction of alumina granules: The stage i-ii transition. *JOURNAL OF THE AMERICAN CERAMIC SOCIETY*, 83(4):685–690.

Kuhn, L., Mcmeeking, R., and Lange, F. (1991). A model for power consolidation. *JOURNAL OF THE AMERICAN CERAMIC SOCIETY*, 74(3):682–685.

Lewis, A. and Simpkin, G. (1994). *Tableting—an industrial viewpoint in Powder Technology and Pharmaceutical Processes, Chulia, D. et al, edts.* Elsevier, Amsterdam.

Lordi (1997). Compaction science and technology. *MATERIAL RESEARCH SOCIETY BULLETIN*, 22(12):34–37.

Miller, R., Ortiz, M., Phillips, R., Shenoy, V., and Tadmor, E. (1998a). Quasicontinuum models of fracture and

plasticity. *ENGINEERING FRACTURE MECHANICS*, 61(3-4):427–444.

Miller, R., Tadmor, E., Phillips, R., and Ortiz, M. (1998b). Quasicontinuum simulation of fracture at the atomic scale. *MODELLING AND SIMULATION IN MATERIALS SCIENCE AND ENGINEERING*, 6(5):607–638.
Needleman, A. (2000). Computational mechanics at the mesoscale. *ACTA MATERIALIA*, 48(1):105–124.
Rodney, D. and Phillips, R. (1999). Structure and strength of dislocation junctions: An atomic level analysis. *PHYSICAL REVIEW LETTERS*, 82(8):1704–1707.
S., R. J. (1995). *Principles of Ceramics Processing*. J. Willey & Sons, New York.
Shenoy, V., Miller, R., Tadmor, E., Phillips, R., and Ortiz, M. (1998). Quasicontinuum models of interfacial structure and deformation. *PHYSICAL REVIEW LETTERS*, 80(4):742–745.
Shenoy, V., Miller, R., Tadmor, E., Rodney, D., Phillips, R., and Ortiz, M. (1999). An adaptive finite element approach to atomic-scale mechanics - the quasicontinuum method. *JOURNAL OF THE MECHANICS AND PHYSICS OF SOLIDS*, 47(3):611–642.
Shenoy, V., Phillips, R., and Tadmor, E. (2000). Nucleation of dislocations beneath a plane strain indenter. *JOURNAL OF THE MECHANICS AND PHYSICS OF SOLIDS*, 48(4):649–673.
Shinbrot, T., Alexander, A., and Muzzio, F. (1999). Spontaneous chaotic granular mixing. *NATURE*, 397(6721):675– 678.
Skinner, B. J. and Porter, S. C. (1997). *Physical Geology*. J. Willey & Sons, New York.
Sleep, N. and Blanpied, N. (1992). Creep, compaction and the weak rheology of major faults. *NATURE*, 359(6397):687–692.
Smith, G., Tadmor, E., and Kaxiras, E. (2000). Multiscale simulation of loading and electrical resistance in silicon nanoindentation. *PHYSICAL REVIEW LETTERS*, 84(6):1260–1263.
Tadmor, E., Phillips, R., and Ortiz, M. (1996). Mixed atomistic and continuum models of deformation in solids. *LANGMUIR*, 12(19):4529–4534.
Westwood, W. D. (1989). Reactive sputtering. *PHYSICS OF THIN FILMS*, 14:57–79.
Zheng, S., Cuiti˜no, A., and G., G. (2000). Evolution of the rearrangement front during unxial compaction.

318g Characterizing the Influence of Raw Material and Process Variation on Product Quality and Yields

Chris McCready, Process Analytical Technology, Umetrics Inc., 17 Kiel Ave, Kinnelon, NJ 07045

Biotech manufacturing is constantly influenced by variations in raw material qualities, biological performance and process changes. The ability to efficiently track and understanding the influence of these changes on the resulting process performance is essential for maintaining quality and optimizing yields. Multivariate (MV) batch analysis methods are powerful tools for characterizing process performance and linking the connection between raw material quality, the manufacturing environment and the resulting product quality. Often important metrics such as biological activity and product yield can be inferred from existing process measurements. A case study is presented demonstrating the ability of these methods to be trained to recognize the need for increased nutrient addition to assure final yields. The MV models are successfully integrated into a closed loop nutrient addition control system resulting in decreased operational costs and increase in yields.

336a Dynamic Impact Of Water Thermodynamics On Normal Phase Chromatography

David Willcox, Process Development, Abbott Laboratories, 1401 Sheridan Road, Dept R452-R13-2, North Chicago, IL 60091

Water was found to have a significant impact on a normal phase chromatography process. This was studied systematically in order to understand the impact of water on the elution, the thermodynamics of water adsorption in the solvent systems used, and the dynamics of water adsorption / desorption. This information was used to optimize two chromatography processes by dynamically controlling the water throughout the process.

336b Application Of In-Line Transmission Nir For Improved Process Understanding And Control Of A Continuous Mixing Process

Gregory M. Troup[1], John P. Higgins[1], Craig A. Mckelvey[2], Michael Lowinger[2], Jeff S. Ko[3], Luke R. Schenck[3], Robert F. Meyer[3], and Brit L. Rudeen[3]. (1) Process Analytics, Merck & Co. Inc., 770 Sumneytown Pike, West Point, PA 19486, (2) Materials Characterization & Technology Assessment, Merck & Co. Inc., 770 Sumneytown Pike, West Point, PA 19486, (3) Pharmaceutical Commercialization Technology, Merck & Co. Inc., 770 Sumneytown Pike, West Point, PA 19486

This presentation discusses the application of an in-line spectroscopic technique during the process development stage of the production of a pharmaceutical intermediate in a continuous mixer. Specifically, we will show the development and implementation of a transmission NIR measurement system to monitor the composition of a multi-component process stream in real time. These real time data are used during process development to facilitate scale up, and to assess ability of the process to dampen input noises. Additionally during developmental processing this measurement system is used for real time process monitoring and detection of process upsets.

336c Solid State Form Conversion Process

Christopher L. Burcham and Jeffrey T. Vicenzi. Chemical Product Research and Development; Process Definition, Eli Lilly and Company, Lilly Research Laboratories, Indianapolis, IN 46285

A process to dehydrate the dihydrate form of a pharmaceutical compound to a crystalline anhydrate form in the solid-state has been developed. An inert gas stream, semi-saturated with solvent is passed continuously through the bed of dihydrate. The degree of dehydration is monitored by PAT in real time to determine the process end point. The effect of solvent type, solvent concentration in the gas stream, and carrier gas flow rate on the rate of form conversion is presented. In the anhydrous state, two enantiotropic polymorphs exist. Polymorph selectivity will also be presented.

336d Development And Implementation Of A Drying Process For An Amorphous Api Solid

Yao-En (David) Li, Alex Coles, Gerrie Liaw, and Stephen Tyler. Global Pharmaceutical Operations, Manufacturing Science and Technology, Dept 045B, Building R13-2,, Abbott Laboratories, 1401 Sheridan Road, North Chicago, IL 60064

An amorphous solid is a solid in which there is no long-range order of the positions of the atoms. It is normally difficult to remove the residual solvents during drying stage, mainly due to the formation of sticky, agglomerated solids.

In this paper, authors will discuss in detail on the development of a drying process based upon its glass transition temperature for amorphous API solids in an agitated dryer. This amorphous API compound is characterized by its low thermal degradation temperature, and the drying temperature of removing trapped solvents is very close to its thermal degradation temperature. There are several operational parameters that strongly effect the glass transition temperature during drying. The authors will present an industrial agitated drying model for such an amouphous compound in a production scale.

336e New Spectrophotometric Method for the Determination of Ambroxol Hydrochloride Using Artificial Neural Networks

Asha Immanuel Raju Chaduvula, Chemical Engineering, Andhra University, Andhra University, Visakhapatnam - 530 003, AP, INDIA, Visakhapatnam, 530003, India and Kiran Babu Uppuluri, Nagarjuna University, Bapatla Engineering College, Bapatla, India.

Simple, accurate, rapid and sensitive spectrophotometric methods for the determination of Ambroxol in pure and dosage forms were proposed. The drug forms (1) A stable green colour with ferric chloride and potassium ferricyanide, exhibiting maximum absorbance at 730 nm. (2) A light blue colour with F.C. reagent and sodium carbonate, exhibiting a maximum absorbance at 780 nm against reagent blanks. MATLAB software was applied for data treatment as a multivariate calibration tool in the spectrophotometric procedure. The artificial neural network (ANN) trained by the back-propagation learning was used for modeling. The ANN used can be used to proceed the data with an average relative error of less than 1.50 %. Furthermore, the estimation of Ambroxol Hydrochloride by ANN was compared with gas chromatographic results of the same sample. The percent difference between predicted and gas chromatographic results were found to be less than 2.00 % for estimation of Ambroxol Hydrochloride

336f The Use Of Pat In The Laboratory To Define And Optimize Api Manufacturing Processes

Mark A. LaPack, Chemical Process R&D, Eli Lilly and Company, Lilly Research Labs South, Indianapolis, IN 46285

Text Not Available.

336g Batch Process End Point Using Pca And Dpca Model Based On Ftir And Calorimetry Data Measurements

Shan Lin[1], Shekhar Viswanath[2], James Marek[1], Steve Richter[1], Zhi Wang[1], and Clifford Mitchell[3]. (1) GPRD Engineering, Abbott Laboratories, 1401 Sheridan Rd, North Chicago, IL 60064-6292, (2) Process R&D, Abbott Laboratories, Dept R452, Bldg R8-3, 1401 Sheridan Rd., North Chicago, IL 60064, (3) GPRD process analytical, Abbott Laboratories, 1401 Sheridan Rd, North Chicago, IL 60064-6292

Batch process is essential to today's drug production in pharmaceutical industries. Batch consistency and end-point timing lead to production quality and efficiency. In this paper, mid-infrared (FTIR) and calorimetry based principal component analysis (PCA) and dynamic principal component analysis (DPCA) were used for identification of batch reactions. We studied two types of triflation reactions and performed many batch processes including six production batches and acquired thousands of (FTIR) measurements directly from each batch. PCA were able to reduce this multi-dimensional data matrix into a few dimensional score plots for visualizing batch concentrations, end point and other batch variations with only limited calibration efforts. The prediction sensitivity was as low as < 1 PA% of starting R-OH residues. Combining process calorimetry data together with FTIR data, we were able to make PCA models more robust. This paper will also briefly discuss the risks and benefits of using PCA monitoring batch process operated in an inconsistent or a faulty mode.

Key word: Batch process, PCA, DPCA, MIR, Calorimetry

Numerical bifurcation analysis of periodic solutions of population balance models

M. Krasnyk[1]*, M. Mangold[1] and A. Kienle[1,2]

[1] *Max-Planck-Institut für Dynamik komplexer technischer Systeme, Sandtorstraße 1, D-39106 Magdeburg, Germany*
[2] *Otto-von-Guericke-Universität Magdeburg, Lehrstuhl für Automatisierungstechnik und Modellbildung, Universitätsplatz 2, D-39106 Magdeburg, Germany*

Theme: 03A00 Population Balance Modeling for Particle Formation Processes: Nucleation, Aggregation and Breakage Kernels

Keywords: nonlinear dynamics, periodic solutions, population balance, crystallization.

1 Introduction

Periodic oscillations in crystallization processes have been found by various authors in theoretical and experimental studies. Such results are important in applications, because the oscillations should be avoided for a better product quality. Conversely in special cases the periodic oscillations may be useful for industrial applications [3, 4].

The main aim of the contribution is to develop a numerical tool for the nonlinear analysis of periodic solutions for particulate processes. A continuous mixed suspension mixed product removal (MSMPR) crystallization process is considered as an example. The process is described by a population balance model including fines dissolution and classified product removal, which are possible sources of nonlinear behavior [2]. Model equations and derivation of a simulation model will be presented in section 2.

The nonlinear analysis of the discretized population balance model is difficult due to the high order and integral terms in the model. The problem is solved by applying the Recursive Projection Method (RPM) [5, 1]. The section 3 highlights the problem of the continuation of periodic solutions for a DAE system:

$$f(x, \dot{x}, \nu) = 0, \qquad f : \mathbb{R}^n \times \mathbb{R}^n \times \mathbb{R} \to \mathbb{R}^n \tag{1}$$

where f is sufficiently smooth, x is a state vector and ν is a parameter. A dynamic flow $\varphi(x_0, t, \nu)$ such that $f(\varphi, \dot{\varphi}, \nu) \equiv 0$ and $\varphi(x_0, 0, \nu) = x_0$ is considered as known. Also the section introduces the RPM and a numerical algorithm that realizes the continuation of periodic solutions with the projection method.

The last chapter presents continuation results for the MSMPR model.

Throughout the text $\langle .,. \rangle$ stands for the usual Euclidean inner product, $\dot{x}$ is a time derivative and F_x, F_ν are derivatives with respect to the state variables x and the parameter ν.

2 MSMPR Crystallizer Model [2]

A crystallizer in continuous mode of operation with fines dissolution and classified product removal is shown in Figure 1. The crystallization process is modelled under the following assumptions:

*Corresponding author. Tel.: +49 391 6110 377, mailto:miha@mpi-magdeburg.mpg.de

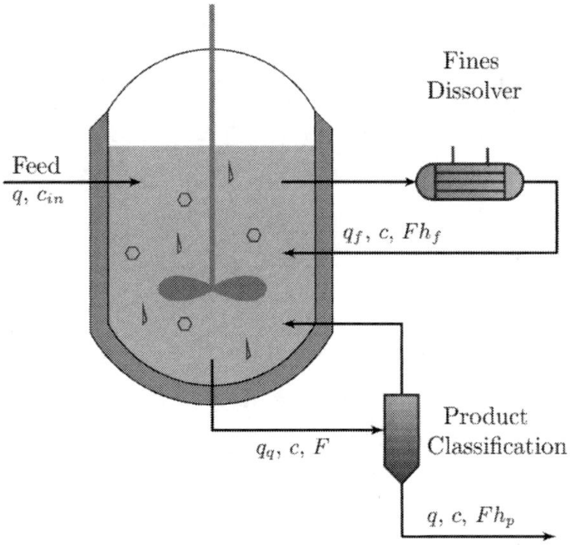

Figure 1: Continuous crystallizer with fines dissolution and classified product removal.

- ideal mixing,
- isothermal operation,
- constant overall volume (liquid+solid),
- nucleation of crystals of negligible size.
- size-independent growth rate,
- no particle breakage, attrition or agglomeration.

The population balance equation for the particle size distribution $F(L,t)$ is

$$\frac{\partial F}{\partial t} = -\frac{\partial (GF)}{\partial L} - \frac{q}{V}(h_f(L) + h_p(L))F \tag{2}$$

with the boundary condition

$$F(0,t) = \frac{B(c,t)}{G(c,t)} = \frac{k_b(c(t) - c_{\text{sat}})^b}{k_g(c(t) - c_{\text{sat}})^g}. \tag{3}$$

The classification functions specifying the fines dissolution and product classification are given by

$$\begin{aligned} h_f(L) &= R_1(1 - h(L - L_f)), \\ h_p(L) &= 1 + R_2 h(L - L_p), \end{aligned} \tag{4}$$

where $h(L)$ is a unit step function and $R_1 = q_f/q$ and $R_2 = (q_q - q)/q$ are the recycle ratios.

The mass balance of solute is

$$M_A \frac{\mathrm{d}c}{\mathrm{d}t} = \frac{q(\rho - M_A c)}{V} + \frac{\rho - M_A c}{\varepsilon}\frac{\mathrm{d}\varepsilon}{\mathrm{d}t} + \frac{q M_A c_{in}}{V\varepsilon} - \frac{q\rho}{V\varepsilon}\left(1 + k_v \int_0^\infty (h_p(L) - 1) F L^3 \mathrm{d}L\right) \tag{5}$$

where ε is the void fraction which is given by

$$\varepsilon = 1 - k_v \int_0^\infty F L^3 \mathrm{d}L.$$

For the numerical simulation the population balance model (2), (5) has to be converted to a set of differential algebraic equations. For the population balance of the dispersed phase a simple finite difference approximation on a non-equidistant but fixed grid is used according to

$$z_i = L_{max}\left(\frac{i}{N}\right)^3, \quad i = 0, 1, \ldots, N$$

with maximal length of crystals L_{max} and a total number of grid points N.

3 Numerical methods

3.1 Continuation of Periodic Solutions

A natural approach to compute stable periodic solutions is straightforward time integration of (1). The time integration is equivalent to a fixed point iteration

$$x^{(i+1)} = \varphi(x^{(i)}, T(x^{(i)}), \nu). \tag{6}$$

The convergence of (6) is determined by the eigenvalues of the linearization of φ at the fixed point x^*. These eigenvalues are Floquet multipliers μ_k of $M^* = \varphi_x(x^*, T^*, \nu)$. If the limit cycle is stable with $|\mu_k| < 1$, except for the trivial Floquet multiplier 1, the fixed point iteration (6) is stable and convergence rate is determined by the multiplier with the largest magnitude.

The solution of this problem is not unique. Indeed, every point on the limit cycle can be chosen as a fixed point x^*. To avoid ambiguity a phase condition equation is used. For example, the integral phase condition

$$s(x, T, \nu) := \int_0^T \langle \varphi(x, t, \nu), \dot{\tilde{\varphi}}(t) \rangle \mathrm{d}t \stackrel{!}{=} 0$$

minimizes the L_2-norm

$$D(\tau) := \int_0^T ||\varphi(x, t + \tau, \nu) - \tilde{\varphi}(t)||_2^2 \mathrm{d}t$$

of the difference between the solution φ and the previously calculated one $\tilde{\varphi}$ with respect to the time shift τ.

In continuation, the parameter ν also may vary and an additional parameterisation equation is required. In this work the pseudo-arclength parameterisation is used

$$n(x, T, \nu) := \langle x - \tilde{x}, \tilde{x}_t \rangle + (T - \tilde{T})\tilde{T}_t + (\nu - \tilde{\nu})\tilde{\nu}_t - \sigma \stackrel{!}{=} 0,$$

where $(\tilde{x}, \tilde{T}, \tilde{\nu})$ is a solution calculated in the previous continuation step, $(\tilde{x}_t, \tilde{T}_t, \tilde{\nu}_t)$ is a tangent vector at the previous solution point and σ is a continuation step size.

3.2 Recursive Projection Method

The Recursive Projection Method [5, 1] addresses the problem, that (6) can show slow convergence or even divergence of the fixed point iteration. The idea of the method is to split the solution space into a "stable" or fast converging subspace and orthogonal an "unstable" one, where the solution has slow convergence or diverges. The splitting allows to apply different numerical algorithms in each subspace. In this work, the Picard iteration in the "stable" subspace and the Newton iteration in the "unstable" one will be used.

Let the initial task be defined in terms of previous section as

$$\begin{aligned} F(x^{(i)}, \gamma) &:= \varphi(x^{(i)}, T, \nu), & x^{(0)} = x_0 \in \mathbb{R}^n, \\ G(x^{(i)}, \gamma) &:= \left\{ \begin{array}{c} s(x^{(i)}, T, \nu) \\ n(x^{(i)}, T, \nu) \end{array} \right\} & \gamma = \{T, \nu\} \in \mathbb{R}^2. \end{aligned}$$

And it is required to find a fixed point $\{x^* \in \mathbb{R}^n, \gamma^* \in \mathbb{R}^2\}$ of a mapping F with a nonlinear constraint G

$$\begin{aligned} x^{(i+1)} &= F(x^{(i)}, \gamma), & F &: \mathbb{R}^n \times \mathbb{R}^2 \to \mathbb{R}^n \in C^\infty, & n \gg 1, \\ 0 &= G(x^{(i)}, \gamma), & G &: \mathbb{R}^n \times \mathbb{R}^2 \to \mathbb{R}^m \in C^\infty, \end{aligned}$$

such that $x^* = F(x^*, \gamma^*)$ and $G(x^*, \gamma^*) = 0$.

The Newton iteration can be used to solve the above task

$$\left. \begin{aligned} r := F(x,\gamma) - x &= 0 \\ G(x,\gamma) &= 0 \end{aligned} \right\} \to \begin{bmatrix} F_x^{(i)} - I & F_\gamma^{(i)} \\ G_x^{(i)} & G_\gamma^{(i)} \end{bmatrix} \begin{bmatrix} \Delta x^{(i)} \\ \Delta \gamma^{(i)} \end{bmatrix} = - \begin{bmatrix} r^{(i)} \\ G^{(i)} \end{bmatrix} \tag{7}$$

with the iteration update $x^{(i+1)} = x^{(i)} + \Delta x^{(i)}$, $\gamma^{(i+1)} = \gamma^{(i)} + \Delta \gamma^{(i)}$. This numerical approach is almost is very expensive in many cases, when the calculation of the first order derivative matrix $F_x^{(i)}$ is the time-consuming part.

To avoid complete computation of the matrix $F_x^{(i)}$ the stabilization procedure is proposed [5]. Let $M := F_x^*$ have eigenvalues $\{\mu_k\}_1^n$ that for some $\delta > 0$ are ordered as

$$|\mu_1| \geqslant \cdots \geqslant |\mu_{n_p}| > 1 - \delta \geqslant |\mu_{n_p+1}| \geqslant \cdots \geqslant |\mu_n| \tag{8}$$

and $n_p \ll n$. Then it is possible to define the maximal invariant subspace $\mathcal{U}$ of M belonging to $\{\mu_k\}_1^{n_p}$ eigenvalues with projectors P and $Q := I - P$ that induce an orthogonal direct sum decomposition

$$\mathbb{R}^n = \mathcal{U} \oplus \mathcal{U}^\perp = P\mathbb{R}^n \oplus Q\mathbb{R}^n.$$

A subspace decomposition can be introduced for the mapping F

$$\begin{aligned} 0 &\stackrel{!}{=} Q(F(x,\gamma) - x) &= QF(p+q,\gamma) - q, \quad & q \equiv Qx \in \mathcal{U}^\perp, \\ 0 &\stackrel{!}{=} P(F(x,\gamma) - x) &= PF(p+q,\gamma) - p, \quad & p \equiv Px \in \mathcal{U}, \end{aligned} \tag{9}$$

so the first mapping has eigenvalues $|\lambda(QF_x^*Q)| \leqslant 1 - \delta$ with magnitude less than one and is stable for the Picard iteration.

Let $V_p \in \mathbb{R}^{n \times n_p}$ define an orthonormal basis for $\mathcal{U}$ that is spanned by the eigenvectors of M corresponding to $\{\mu_k\}_1^{n_p}$ and $V_q \perp V_p$ so

$$\begin{aligned} P &= V_p V_p^T, & & \bar{p} = V_p^T x \in \mathbb{R}^{n_p}, & p = V_p \bar{p}, \\ Q &= V_q V_q^T &= I - V_p V_p^T, \quad & \bar{q} = V_q^T x \in \mathbb{R}^{n - n_p}, & q = V_q \bar{q}, \end{aligned}$$

where $\bar{p}$ and $\bar{q}$ solution coordinates in "unstable" and "stable" subspaces. With the projectors P and Q the original task (7) can be rewritten as

$$\begin{aligned} \bar{r}_q &= V_q^T r &= V_q^T(QF(V_p\bar{p} + V_q\bar{q}, \gamma) - V_q\bar{q}) &= 0 \\ \bar{r}_p &= V_p^T r &= V_p^T(PF(V_p\bar{p} + V_q\bar{q}, \gamma) - V_p\bar{p}) &= 0 \\ & & G(V_p\bar{p} + V_q\bar{q}, \gamma) &= 0 \end{aligned}$$

and the Newton iteration is transformed to

$$\begin{bmatrix} V_q^T(F_x^{(i)} - I)V_q & V_q^T(F_x^{(i)} - I)V_p & V_q^T F_\gamma^{(i)} \\ V_p^T(F_x^{(i)} - I)V_q & V_p^T(F_x^{(i)} - I)V_p & V_p^T F_\gamma^{(i)} \\ G_x^{(i)} V_q & G_x^{(i)} V_p & G_\gamma^{(i)} \end{bmatrix} \begin{bmatrix} \Delta\bar{q}^{(i)} \\ \Delta\bar{p}^{(i)} \\ \Delta\gamma^{(i)} \end{bmatrix} = - \begin{bmatrix} V_q^T r^{(i)} \\ V_p^T r^{(i)} \\ G^{(i)} \end{bmatrix}.$$

Using the fact that $V_p^T V_p = I_p$, $V_q^T V_q = I_q$ and $V_p \perp V_q$ the previous system can be separated into

$$\begin{bmatrix} V_q^T F_x^{(i)} V_q - I_q & 0 & 0 \\ 0 & V_p^T F_x^{(i)} V_p - I_p & V_p^T F_\gamma^{(i)} \\ 0 & G_x^{(i)} V_p & G_\gamma^{(i)} \end{bmatrix} \begin{bmatrix} \Delta\bar{q}^{(i)} \\ \Delta\bar{p}^{(i)} \\ \Delta\gamma^{(i)} \end{bmatrix} = - \begin{bmatrix} V_q^T(r^{(i)} + F_\gamma^{(i)}\Delta\gamma^{(i)}) \\ V_p^T(r^{(i)} + F_x^{(i)} V_q \Delta\bar{q}^{(i)}) \\ G^{(i)} + G_x^{(i)} V_q \Delta\bar{q}^{(i)} \end{bmatrix} \tag{10}$$

The resulting system (10) has the properties:

- $|\lambda(V_q^T F_x^{(i)} V_q)| \leqslant 1 - \delta$;
- relatively small size of the bottom right part $(n_p + m \times n_p + m)$.

3.3 Numerical Algorithm

The algorithm includes four steps

- The first step is the subspace iteration that approximates dominant $n_p + n_{pe}$ eigenvectors of $F_x^{(i)}$

  Input: $V_e^{[0]}$
  Output: V_p with $\mathcal{U} \approx \text{span}\{V_p\}$
  begin
  $W = V_e^{[0]}$
  repeat
  $V_e \leftarrow \text{orth}(W);\ W \leftarrow F_x^{(i)} V_e$
  $[S, Y] \leftarrow \text{schur}(V_e^T W)$
  $\mu \leftarrow \text{diag}(S)$
  $V_e \leftarrow V_e Y;\ W \leftarrow WY$
  until *convergence of* $\mu(1 : n_p)$;
  $V_p \leftarrow V_e(1 : n_p)$
  end

- In the second step the solution with an iterative method in Q subspace is computed

$$\left[V_q^T(F_x - I)V_q\right]\Delta\bar{q} = -V_q^T(r + F_\gamma \Delta\gamma).$$

  With $|\lambda(V_q^T F_x V_q)| < 1$ the Picard iteration

$$\Delta\bar{q}^{(k+1)} = V_q^T F_x V_q \Delta\bar{q}^{(k)} + V_q^T(r + F_\gamma \Delta\gamma)$$

  is convergent.

- In the third step the solution in P subspace with a direct method is implemented

$$\begin{bmatrix} V_p^T(F_x - I)V_p & V_p^T(F_\gamma + F_x \Delta q_\gamma) \\ G_x V_p & G_\gamma + G_x \Delta q_\gamma \end{bmatrix} \begin{bmatrix} \Delta\bar{p} \\ \Delta\gamma \end{bmatrix} = -\begin{bmatrix} V_p^T(r + F_x \Delta q_r) \\ G + G_x \Delta q_r \end{bmatrix}.$$

- In the last step the solution vector is updated with a vector

$$\Delta x^{(i)} = V_p \Delta\bar{p} + \Delta q_r + \Delta q_\gamma \Delta\gamma,$$

  so the next iteration will be

$$\begin{aligned} x^{(i+1)} &= x^{(i)} + \Delta x^{(i)} \\ \gamma^{(i+1)} &= \gamma^{(i)} + \Delta\gamma^{(i)}. \end{aligned}$$

4 Simulation Results

To present the simulation results the model that is described in section 2 is used. The model parameters can be found in [2]. The population balance equation (2) of the model is discretized with 200 grid points.

To locate a region with periodic solutions the Hopf point continuation is performed. The results are shown in Figure 2. The shaded region presents parameter domain in the R_1/b plane with oscillatory behavior. As shown in the figure for a fixed reflux ratio R_1 it is possible to have domains without oscillations, and domains with one or two oscillatory regions. In figure 3 periodic continuation diagrams computed with the RPM are presented. Four subfigures in figure 3 correspond to the dashed lines on the Hopf points diagram for different reflux ratios R_1.

In figure 4 corresponding diagrams with the "slowly convergent" Floquet multipliers are presented. For $\delta = 0.9$ there are only two such multipliers in the sense of ordering (8). Consequently the restricted monodromy matrix $V_p^T M V_p$ has low dimension (2×2). Treating this small subspace by Newton iteration is of course much faster then applying Newton iteration to the whole 200 dimensional model.

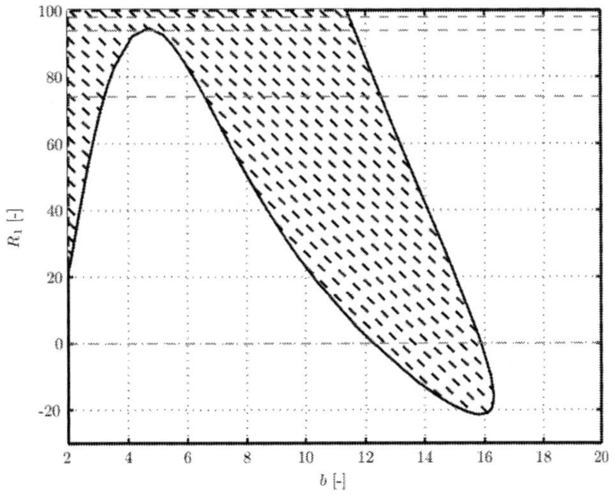

Figure 2: Hopf points continuation and instability region in the R_1/b plane (results for unphysical negative values of R_1 are added for the sake of completeness).

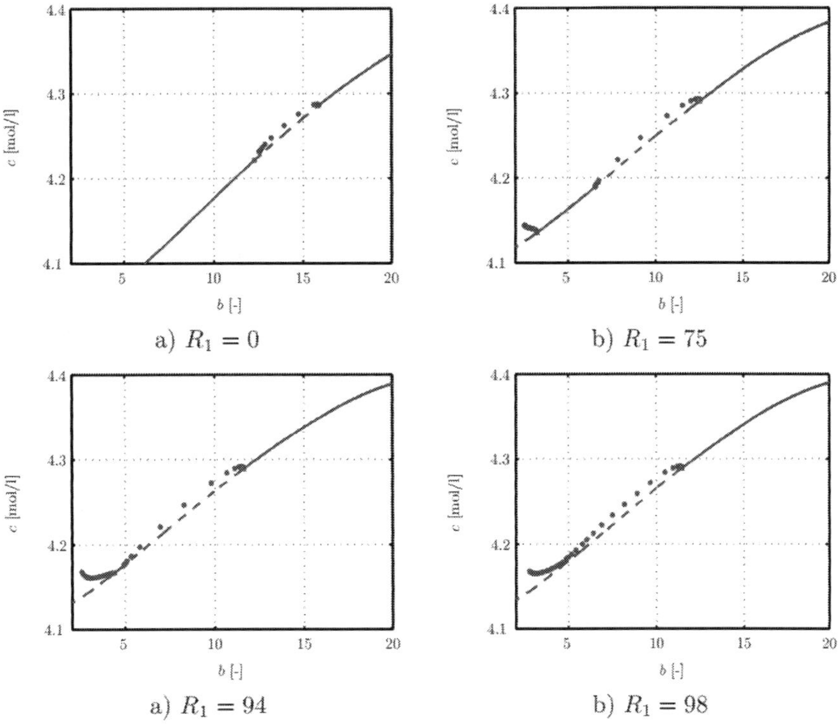

Figure 3: Bifurcation diagrams for different values R_1 in Fig. 2.

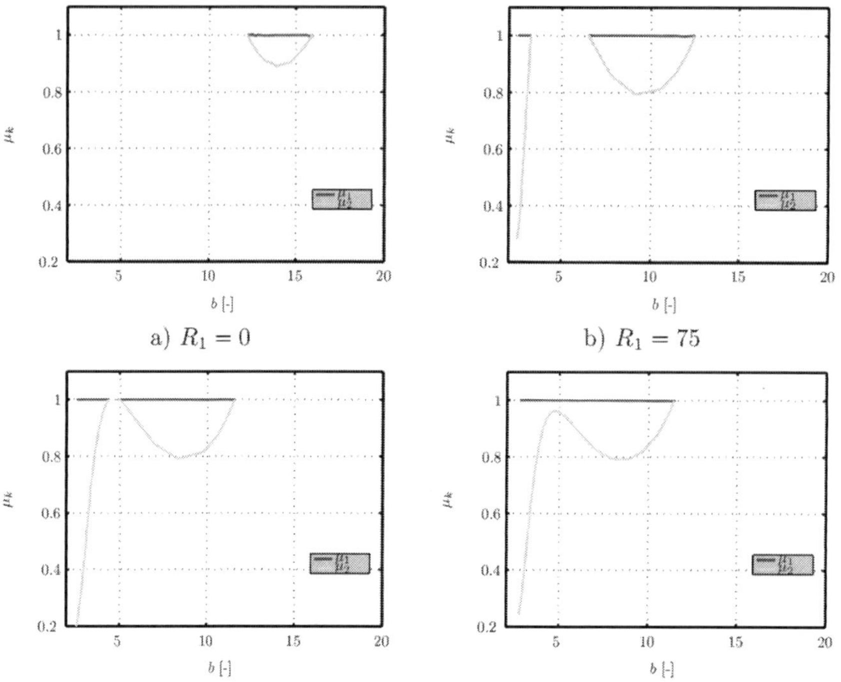

a) $R_1 = 0$ b) $R_1 = 75$

a) $R_1 = 94$ b) $R_1 = 98$

Figure 4: Corresponding Floquet multipliers to the diagrams in Fig. 3.

5 Conclusion

In this contribution an application of the Recursive Projection Method to the numerical bifurcation analysis is presented. Focus is on efficient computation of periodic solutions of a large nonlinear system. The presented results show that the analysis of periodic solutions in population balance models can be done efficiently by the RPM in combination with conventional numerical algorithms for the nonlinear analysis.

The Recursive Projection Method can be used to calculate steady-state and periodic solutions with a "black-box" DAE solver, and has good convergence properties.

Further possible enhancements of the numerical algorithm may be:

- implementation of the Krylov subspace iterations
  - for the calculation of the basis V_p (Arnoldi iteration),
  - for a solution in the Q subspace (GMRES method);
- implementation of advanced methods for the periodic boundary value problem
  - multiple shooting,
  - collocation method;
- improvement of the matrix-vector multiplication $F_x V$ for dynamic systems.

References

[1] K. Lust. *Numerical bifurcation analysis of periodic solutions of partial differential equations.* PhD thesis, Katholieke Universiteit Leuven, Leuven, Belgium, 1997.

[2] P. K. Pathath and A. Kienle. A numerical bifurcation analysis of nonlinear oscillations in crystallization processes. *Chemical Engineering Science*, 57(10):4391–4399, 2002.

[3] R. Radichkov, T. Müller, A. Kienle, S. Heinrich, M. Peglow, and L. Mörl. A numerical bifurcation analysis of continuous fluidized bed spray granulation with external product classification. *Chemical Engineering and Processing*, 45(10):826–837, Oct. 2006.

[4] R. Schütte, A. Ruhs, I. Pelgrims, C.-J. Klasen, and L. Kaiser. Verfahren zur Herstellung von Granulat mit periodisch oszillierender Korngrößenverteilung und Vorrichtung zu seiner Durchführung, 1998. Patent Application: DE 19639579 C1.

[5] G. M. Shroff and H. B. Keller. Stabilization of Unstable Procedures: The Recursive Projection Method. *SIAM Journal on Numerical Analysis*, 30(4):1099–1120, Aug. 1993.

373b Modeling Soot Formation in Turbulent Flames by Direct Quadrature Method of Moments

Federica Furcas, Alessandro Zucca, Daniele Marchisio, and Antonello Barresi. Scienza dei Materiali e Ingegneria Chimica, Politecnico di Torino, C.so Duca degli Abruzzi 24, Torino, 10129, Italy

The effect of turbulence on chemical reactions is known to be important in many practical combustion systems, in fact, it strongly influences concentration profiles of fuel, oxidiser, secondary and intermediate products, as well as of some important pollutants. The chemical pathways leading to the formation of the major air-pollutants are known with acceptable accuracy, but the formation of soot is the most complex and least understood phenomenon. Soot emissions are considered to have significant negative effects on human health and global warming. Soot formation is caused by incomplete oxidation of the species that constitute the fuel, with the production of benzene molecules, which grow to give polycyclic aromatic hydrocarbons (PAH), considered key compounds in the reactions involved in the first stages of the process. The objective of the present study is to develop a fully predictive numerical model for the prediction of soot formation and evolution based on Population Balance Equations (PBE). In this first part of the study the different approaches available (i.e., in terms of turbulence models, micro-mixing approaches, soot particle nucleation, aggregation, surface growth and oxidation models) are tested and results are compared with experimental data [1] for a first screening. The investigation carried out in this work is limited to the Reynolds-averaged Navier-Stokes approach (RANS). The closure problem is then solved with the k-• turbulence model, which has been shown to give results in good agreement with experimental data for non-swirling flames [2]. Moreover, two different micro-mixing models have been tested (i.e., the beta-pdf and the multiple laminar flamelet approach) coupled with the same kinetic scheme involving twenty different chemical species [3]. The core of the mathematical model is the PBE that tracks the evolution of soot particles. Particle nucleation, aggregation, surface molecular growth, oxidation and restructuring were taken into account into the model, the rate of each process being evaluated by simplified kinetic expressions available in the literature [4,5,6,7]. Aggregation was instead described in terms of the Fuchs theory [8], that interpolates between the continuum and the free molecule regimes. All the rate expressions were adapted to take into account the fractal nature of the particle; the simplest choice is to fix an unique value for all the particles but it is known that aged soot aggregates might undergo some restructuring processes that compact the aggregates increasing their final fractal dimension [9]. The radiative transfer due to soot particles was taken into account by introducing an additional source term for the energy equation and an effective absorption coefficient for soot was determined. Since the presence of soot particles, through radiation, has a strong effect on the flame structure, the PBE must be solved together with the continuity, Navier-Stokes, enthalpy balance, and species transport equations. A convenient approach for the solution of the PBE is based on the use of methods based on quadrature approximations. In fact, it is possible to show that, given 2Nd moments of the Number Density Function (NDF) one can calculate the Nd values of the weights and the Nd values of the abscissas of the quadrature approximation, directly solving transport equations of weights and abscissas themselves [10]. In this work a commercial Computational Fluid Dynamics (CFD) code (Fluent 6.2) is used and the PBE is directly implemented in the CFD code by resorting to the so-called the Direct Quadrature Method of Moments (DQMOM).

In the figure model predictions for soot volume fraction (fv) and total soot particle number density (m0) along the axial coordinate of the flame are reported and compared with experimental data. Experimental data taken from the literature [1] are indicated in the Figure with filled symbols whereas the different continuous lines represents model predictions obtained by using different choices of micro-mixing models and kinetics expressions for soot particle nucleation, surface growth, and oxidation. As it is seen the agreement with experimental data is acceptable showing that this quite simple modelling approach is suitable for predicting soot properties with sufficient accuracy. Moreover in this work the issue of the number of the internal coordinates is also addresses. In fact, by adopting the very same kinetics expressions DQMOM in a pseudo-monovariate and in a truly bi-variate formuations are compared and the advantage of using two internal coordinates instead than one (at least for this particular case) is eventually assessed.

References 1. Hu B.,Yang B., Koylu U. O.: Combustion and Flame 134: 93 (2003). 2. Kent J.H., Honnery D.: Combustion Science and Technology, 54:383 (1987). 3. Peeters. T.: PhD thesis, Delft Technical University, Delft, The Netherlands, (1995). 4. Fairweather M., Jones W. P., Lindstedt R. P.:Combustion and Flame, 89: 45 (1992). 5. Moss J.B., Stewart C.D., Young K. J.: Combustion and Flame, 101: 491(1995). 6. Park S. H., Rogak S. N.: Aerosol Science and Technology, 37:947 (2003). 7. Lee K., Thring M. W., Beer, J.: Combustion and Flame, 6:.137 (1962). 8. Fuchs N. A.: Mechanics of Aerosols, Pergamon, New York, U.S.A., (1964). 9. Di Stasio S., Kostandopoulos A.G., Kostoglou M. Journal of Colloid and Interface Science, 247, 33-46.(2002). 10. Marchisio D.L., Fox R. O., Journal of Aerosol Science, 36, 43 (2005).

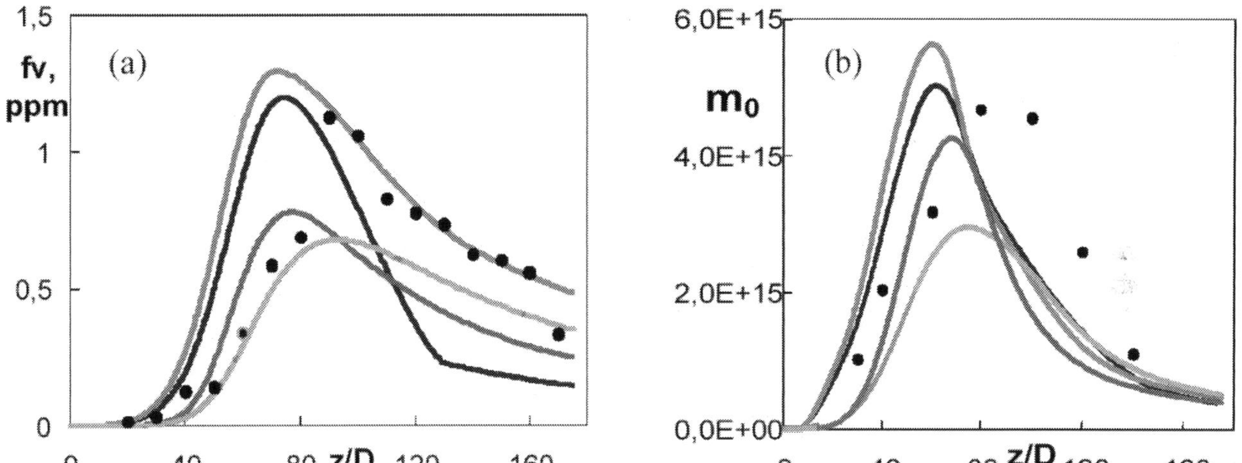

A Mechanistic Model for Nucleation and Aggregation in Population Balances of Granulation: Batch Characterisation Studies and Experimental Validation

Jonathan M.-H. Poon[1], Rohit Ramachandran[1], Constantijn F. W. Sanders[2], Thomas Glaser[2,5], Francis J. Doyle III[2], James D. Litster[3,4], Frantisek Stepanek[1], Fu-Yang Wang[3], Ian T. Cameron[3] and Charles D. Immanuel[1]

[1] *Imperial College London, London, UK*
[2] *University of California, Santa Barbara, CA, USA*
[3] *University of Queensland, St. Lucia, Australia*
[4] *Purdue University, West Lafayette, IN, USA*
[5] *Universität Stuttgart, Stuttgart, Germany*

Abstract

A dynamic model is presented for granulation processes, employing a three-dimensional population balance framework. As a first attempt to account for the multi-scale character of the process, the kernels used in the population balance model are derived using mechanistic representations of the underlying particle physics such as wetting kinetics and energy dissipation effects. An integrated population balance incorporating three dominant granulation rate processes is validated against experimental data using a calcite/PVOH-H_2O recipe performed on a lab-scale drum granulator for granule size, fractional binder content and porosity. Results obtained validate the model quantitatively as well as qualitatively.

Introduction

Granulation is a particle production process of converting fine powdery solids into larger free-flowing agglomerates. It finds application in a wide range of industries (*e.g.* pharmaceuticals, fertilisers and minerals). Granulated products often have notable improvements compared to their ungranulated form and some of these include increased bulk density, improved flow properties and uniformity in the distribution of multiple solid components. Granulation processes have been ubiquitous in the industry for many years with significant research undertaken to gain further insight into the underlying phenomena occurring during the process. However, industrial granulation processes are by-and-large operated in a highly inefficient manner with large recycle ratios within the process (3-4:1, recycle/product) [1]. As will be described next, an integrated systems modelling approach will be a crucial aid to mitigate this situation [1,2,3,4]. Furthermore, a comprehensive model of the process will enable an analysis of the system dynamics and the formulation of a suitable control strategy, which will contribute to a more efficient operation of the process [1,3].

Granulation has traditionally been described via a number of different mechanisms (namely, nucleation, layering, aggregation, abrasion transfer, and crushing). However, it is becoming more common to view granulation as a combination of three phenomenological rate processes: wetting and nucleation, consolidation and growth and attrition and breakage [5]. The formation of granules is first initiated by the nucleation of fine powder (primary particles). This involves the distribution of liquid binder among the powder, followed by the penetration of the droplet into the powder wherein the binder droplet will capture the particles surrounding it thereby converting it into a nucleus. The nuclei particles will continue to grow via aggregation and consolidation as they collide with other nuclei particles and the walls of the drum. Granule breakage may also result due to the collisions. In this study, minimal breakage occurs due to the low shear fields encountered in the drum granulator. The impetus behind this study is to validate a three-dimensional population balance model incorporating mechanistic kernels for the nucleation and aggregation phenomena whilst employing a simple empirical correlation for the consolidation phenomenon. This is carried out by performing batch granulation experiments using limestone (calcite) as the powder and a solution of poly vinyl alcohol (PVOH) as the liquid binder. The granules are characterised according to the granule size distribution (GSD), fractional binder content and porosity.

Mechanistic Model

The granulation process is usually modelled using population balance equations (PBEs) accounting for the individual rate processes [7,8]. One major challenge in developing these population balance models is the identification of suitable kernels (rate laws) for the sub-processes (e.g. aggregation). The aggregation kernel is an important component in the population balance model, as it provides the functional dependency of the growth rate on the process and material variables [9]. Due to the limited knowledge of the mechanisms that underlie the granulation process, most of the proposed aggregation kernels are empirical or semi-empirical in nature and have been largely formulated based on experimental observations. This involves fitting the coefficients in these kernels to plant or laboratory data. Such kernels have limited validity and utility in the broader perspective. Other studies have also considered physically based kernels and whilst these aggregation rate constants are more meaningful, the rate constants are still fitted from experimental data and therefore may not be completely predictive [10]. Therefore, in order to further improve the predictive capabilities of the aggregation models and to extend their region of validity, one has to directly incorporate the mechanistic features of the process and develop a kernel strongly based on first principles.

The three-dimensional population balance model employed in a previous study for modelling the granulation process accounting for aggregation and consolidation is employed in this work [11]. In addition, a mechanistic kernel was developed for the nucleation phenomenon and incorporated into the population balance model [12]. In the present model the particle population distribution is tracked according to the particle size, binder content and porosity, which are then recast in terms of the volume of solid (s), volume of liquid (l) and volume of gas (g) to facilitate the formulation of the three-dimensional PBE given in Equation (1). Breakage, layering and the drying effect have not been accounted for in the present model.

$$\frac{\partial}{\partial t}F(s,l,g,t)+\frac{\partial}{\partial g}\left(F(s,l,g,t)\frac{dg}{dt}\right)=\Re_{nuc}+\Re_{aggre} \qquad (1)$$

In Equation (1), $F(s,l,g,t)$ represents the population density function defined such that $F(s,l,g,t)\,ds\,dl\,dg$ is the moles of granules of solid volume between s and $s+ds$, liquid volume between l and $l+dl$, and gas volume between g and $g+dg$. The partial derivative term with respect to g accounts for the consolidation phenomenon. During consolidation the granules will experience a reduction in pore volume and an increase in the liquid saturation as there is only a decrease in the pore volume but the volume of solid and liquid remain unchanged. The consolidation phenomenon accounts for the densification of the granules and an associated reduction in the particle porosity (ε, which is modelled using an empirical relation as shown below where the constant c is dependent upon the mixing parameters in the granulator).

$$\varepsilon = \frac{l+g}{s+l+g} \tag{2}$$

$$\frac{d\varepsilon}{dt} = -c(\varepsilon - \varepsilon_{\min}) \tag{3}$$

$$\frac{dg}{dt} = -c\frac{s+l+g}{s(1-\varepsilon_{\min})}\left[l - \left(\frac{\varepsilon_{\min} s}{1-\varepsilon_{\min}}\right) + g\right] \tag{4}$$

The rate of nucleation is represented by $\Re_{nuc}$ whilst $\Re_{aggre}$ accounts for the formation and depletion of granules during the aggregation phenomenon. These phenomena are discussed in the next two sections.

Nucleation Kernel

The nucleation stage is considered to be crucial during the initial stages of granulation [15]. This is attributed to the fact that the nuclei size distribution obtained will influence the resulting granule size distribution. Hence, the development of a mechanistic kernel for the nucleation phenomenon taking into consideration the wetting kinetics and nucleation equilibria would provide a greater insight into the spatio-temporal behaviour of the nucleation process. In this study, a droplet controlled regime is assumed, where each droplet results in a single nucleus [13] with uniform binder distribution leading to well formed nuclei [14].

The nucleation kernel accounts for the wetting kinetics and nucleation equilibria as described next. The wetting kinetics are explicitly accounted for in determining the rates of nucleation whilst both the wetting kinetics and nucleation equilibria account for the nuclei characteristics and drop penetration time. The underlying theory is based on the studies by [13,15]. The form of the nucleation kernel is derived from the collision/transition state theory in which a minimum energy requirement for the rate process must be overcome for the nucleation event to be successful. This minimum energy requirement is represented by the spreading coefficient, which determines if it is energetically favourable for the liquid binder to spread over the powder resulting in nucleation. This in turn is related to the works of adhesion and cohesion which are defined in Equations (5)-(7).

$$\textit{Work of cohesion for a solid}: W_{CS} = 2\gamma_{SV} \tag{5}$$

$$\textit{Work of cohesion for a liquid}: W_{CL} = 2\gamma_{LV} \tag{6}$$

$$\textit{Work of adhesion for an interface}: W_{A} = \gamma_{SV} + \gamma_{LV} - \gamma_{SL} \tag{7}$$

where γ is the interfacial energy and the subscripts S, L and V represent the solid, liquid and vapour phases respectively. There are two possible scenarios in which spreading may occur (Equation (8)), [16]. The first is the formation of a surface of liquid film when the liquid spreads evenly over a solid and this occurs when the droplet diameter (D_d) is equivalent to the solid particle diameter (D_p). The second is when the solid particles embed themselves onto the droplet surface without the formation of a liquid film or when both the solid and the liquid interfaces have high works of cohesion such that the solid-liquid interfacial strength will be minimised. This occurs when the droplet diameter is significantly greater than the solid particle diameter.

$$\lambda = \begin{cases} W_A - W_{CL}, & D_d \approx D_p \\ W_A - W_{CS}, & D_d >> D_p \end{cases} \tag{8}$$

The net work required for the wetting phase between the solid-liquid combination and the process operating variables are incorporated into the calculation of the nucleation kernel given in Equation (9). In this equation, λ represents the spreading coefficient, A_0 is an adjustable parameter, Q is the binder flow rate, R is the ideal gas constant, and T is the operating temperature.

$$k_{nuc}(s,l,g) = A_o Q e^{\left(\frac{\lambda}{RT}\right)} \tag{9}$$

The rate of formation of nuclei particles and the rate of depletion of the primary particles are given in Equations (10) and (11) under the assumption of a zeroth order dependency on the concentration of the primary particles.

$$\Re_{nuc}(s_{nuc}, l_{nuc}, g_{nuc}) = k_{nuc} \tag{10}$$

$$\Re_{nuc}(s,l,g) = -k_{nuc}(s,l,g) N\, x(s,l,g) \tag{11}$$

where N is the number of primary particles in the nucleus and $x(s,l,g)$ is the fraction of primary particles with characteristics s, l and g that are present in the nucleus. The nucleation phenomenon is depicted in Figure 1.

Aggregation Kernel

The formation and depletion terms associated with the aggregation phenomenon is given by Equations (12)-(14) [11,17]. The model prescribes two conditions under which aggregation can occur. Type I aggregation occurs when a viscous force is able to overcome the initial kinetic energy of the approaching particles, causing them to aggregate prior to the surfaces making contact. Type II aggregation involves two sequential stages – the forward and reverse path. The path with the higher energetics is taken as the rate-determining step. The forward path is identical to the Type I aggregation with the addition that the particles are repelled from each other due to the elastic energy. Depending on the viscous dissipation, particles either rebound or aggregate. If the particles possess sufficient kinetic energy after complete recovery, particle separation occurs and no aggregation takes place. A schematic of the two types of aggregation is depicted in Figure 2 [6]. The steady state forces are then employed for the dynamic calculation for the aggregation rates using the Smoluchowski equation as shown in

Equation (15). The Fuch's stability ratio is defined in Equations (16)-(17) for Type I and Type II aggregation respectively.

$$\Re_{aggre}(s,l,g,t) = \Re_{aggre}^{formation} - \Re_{aggre}^{depletion} \tag{12}$$

$$\Re_{aggre}^{formation} =$$

$$\frac{1}{2}\int_{s'=s_{nuc}}^{s-s_{nuc}}\int_{l'=0}^{l}\int_{g'=0}^{g}\beta(s',s-s',l',l-l',g',g-g')F(s',l',g',t)F(s-s',l-l',g-g',t)ds'dl'dg' \tag{13}$$

$$\Re_{aggre}^{depletion} = F(s,l,g,t)\int_{s'=s_{nuc}}^{s_{\max}}\int_{l'=0}^{l_{\max}}\int_{g'=0}^{g_{\max}}\beta(s',s,l',l,g',g)F(s',l',g',t)\,ds'dl'dg' \tag{14}$$

$$\beta(p_1,p_2) = c_1\frac{4\pi u_0(r_1+r_2)^2}{W} \tag{15}$$

Type I aggregation: $$W(p_1,p_2) = (r_1+r_2)\int_{D=(r_1+r_2)}^{\infty}\frac{\exp\left(\frac{-\psi(p_1,p_2,D)}{kT}\right)}{D^2}d(D) \tag{16}$$

Type II aggregation:

$$\frac{W(p_1,p_2)}{r_1+r_2} = \max\left(\int_{D=(r_1+r_2)}^{\infty}\frac{\exp\left(\frac{-\psi_{forward}(p_1,p_2,D)}{kT}\right)}{D^2}d(D),\ \int_{D=(r_1+r_2)}^{\infty}\frac{\exp\left(\frac{-\psi_{reverse}(p_1,p_2,D)}{kT}\right)}{D^2}d(D)\right) \tag{17}$$

In the above equations p_1 and p_2 are the two particles with their respective radius r_1 and r_2, k is the Boltzmann constant, ψ is the net attractive potential, and u_0 represents the relative collision velocity between the two surface wet granules [5, 6, 11]. Mechanistic representations for the nucleation and the aggregation kernel have been incorporated into the three-dimensional population balance equation.

Numerical Solution

In solving the three-dimensional population balance model, a hierarchical solution strategy is employed [11] , that explicitly casts the population balance in terms of the underlying rate processes (i.e., nucleation, aggregation and consolidation). This technique is based on the discretisation of the three-dimensional population into bins and the recasting of the population into finite volumes along each of the three internal coordinates. In the solution of the finite volume population balance, a decomposition solution framework is used, whereby the various rate processes are accounted for individually at each time step while advancing forward in time. The adopted solution framework allows for efficient computations to be performed, which have been described in detail in [11,18].

Materials and Methods

For the drum granulation experiments, the calcite (Aglime, Landmark, Australia) feed was used as the the primary powder. This material had a mass mean diameter of 130 μm with 95% of the mass smaller than 250 μm. The skeletal density of the material was 2.746 kg/m^3. The PVOH solution (PVOH-H_2O, Elvanol T66, Du Pont, USA) was used as the binder (2.5% concentration by mass). A stainless-steel rotating drum was used with dimensions of 0.3 *m* (diameter) and 0.2 *m* (length), which was operated at 25 r.p.m. For each experiment the drum granulator was loaded with approximately 1.5 *kg* of calcite powder. Samples were taken at distinct intervals by scooping approximately 150 grams of material from 6 fixed locations within the drum. In order to test the reproducibility of the sampling method, one experiment was performed where the granuation process was stopped after 15 *mins* and 6 samples were withdrawn from the drum at the 6 fixed locations as previously described. By comparing the standard deviation across the samples, it was realised that there was minimal sampling error. The sample(s) were subsequently dried in a conventional oven for 2 hours at 110°C. After drying, the GSD was determined via sieve analysis using a $\sqrt[4]{2}$ series, ranging from 250 μm to 5648 μm. Binder content (within each sieve size range) was measured by means of spectrophotometry (Lambda 40, Perkin Elmer Instruments). Brilliant Blue dye (E133, Queen fine foods, Australia) of known concentration (8 *g/L*) was added to the liquid binder and via an absorbance-concentration calibration curve, the binder content was determined. In a similar fashion, the porosity within each sieve size range was determined by means of pycnometry. The absolute density (ρ_a) was measured using a helium pycnometer (Accupyc, Micromeritics) and the bulk density (ρ_b) using an envelope density analyser (Geopyc, Micromeritics). The porosity was calculated using Equation (18). All experiments were performed with a target binder-to-solids ratio of 0.11 and 0.12.

$$\varepsilon = 1 - \left(\frac{\rho_a}{\rho_b} \right) \tag{18}$$

Results and Discussion

The population balance model was validated by comparing the simulated output with experimental measurements. Granule samples were taken out from the drum granulator at intermittent time intervals at time instances of 180 *s*, 300 *s*, 600 *s* and 900 *s*. In the model, the primary particles are monodispersed with an average diameter of 130 μm. Other model inputs such as the initial number of seed particles, duration of binder addtion, binder flow rate, droplet volume and binder-to-solids ratio were set accordingly as per actual experimental conditions. A sensititivy analysis around the nominal operating regime was performed to determine the influence of the model inputs on the final output (particle density). Based on this analysis, sensitive parameters such as the nucleation and aggregation size limits, proportion of liquid binder penetration, *c* (Equation 4) and c_I (Equation 15) were used as tuning parameters for matching the predicted model outputs with experimental measurements for the

GSD, fractional binder content and porosity. The first set of data (binder-to-solids ratio = 0.11) was used to tune the model i.e., estimate the parameters in the model. Subsequently, the tuned model was used to predict the GSD, fractional binder content and porosity for other combinations of operating conditions and formulations. The parameters c (5×10^{-2}), c_1 (4.5×10^{7}) and A_0 (4×10^{-25} $mol\ m^{-3}$) were used for both simulations with binder-to-solids ratio of 0.11 and 0.12. The duration of the binder addition was 112 s and 122 s respectively for each case.

Binder-to-solids ratio = 0.11

Figure 3a depicts the simulated time evolution of the total particles in the system which reduces due to the combined effects of nucleation and aggregation. It can be seen that for the first 20 s there is no change in the particle numbers since the liquid binder is only introduced into the drum granulator after 20 s, then from 20 s to 2 *mins* there is a rapid decrease in particles and from 2 to16 *mins* there is a gradual decline in the rate of decrease of particle numbers. This trend is typical as in the first two minutes, binder is being sprayed which results in a sharp decrease in total particles (sum of granules and fines). Thereafter, the nucleation event stops and only aggregation occurs as a result of the residual binder that coats the surface of the particles. Figures 3b and 3c show the change in average diameter and the fractional binder content with time respectively. The model is able to capture the particle level phenomena of the process resulting in a close match between the simulated profiles and the experimentally measured data. The GSD profiles at each time instance corresponding to the intermittent sampling times during the experiment (i.e., 180 s, 300 s, 600 s and 900 s) are compared with those obtained from sieve analysis as seen in Figure 4. The GSDs are represented as the normalised mass frequency of granules with respect to granule size. A clear progression is observed during the course of granulation whereby particles agglomerate together forming larger sized particles, which is depicted by a gradual reduction in the peak spanning 0-250 μm and a corresponding growth in the peak spanning 700-1000 μm. Figure 5 shows the fractional binder content and Figure 6 shows the porosity distribution at each of the three different sampling times. The fractional binder content reported here represents the averaged quantity for a particular size range using Equation (19).

$$LS(r) = \frac{\sum_{LS} LS \times F(LS,r)}{\sum_{LS} F(LS,r)} \tag{19}$$

where LS is the liquid binder-to-solids ratio and r is the particle size. In a similar fashion, porosity is defined to be the averaged porosity of all the granules in each individual size range. From Figure 4, it can be seen that the model is able to accurately capture the GSD and to a reasonable extent the range of the experimentally measured data for fractional binder content (Figure 5). For porosity (Figure 6) a similar trend is observed between the model data and experimental data. A very good match is obtained in the GSDs as compared to fractional binder content and porosity because more emphasis was placed on the validation of the GSD (often deemed to be the more important granule attribute). Furthermore, due to the complex non-convex nature of the model, an optimisation algorithm is not used for parameter fitting. This implies that the model is tuned by increasing or decreasing one of the sensitive parameters while holding the others constant. Figures 7(a)-(f) depict the 2D distribution of total particles with respect to particle diameter and fractional binder content. As seen in the plots, the particles in the initial seed undergo nucleation and aggregation and form a larger mode of particles, thereby resulting in a bimodal distribution.

Binder-to-solids ratio = 0.12

In this section, the binder-to-solids ratio was increased to 0.12. Intuitively, this tells us that the mean diameter of granules and the fractional binder content over time should increase. This is consistent with the theory that a higher binder-to-solids ratio (provided it does not exceed a certain upper bound) promotes nucleation and aggregation resulting in lesser particles. Similar to the previous section, the predicted GSDs, fractional binder content and porosity distribution from the model are compared with those that are experimentally obtained. In Figure 8 it can be seen that the model output adequately captures the bimodal nature of the experimentally measured data for the GSD. A comparison between the predicted output and experimetnal measurements for the fractional binder content is also shown in Figure 9. It can be seen that the model is able to suitably capture the range of experimental measurements. The model output for porosity is compared with that which is experimentally measured as shown in Figure 10. It is observed that the trend of decreasing porosity (across the size fractions) is similar for both the model and experiment. Overall, for both binder-to-solids ratio (0.11 and 0.12) a good agreement was obtained between the simulated output and experimental data for the GSDs. For the fractional binder content and porosity across the size fractions, the model was adequately able to capture the range and trend of the measured data.

Conclusions

In this study, the three-dimensional population balance model employing mechanistic representations for both the nucleation and aggregation phenomena was validated for a calcite/PVOH-H_2O granulation recipe. Lab-scale experiments and data analyses were performed to obtain distributions of size, binder content and porosity data, which are an extension to typical data aquired (size and average binder content). Validation of the GSDs between model predictions and those obtained in the experiments showed good agreement. The model prediction for porosity also showed reasonable agreement particularly in matching the range and trend of the experimental data. Experimental measurements for the binder content with respect to each sieve size range were also compared with the model predictions. The simulated output for the binder content was able to capture the range, which may prove to be an added utility in anticipating an expected binder composition of the granules. This may facilitate the operation of the graulation process to ensure granule characteristics do not deviate too far from the desired values. These results are promising for the long term step toward a comprehensive integrated population balance formulation enabling the study of complex nonlinear processes, with granulation being a prime example. Future work will encompass validation of the three granule attributes at varying operating conditions and formulations to fully capture the richness of mechanistic modelling.

Acknowledgements

The authors would like to acknowledge the funding provided by EPSRC, UK, the department training allocation fund, International Fine Particles Research Institute (IFPRI) and the Australian Research Council Discovery Grant DP0345777.

References

1. P.R. Mort (2005) "Scaleup of binder agglomeration processes," *Powder Technology*, January, pp. 86-103.

2. J.D. Litster (2003) "Scaleup of wet granulation processes: science not art," *Powder Technology*, February, pp. 35-40.

3. M. Bardin, P.C. Knight and J.P.K. Seville (2004) "On control of particle size distribution in granulation using high shear mixers," *Powder Technology*, April, pp. 169-175.

4. P.C. Knight (2004) "Challenges in granulation technology," *Powder Technology*, March, pp.156-162.

5. S.M. Iveson, J.D. Litster, K. Hapgood and B.J. Ennis (2001), "Nucleation, growth and breakage phenomena in agitated wet granulation processes: a review," *Powder Technology*, June, pp. 3-39.

6. L.X. Liu, J.D. Litster, S.M. Iveson and B.J. Ennis (2000) "Coalescence of Deformable Granules in Wet Granulation Processes," *AIChE Journal*, March, pp. 529-539.

7. A.A. Adetayo, J.D. Litster, S.E. Pratsinis and B.J. Ennis (1995) "Population balance modelling of drum granulation of materials with wide size distribution," *Powder Technology*, August, pp. 37-49.

8. D. Verkoeijen, G.A. Pouw, G.M.H. Meesters and B. Scarlett (2002) "Population balances of particulate processes–a volume approach," *Chemical Engineering Science*, May, pp. 2287-2303.

9. T. Abberger (2001) "The effect of powder type, free moisture and deformation behaviour of granules on the kinetics of fluid-bed granulation," *European Journal of Pharmaceutics and Biopharmaceutics*, November, pp. 327-336.

10. L.X. Liu and J.D. Litster (2002) "Population balance modelling of granulation with a physically based coalescence kernel," *Chemical Engineering Science*, June, pp. 2183-2191.

11. C.D. Immanuel and F.J. Doyle III (2005) "Solution technique for a multi-dimensional population balance model describing granulation processes," *Powder Technology*, August, pp. 213-225.

12. J.M.-H. Poon, C.D. Immanuel, F.J. Doyle III and J.D. Litster (To appear) "A three-dimensional population balance model of granulation with a mechanistic representation of the nucleation and aggregation phenomena," *Chemical Engineering Science*.

13. K.P. Hapgood, J.D. Litster, S.R. Briggs and T. Howes (2002) "Drop Penetration into Porous Powder Beds," *Journal of Colloid and Interface Science*, September, pp. 353-366.

14. S.M. Iveson and J.D. Litster (1998) "Growth regime map for liquid-bound granules," *AIChE Journal*, July, pp. 1510-1518.

15. K. Hapgood (2002) "Nucleation and binder dispersion in wet granulation," Ph.D. thesis, Department of Chemical Engineering, University of Queensland.

16. B.C. Hancock, P. York and R.C. Rowe (1994) "An assessment of substrate-binder interactions in model wet masses. 1: Mixer torque rheometry," *International Journal of Pharmaceutics*, February, pp. 167-176.

17. D. Ramkrishna (2000) "Population Balances," Academic press, San Diego.

18. C.D. Immanuel and F.J. Doyle III (2003) "Computationally efficient solution of population balance models incorporating nucleation, growth and coagulation," *Chemical Engineering Science*, August, pp. 3681-3698.

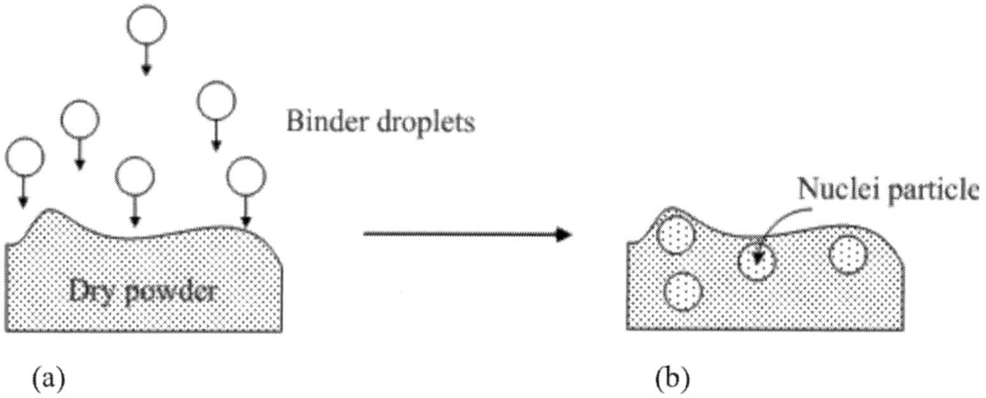

Figure 1. Schematic of the (a) wetting and (b) nucleation phenomenon

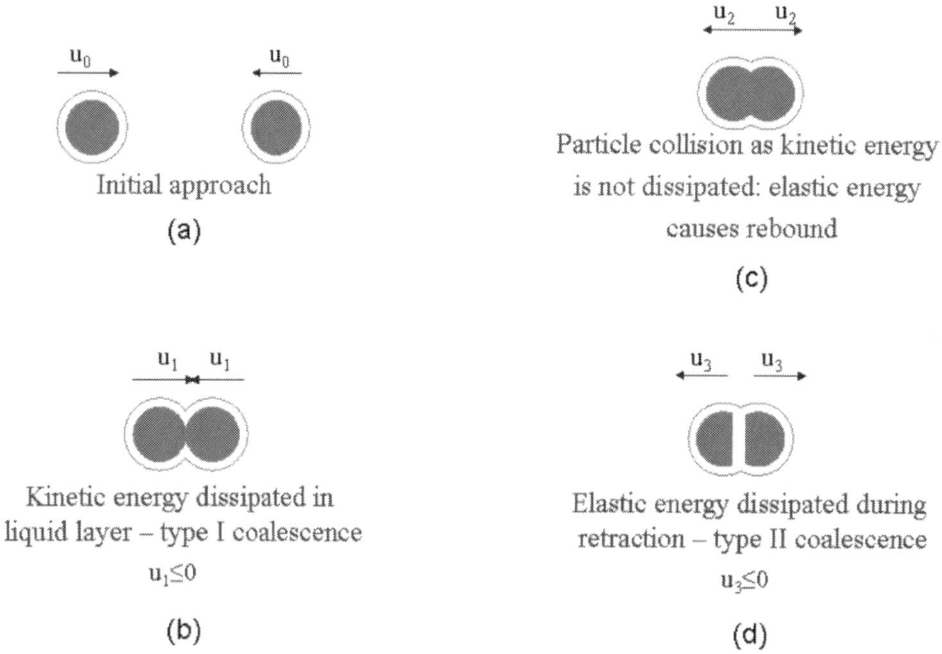

Figure 2. Schematic representation of the two types of aggregation phenomena [6]. The velocities depicted in the schematics indicate the relative velocities of the two particles. u_0 is the approach velocity at infinite separation; u_1 is the velocity upon collision; u_2 is the initial rebound velocity; u_3 is the velocity at the point when the binder liquid layers separate.

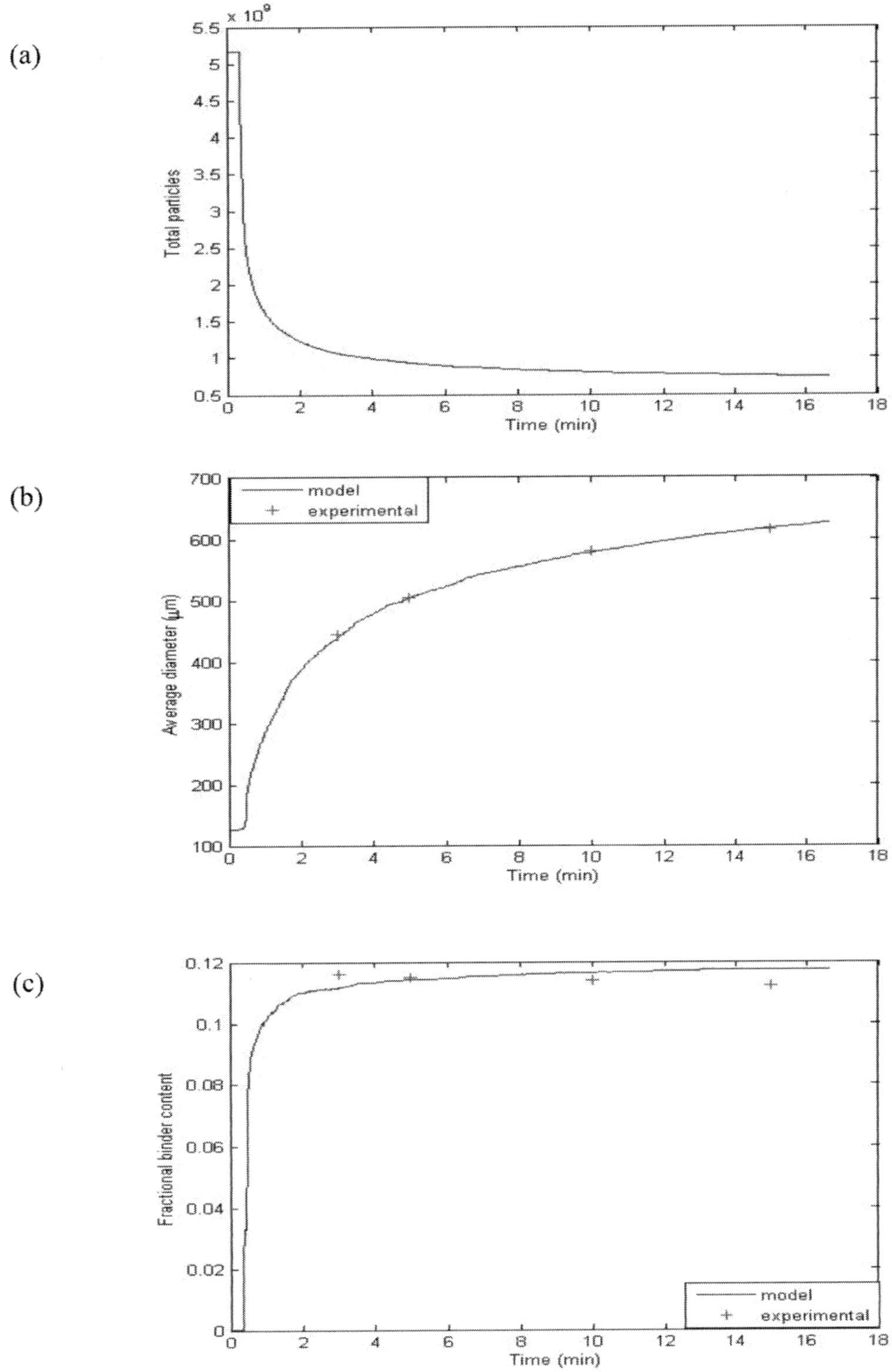

Figure 3. Comparison between simulated outputs with experimental data of time profiles for (a) total particles, (b) average particle size and (c) fractional binder content for binder-to-solids ratio of 0.11.

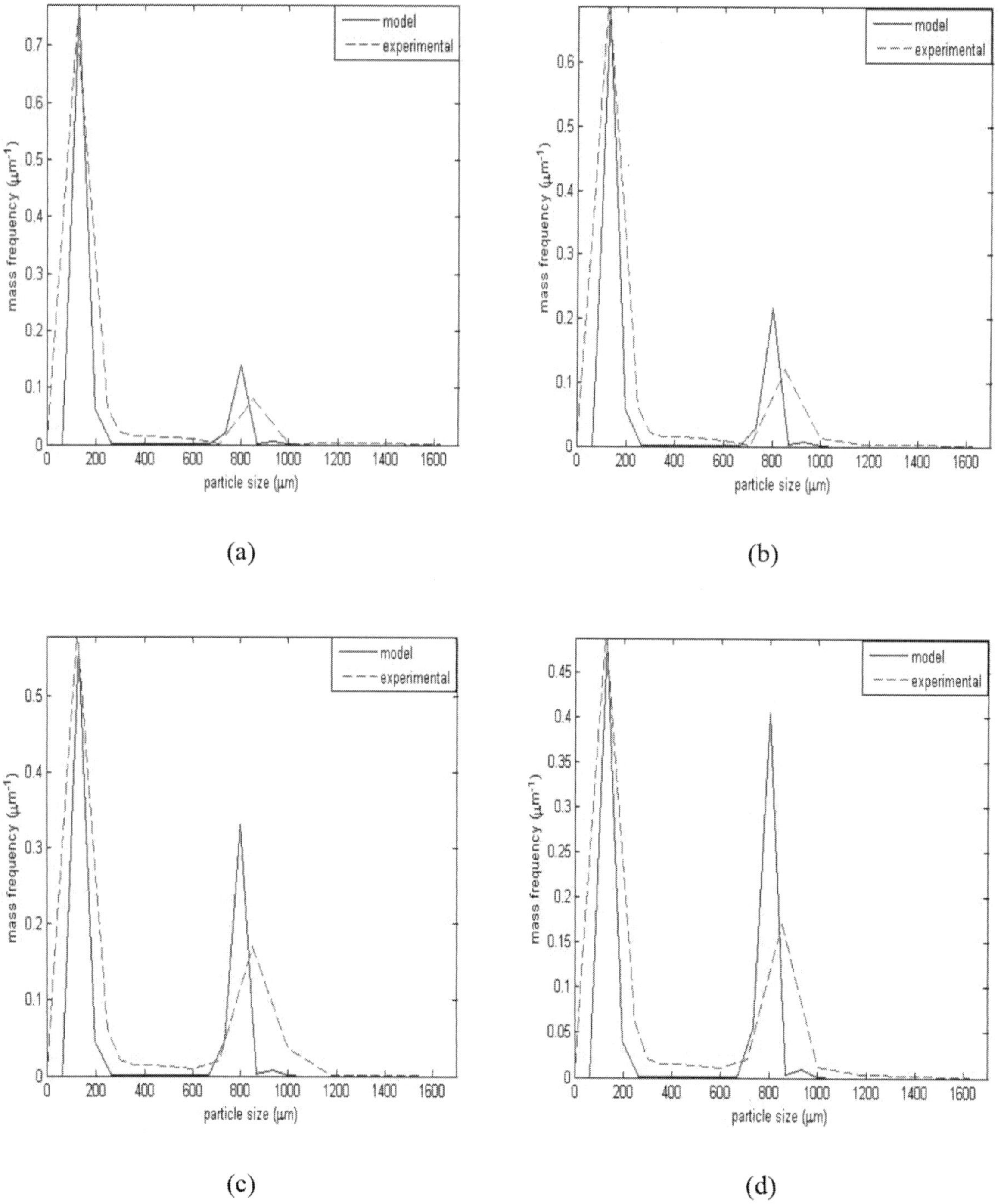

Figure 4. Comparison between simulations and experimental data for time profiles of granule size distributions at (a) t = 180 *s*, (b) t = 300 *s* , (c) t = 600 *s* and (d) t = 900 *s* for binder-to-solids ratio of 0.11.

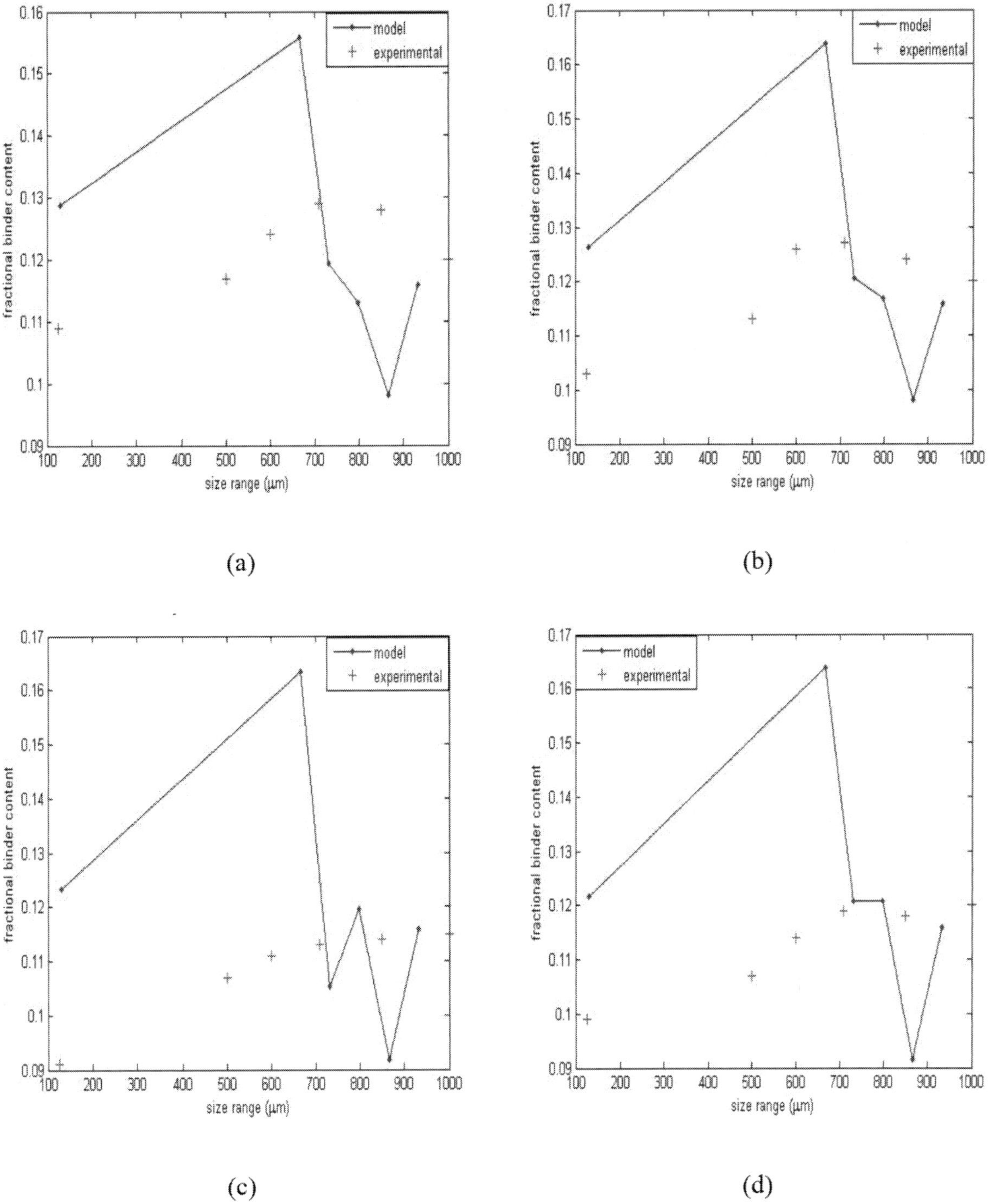

Figure 5. Time profiles of the fractional binder content for both model outputs and experimental data at (a) t = 180 *s*, (b) t = 300 *s*, (c) t = 600 *s* and (d) t = 900 *s* are compared, for binder-to-solids ratio of 0.11.

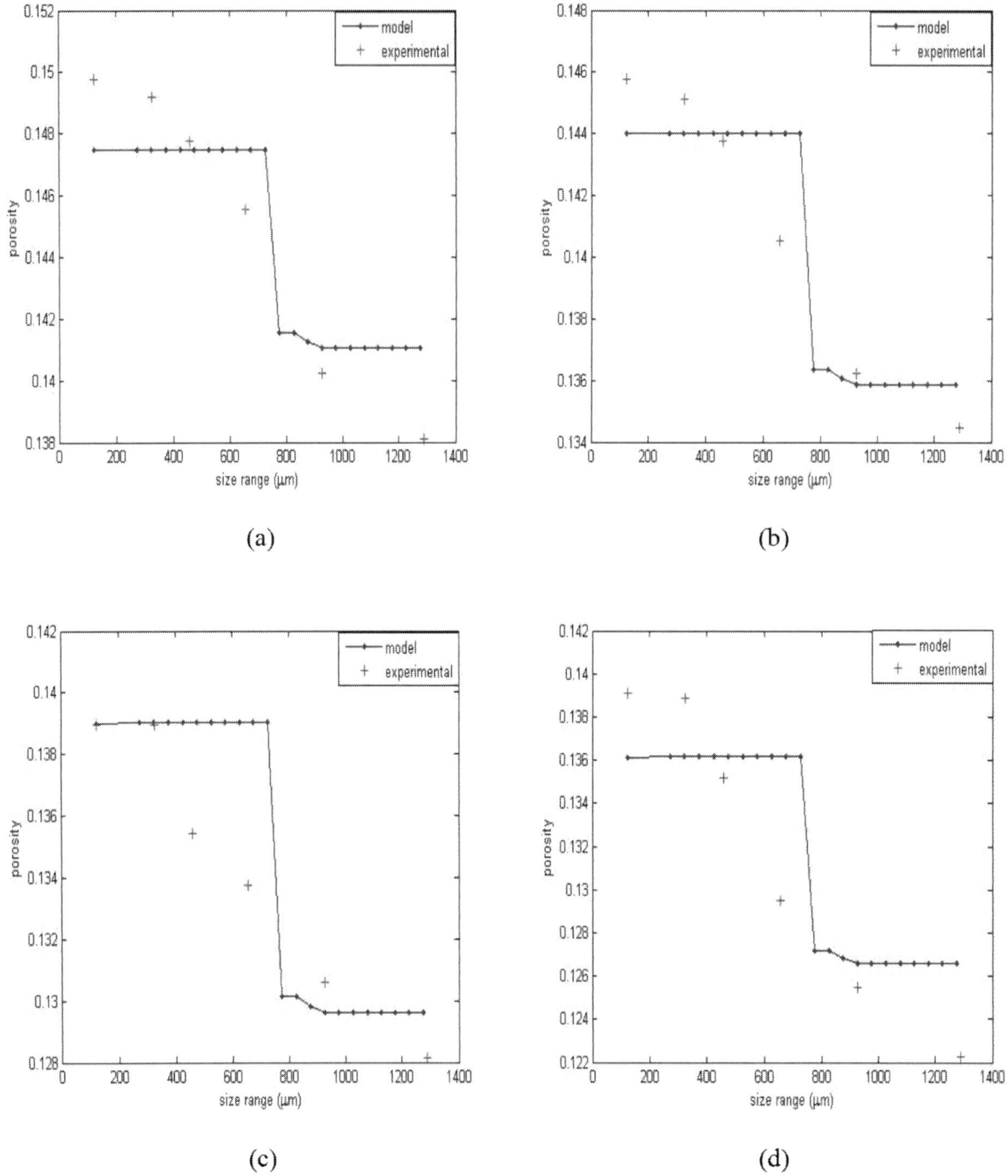

Figure 6. Distribution of the granule porosity with particle size at time
(a) t = 180 *s*, (b) t = 300 *s* , (c) t = 600 *s* and (d) t = 900 *s* are compared between simulated outputs and experimental data, for binder-to-solids ratio of 0.11.

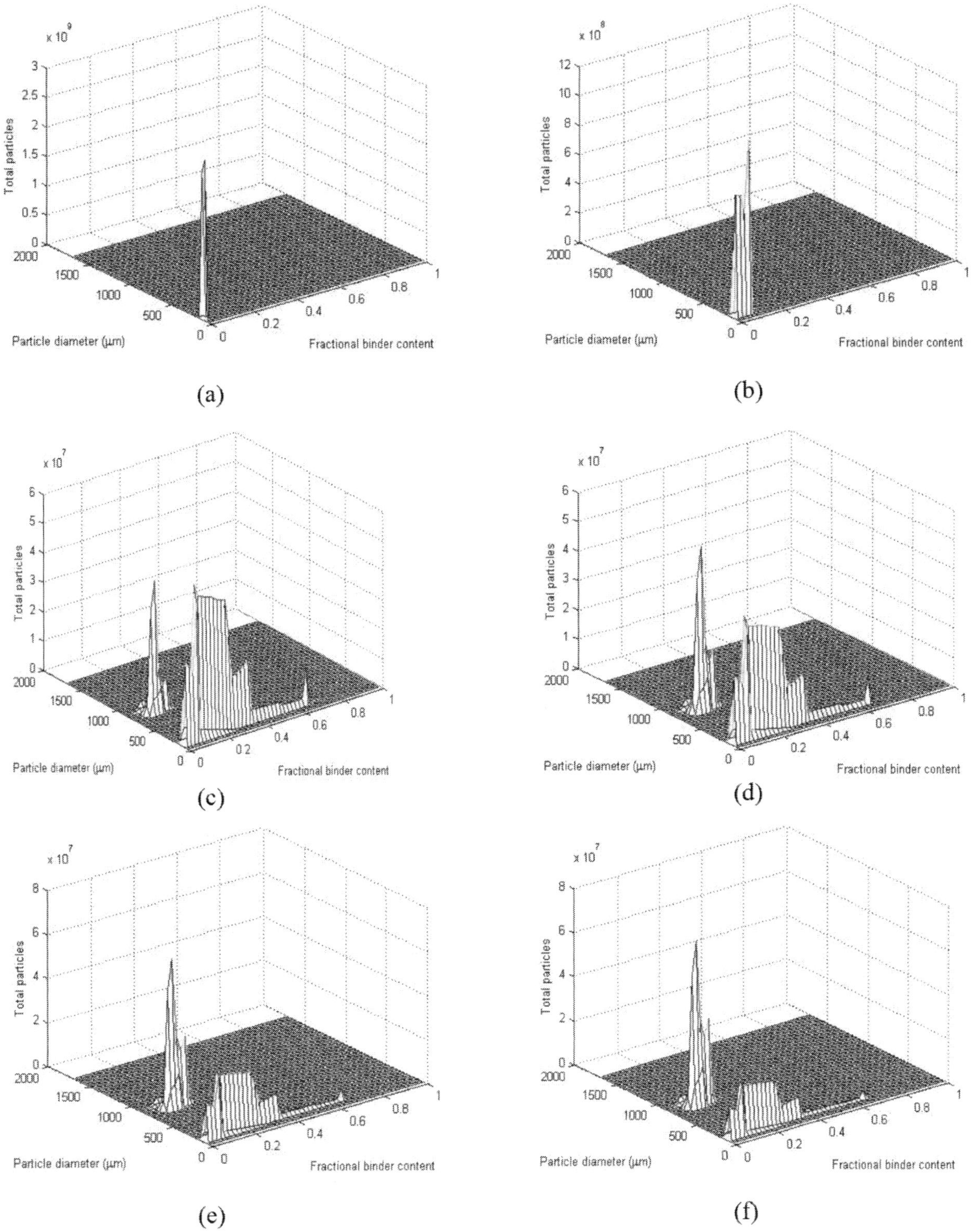

Figure 7. Evolution of the granule size distributions under combined nucleation, consolidation and aggregation at various time instances (a) 0 *s*, (b) 20.01 *s*, (c) 180 *s*,
(d) 300 *s*, (e) 600 *s* and (f) 900 *s*, for binder-to-solids ratio of 0.11.

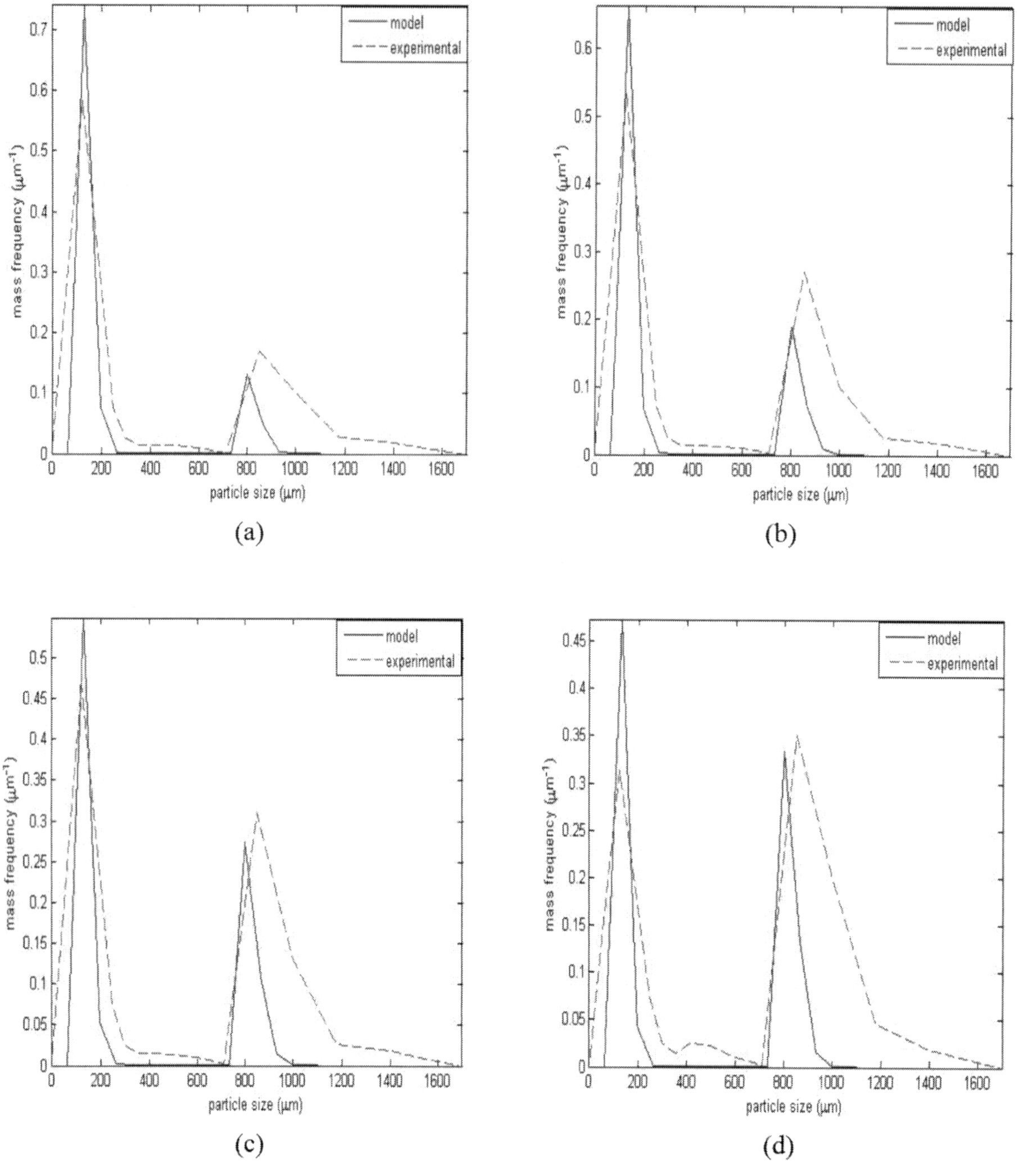

Figure 8. Comparison between simulations and experimental data for time profiles of granule size distributions at (a) t = 180 *s*, (b) t = 300 *s* , (c) t = 600 *s* and (d) t = 900 *s* for binder-to-solids ratio of 0.12.

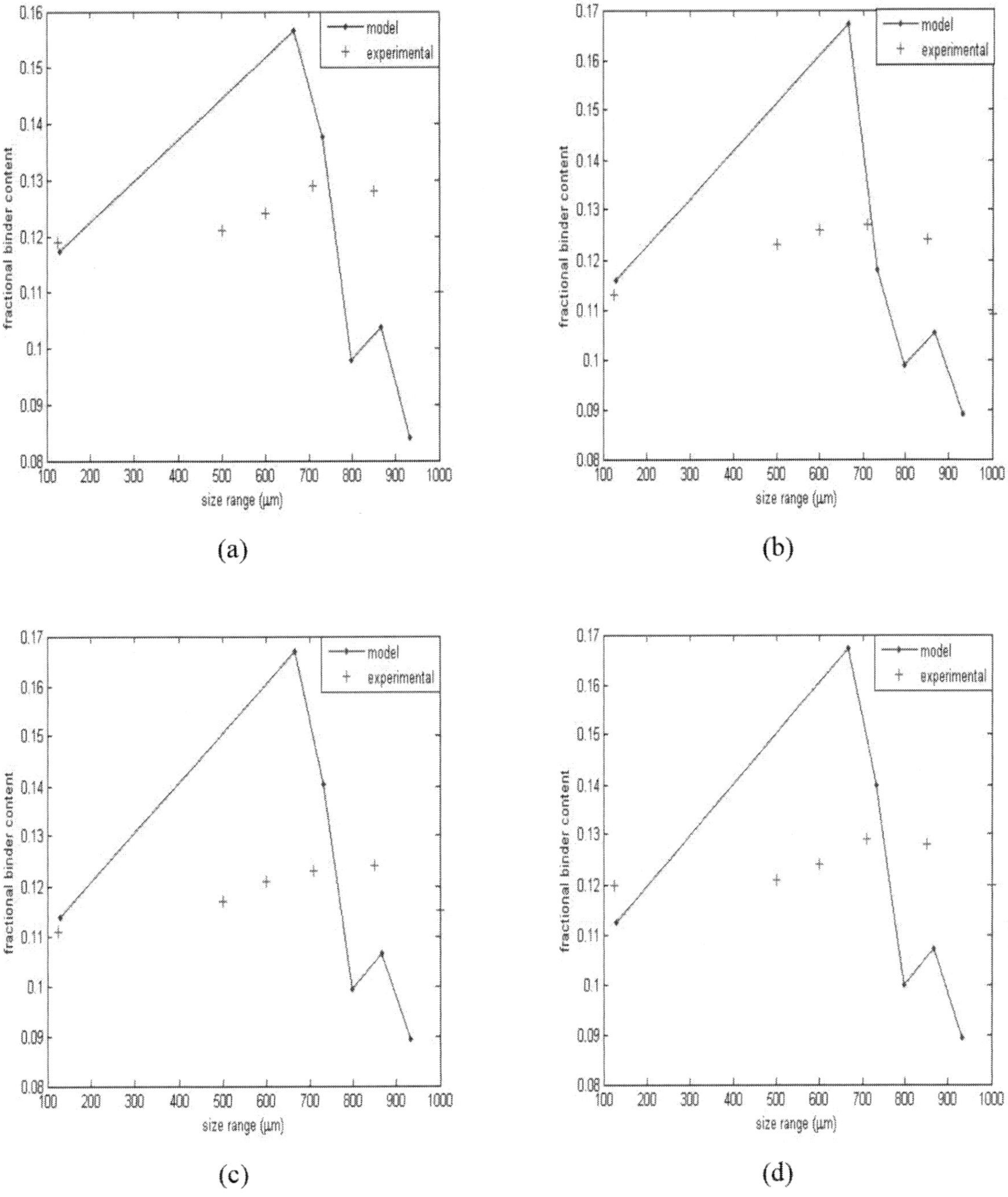

Figure 9. Time profiles of the fractional binder content for both model outputs and experimental data at (a) t = 180 *s*, (b) t = 300 *s* , (c) t = 600 *s* and (d) t = 900 *s* are compared, for binder-to-solids ratio of 0.12.

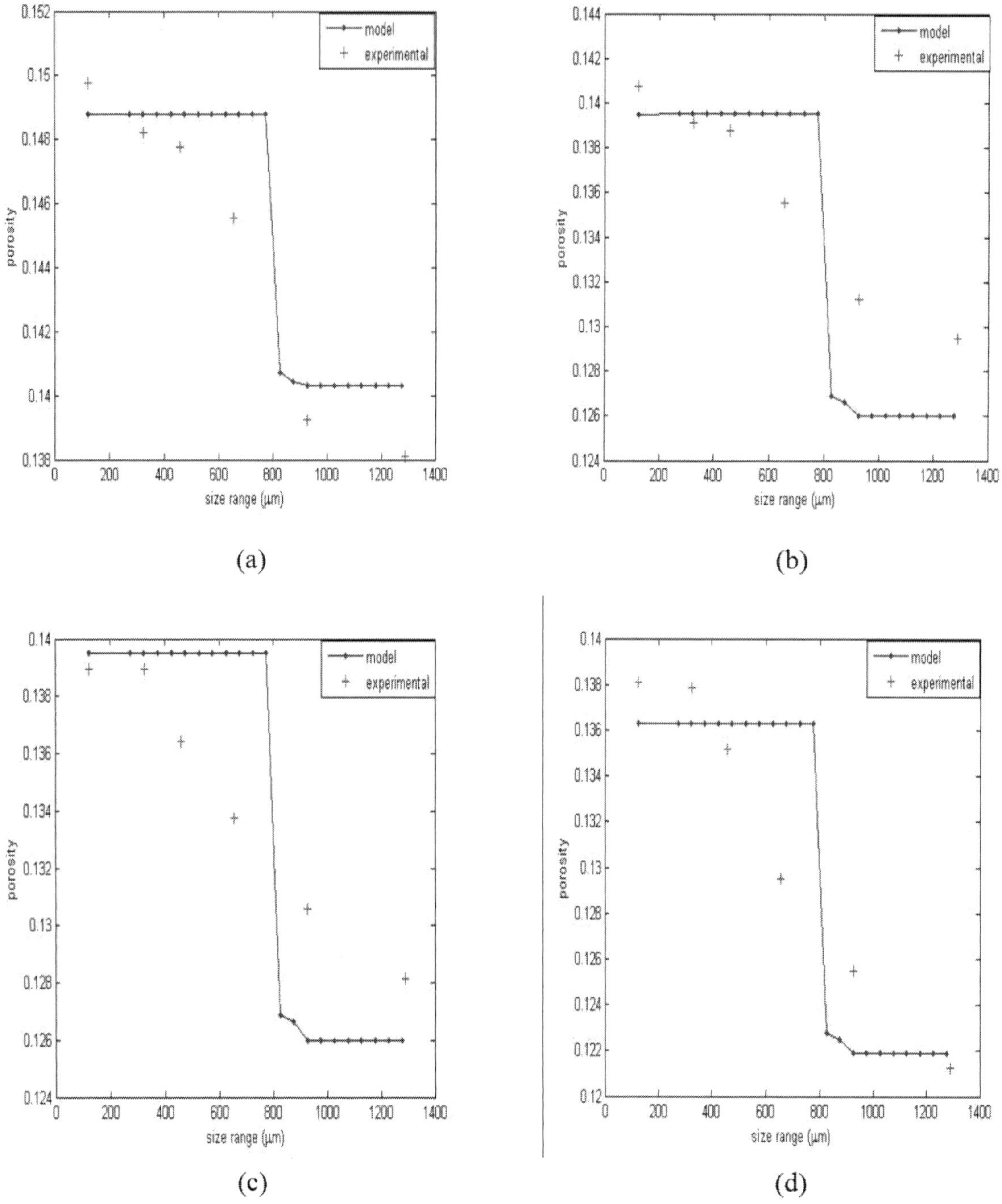

Figure 10. Distribution of the granule porosity with particle size at time (a) t = 180 *s*, (b) t = 300 *s* , (c) t = 600 *s* and (d) t = 900 *s* are compared between simulated outputs and experimental data, for binder-to-solids ratio of 0.11.

373d Numerical Solution Of Multi-Variable Cell Population Balance Models With A Free Boundary Algorithm

Mihail E. Kavousanakis, School of Chemical Engineering, National Technical University of Athens, Athens, Greece, Andreas G. Boudouvis, School of Chemical Engineering, National Technical University of Athens (NTUA), GR-15780, Athens, Greece, and Nikos V. Mantzaris, Chemical and Biomolecular Engineering, Rice University, Houston, TX 77005.

The complex behavior of heterogeneous cell populations can be accurately described by a class of mathematical models widely known as cell population balance models. They are first order partial integro-differential equations, whose formulation is fully defined by the single-cell reaction and division rates and the partition probability density function for the description of the mechanism which defines the distribution of cellular material amongst the daughter cells at cell division. Numerical algorithms have been developed for the numerical solution of cell population balance models due to the difficulty in deriving analytical solutions. However, their efficient solution still remains a challenging task; the main challenge comes from the fact that the boundaries of the intracellular state space are typically not known a priori. A novel free boundary algorithm has been recently developed treating with success the challenges of unknown boundaries and the co-existence of solutions differing by orders of magnitude, based on a transformation of the independent variable with respect to the mean expression level of the intracellular content. We present an extension of this free boundary algorithm in systems with two intracellular species. As a test bed of our algorithm we considered the case of linear single-cell reaction rates with respect to both species. Equal partitioning of cellular material at cell division exhibits oscillatory behavior for the number density function normalized around the mean intracellular content. The period of oscillations depends on the sharpness of the division mechanism, while unequal partitioning at cell division destroys the oscillations. We also present a parametric study of steady state solutions in E-Coli populations with a genetic toggle switch. Our target is to elucidate the effect of heterogeneity in the extent of the bistability region which appears at the single-cell level.

Fast Monte Carlo algorithms for Modeling and Optimization of Complex Particulate Processes: PEMC and τ-PEMC

Roberto Irizarry
DuPont Electronic Technologies.
14 T.W. Alexander Drive
P.O. Box 13999
Research Triangle Park, NC 27709-3999
919-248-5246
E-mail: Roberto.Irizarry@usa.dupont.com

1. Introduction

Solution of population balance equations by Monte Carlo method consists of an artificial realization of the population dynamics of a finite system. In this work, new tools are presented for the simulation and optimization of complex particulate systems modeled by multi-dimensional population balance equations. New fast Monte Carlo methodologies, PEMC and τ-PEMC ([1], [2]) reduce the computational load of the MC simulation by orders of magnitude with negligible accuracy losses, when compared with exact MC algorithms. This improvement in simulation speed makes process optimization feasible ([3], [4]). The PEMC method is applied to problems of industrial importance, in particular, the determination of optimal nucleation profile of a nucleating-coalescing system including DLVO interactions.

2. Point ensemble Monte Carlo.

The point ensemble Monte Carlo, PEMC, method was introduced in [1]. The method consists in the stochastic simulation algorithm, SSA, solution of the Jump Markov model called Particle Ensemble Random product model, PERP. This model is based on three ideas. First, the total population is "discretized" in subpopulations of particles with size in a specified interval. Each subpopulation is viewed as a "chemical species" with the number of particles in the subpopulation representing the concentration of that species. Second, the inter-particle interactions (i.e. coalescence, nucleation, breakage) are viewed as a set of special type of reaction, in which the reaction products are allocated stochastically to the existing species using probability functions that are mass conserving on average. Third, The original set of subpopulations is coupled with the system of "chemical species".

The Markov model, PERP, is summarized as follows. The state vector consists of the number of particles and the point ensembles in each subpopulation, $N=(N_1,N_2...N_M)$, $\Phi=\{\Phi_1,...,\Phi_M\}$, associated with chemical pseudo-species, $S=(S_1,...,S_M)$. The ensemble Φ includes all properties that fully describe the state of individual particles (volume, surface area, etc.). The state of the system is modified by a set of special type of reaction channels called Random product channels. Figure 1 shows one example in which two particles from sub-ensembles 2 and 4 coalesce to form a new particle that is allocated stochastically either to ensemble 4 or 5. Each possible event is described by a pseudo-reaction as explained in [1]. For example,a coalescence between particle *i* and *j* will produce a new particle with volume $v_{new}=v_i+v_j$ which falls in the interval $[v_k, v_{k+1}]$. After a coagulation event the new particle is assigned to species S_k with probability P_s or to species S_{k+1} with probability $1-P_s$.

$$E_s : S_i + S_j \rightarrow \begin{cases} \overset{P_s}{\rightarrow} & S_k \\ \overset{1-P_s}{\rightarrow} & S_{k+1} \end{cases} \tag{9}$$

The product probability parameter, P_s, of the random product channels are calculated using average mass conservation criteria. The details of these reaction channels and the product probability parameter are given in [1].

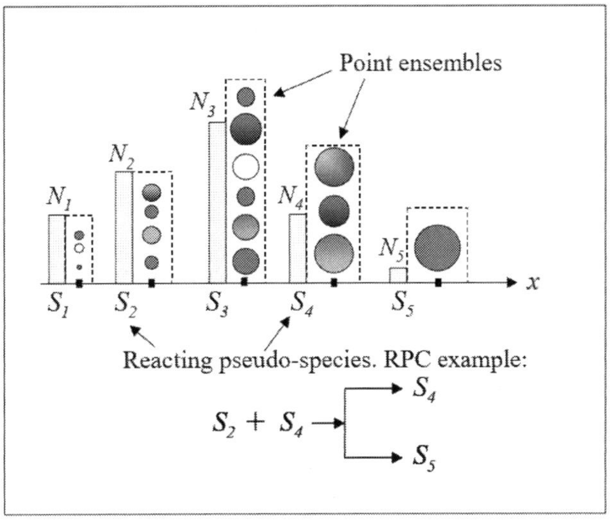

Figure 1. The reacting system of pseudo-species and the point ensemble are coupled to complete the point ensemble Monte Carlo method.

The algorithm is described next in details.

3. Point Ensemble Monte Carlo algorithm (PEMC).

The PEMC algorithm is described next. The method consists in solving the PERP jump Markov model constructed in the last section using SSA steps:

Algorithm. *Point Ensemble Dynamic Monte Carlo (PEMC)*

Input Parameters. The problem input parameters are the initial distribution of particles properties, and, the initial concentration of particles, c_{par} (number of particles/unit volume) The simulation parameters are the total number of particles to be used in the simulation, N_p, and the set of M mass points discretizing the volume (v_i, $i=1,M$). The simulation volume is then calculated ($V_s=N_p/c_{par}$). Enumerate all possible random product channels (in any order), E_s, $s=1,T$, where T is the total number of possible random product channels (events). For each possible event E_s, calculate and store the probability P_s ($s=1,T$) using the volume conservation in Eq. (12). Calculate and store all problem kernels evaluated at the grid points ($J(v_i)$, $q(v_i,v_j)$, $\Gamma(v_i)$, $b(v_i,v_j)$ $i,j=1,\ldots M$).

Initialization. The point ensembles are initialized generating N_p random particles from the prescribed initial density function of particle properties. Each particle, p, is allocated to the corresponding sub-ensemble Φ_i, using the criteria that $x_p \in [(v_{i-1}+v_i)/2,(v_i+v_{i+1})/2]$. Let N_i be the number of particles in sub-ensemble Φ_i. Set the time equal to zero.

Step 1. *Next event time* (SSA step)

$$\langle\tau\rangle = 1/\sum_{i=1}^{T} R(E_i) \qquad (16)$$

Set time to:

$$t = t + \langle\tau\rangle \qquad (17)$$

Step 2. If there is a complex intra-particle dynamics (chemical reactions inside the particles etc.), then, for all particles in all point ensembles integrate the intra-particle model dynamics from $t-\langle\tau\rangle$ to t. Particle growth is considered as part of the intra-particle dynamics (see subsection on handling growth mechanism).

Step 3: *Find the PERP event to be fired next* solving the following equation (SSA step):

$$\sum_{i=1}^{f} R(E_i) < r\sum_{i=1}^{T} R(E_i) \le \sum_{i=1}^{f+1} R(E_i) \qquad (18)$$

The selected event is E_f.

Step 4: *Fire PERP event E_f:*

(I) If the event is a coalescence event: $E_f = S_i + S_j \rightarrow v_{new}$, where $v_{new} = v_i + v_j \in [v_k, v_{k+1}]$.

a) Select a random particle, *n*, from point ensemble $_i$.
b) Eliminate particle from Φ_i and set $N_i \leftarrow N_i - 1$.
c) Select a random particle, *m*, from point ensemble $_j$.
d) Eliminate particle from φ_j and set $N_j \leftarrow N_j - 1$.
e) Form a new particle from mother particles.
f) Select a random number, *r*, from uniform probability distribution in (0,1). If $r<P_f$, add the new particle to ensemble $_k$ and set $N_k \leftarrow N_k + 1$. Otherwise add the new particle to ensemble Φ_{k+1} and set $N_{k+1} \leftarrow N_{k+1} + 1$.

(II) If the event is a breakage event: $E_f = S_i \rightarrow 2v_{new}$ where, $v_{new} = (v_i / 2) \in [v_k, v_{k+1}]$.

a) Select a random particle, *n*, from point ensemble $_i$.
b) Eliminate particle from Φ_i and set $N_i \leftarrow N_i - 1$.
c) Form two new twin particles from mother particle.
d) For each new particle, select a random number, *r*, from uniform probability distribution in (0,1). If $r<P_f$, add the new particle to ensemble Φ_k and set $N_k \leftarrow N_k + 1$. Otherwise add the new particle to ensemble Φ_{k+1} and set $N_{k+1} \leftarrow N_{k+1} + 1$.

(III) If the event is a nucleation event: $E_f = \varnothing \rightarrow v_{new} = v_o \in [v_k, v_{k+1}]$.

a) Select a random number, *r*, from uniform probability distribution in (0,1). If $r<P_f$, add the new particle to ensemble $_k$ and set $N_k \leftarrow N_k + 1$. Otherwise add the new particle to ensemble Φ_{k+1} and set $N_{k+1} \leftarrow N_{k+1} + 1$.

Step 5. If the final time is reached, end the instance; otherwise go to step 1. Check for particle control (See subsection on controlling number of particles during the simulation)

---

4. Determination of optimal nucleation profile of complex coagulation kernels

Size control of nano-particles formed by the nucleation/coalescence mechanism was considered in [3]. In this section, the same problem is solved using PEMC. This problem is a good model system to test PEMC because direct numerical solutions are available as bench mark solutions. Furthermore, standard MC solutions are computationally intensive due to large nucleation rate and complex coagulation kernel including colloidal interactions (DLVO theory). This will limit the use of standard MC simulation when solving optimization problems.

Initially the reactor is charged with particles of volume v_r. While the initial particle population is coalescing, more nuclei of volume v_o are added to the reactor at a given rate to control the final particle size distribution (PSD). In this section, the problem is postulated as dynamic optimization problem to find the optimal schedule for nuclei addition for a target size with a minimized standard deviation. Two operational constraints are added in the analysis. First, the addition rate can have only two values: no addition or a specified constant addition rate. The second operational constraint is that the number of intervals for additions cannot exceed a determined value. The details of the dynamic optimization problem formulation are given in [3].

For the collision kernel, it is assumed that the dominant mechanism is Brownian motion given by:

$$q_B(v,v') = \frac{8kT}{3\mu}\left(v^{1/3} + v'^{1/3}\right)\left(v^{-1/3} + v'^{-1/3}\right) \tag{34}$$

where k is Boltzmann's constant, μ is the viscosity and T is the reactor temperature. The Brownian collision kernel is corrected with the so-called stability factor, using standard DLVO theory to account for the attractive and repulsive forces between particles ([5], [6]):

$$q(v,v') = \frac{q_B}{2\int_0^{\infty} \frac{\exp\left(V_T(x,v,v')/kT\right)}{(x+2)^2} dx} \tag{35}$$

where x is the shortest distance between particle surfaces scaled by the average radius between the two particles. The interaction energy V_T is the sum of all forces between particles. In particular, the effects of van der Waals attractive forces and electrostatic repulsive forces are considered ($V_T = V_{vdw} + V_{elec}$). The van der Waals potential is given by:

$$V_{vdw} = -\frac{A_H}{6}\left[\frac{2a_1a_2}{R^2-(a_1-a_2)^2} + \frac{2a_1a_2}{R^2-(a_1+a_2)^2} + \ln\frac{R^2-(a_1+a_2)^2}{R^2-(a_1-a_2)^2}\right] \tag{36}$$

There are several formulations for the electrostatic interaction. To simplify the problem, the following expression is used,

$$V_{elec} = 4\pi\varepsilon\psi_0^2 \frac{a_1a_2}{a_1+a_2}\ln\left[1+\exp(-kx)\right] \tag{37}$$

where a is the particle radius and R is the distance between two particle centers; all other parameters and values are defined in Table 1.

All the parameters in the MC simulation are in dimensionless forms using the following scales: the time is scaled with $1/(q_o f_o)$, the particle density function is scaled with f_o, the nuclei addition rate with (q_o f_o f_o) and the particle volume with v_o, where $q_o=8kT/3\upsilon$, f_o is the initial particle density in the reactor and v_o is the nuclei volume. In the simulation 50 volume nodes were used ($v_1=v_o$, $v_i=1.3v_{i-1}$, $i=2,10$, $v_i=1.5v_{i-1}$, $i=11,50$). All simulations were executed from random initial profile and solved with a termination criterion of 10,000 function evaluations. The simulation box was initiated with 50,000 particles.

Case study: The starting particles are 26 nm in radius and the nuclei have a radius of 2 nm. The target is to grow 180-nm particles with the same initial particle concentration as in previous cases. The dimensionless addition rate is J_{set}=1.0 with no constraints on the number of addition intervals. The DLVO parameters are shown in Table 1. Figure 2 shows the optimal profile of with 5 combinatorial elements [3] δ_i={1, 2}={u(t)= 0, u(t)= J_{set}}. Figure 3 shows the volume-cumulative function of the two profiles and the volume-cumulative function of two suboptimal solutions (no addition, constant addition). In this case, the PEMC solution is almost identical to the solution obtained by direct numerical solution of the population balance equation.

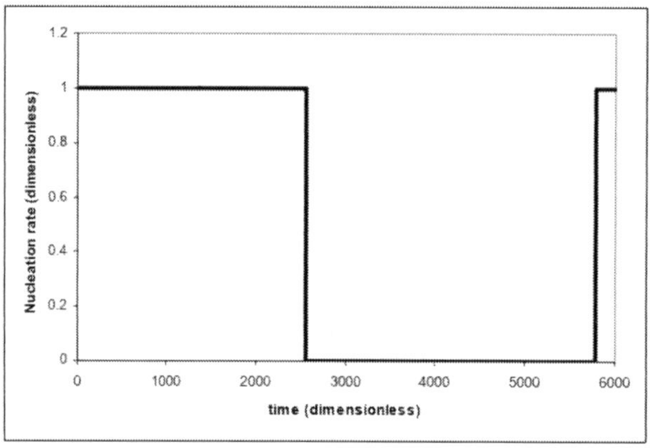

Figure 2. Optimal profile of the nano-particle synthesis with operational constraints (DLVO case, v_o=2 nm, v_r=26 nm).

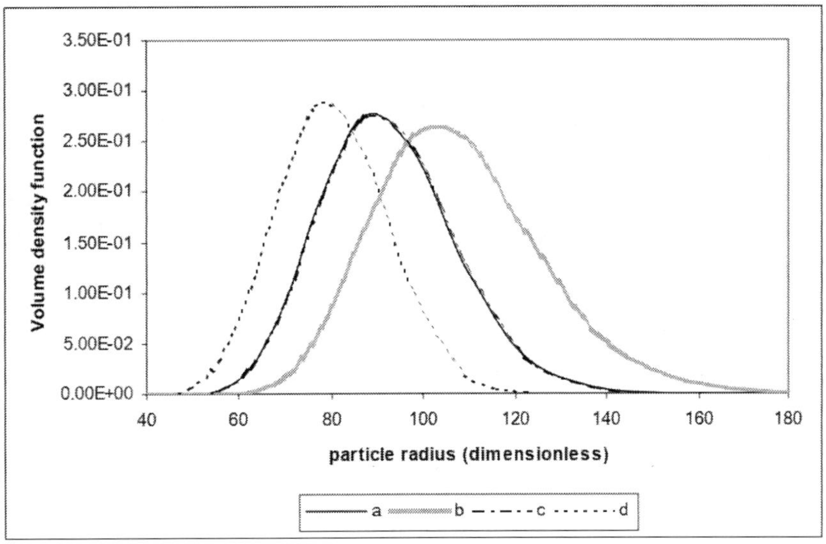

Figure 3. Volume fraction distribution vs. dimensionless particle radius: (a) optimal profile of unconstrained problem, (b) optimal profile of constrained problem, (c) sub-optimal constant nuclei addition and (d) sub-optimal no nuclei addition.

Table 1. DLVO parameters.

Variable	Description	Value
A_H	Hamaker constant	1×10^{-19} J
T	Temperature	300 K
μ	Viscosity	0.001 Pa.s
$\varepsilon\varepsilon_o$	Dielectric constant of medium	2.17×10^{-10} C/(V m)
κ	Inverse Debye length	2×10^{8} (1/m)
ψ_o	Surface potential	50 mV
a_o	Seed particle radius	2 nm

a_r	Initial population particle radius	40 nm

5. Discussion and future work

In this work, the feasibility of solving dynamic optimization problems of systems with complex coagulation kernels was demonstrated. The method is shown to reduce the CPU time without sacrificing the simulation accuracy. The algorithms also have been applied to complex multidimensional problems [1], [2]. This algorithm has also been adapted to reaction precipitation in reverse micelles [3].

Nomenclature

b	Number of daughter particles.
c_{par}	Particle concentration (number of particles/unit volume).
E	Event that changes the state vector (i.e. reaction channel).
G	Growth rate.
J	Nuclei addition rate.
M	Number of species or pseudo-species in the system.
N	State vector representing the number of particles in a point ensemble.
N_i	Number of particles in point ensemble i.
N_p	Number of particles in simulation volume.
P_i	Product probability parameter for event E_i.
q	Aggregation rate.
$R(E_i)$	Propensity function of event E_i.
r, r_1, r_2	Random number from uniform probability distribution.
S	Vector of species in reacting system. Vector of Pseudo-species in PEMC.
s	Factor used in generating the mass point ($v_{i+1}=sv_i$).
S_i	Specie or Pseudo specie i.
T	Total number of reactions or total number of random product channels.
t	Time.
u, v	Dummy variables used for volume.
v_i	Volume of mass points i.
v_{new}	v_i+v_j for coalescence, $v_i/2$ for symmetric binary breakage, and v_o for nucleation.
V_s	Simulation volume.
x_{avg}	Ensemble average volume.
x_j	Volume of particle j.
Φ_i	Point ensemble associated with pseudo-specie S_i.

Γ	Breakage frequency.
τ_{micro}	Stochastic time for next event in SSA algorithm.

References

1. Irizarry, R., Fast Monte Carlo Methodology for Multivariate Particulate Systems Point Ensemble Monte Carlo, *Accepted in Chemical Engineering Science*.
2. Irizarry, R, Fast Monte Carlo Methodology for Multivariate Particulate Systems-II: τ-PEMC, *accepted in Chemical Engineering Science.*
3. Irizarry, R. (2005), Ageneralized framework for solving dynamic optimization problems using the artificial chemical process paradigm: Applications to particulate processes and discrete dynamic systems, *Chemical Engineering Science*, **60**, 5663-5681.
4. Irizarry, R. (2006), Hybrid Dynamic Optimization using Artificial Chemical Process Method: Extended LARES-PR, *Ind. Eng. Chem. Res.*, **45**, 8400-8412.
5. Russel W.B., Saville D.A.., and Schowalter W.R. (1989), Colloidal dispersions. Cambridge Uuniversity Press, Cambridge, U.K.
6. Schwarzer H.C. and W. Peukert (2005), Prediction of aggregation kinetics based on surface properties of nanopartcles, *Chemical Engineering Science*, **60**, 11-25.
7. Irizarry, R, Fast Monte Carlo Simulation for reaction precipitation in reverse micelles, *to be submitted.*

373f A New Simulation Method For Population Balances Using Constant-Number Monte Carlo With Adjustable Time Step

Samira Khalili[1], Antonios Armaou[1], and Themis Matsoukas[2]. (1) Chemical Engineering, The Pennsylvania State University, University Park, PA 16802, (2) Chemical Engineering, Penn State, 150 Fenske Laboratory, University Park, PA 16802

Monte Carlo (MC) methodologies have become mainstream numerical tools for the simulation of particulate processes that require the solution of the population balance equation. Such methodologies have been shown to be very efficient in simulations of well-mixed systems. Implementing population balance strategies within a larger 2D or 3D simulator remains a challenge, however, because the computational cost of solving the population balance equation in multiple dimensions is high. Standard simulators advance time and then compute changes within that interval and for this reason, time-driven MC has been proposed as the most appropriate MC variant for simulations that combine population balances with standard numerical integrators. However, *time-driven* MC as a stand-alone algorithm is slower than standard event-driven MC with the same number of particles; when interfaced with 2- or 3D effects (e.g. fluid dynamics) it is rather slow for practical implementations. Event-driven MC, while generally faster, operates in the opposite way of standard integrators: it effects a change in the population and then advances time by a corresponding amount. Thus the time step is not directly controlled.

Here we present a new hybrid algorithm, which (i) utilizes a fixed (predefined) number of simulation particles, N, and (ii) which is capable of transforming a sample of N simulation particles at t to a corresponding sample of N particles at time t+Δt, where Δt is a predefined time increment. This is internally accomplished by temporarily inflating the number of simulation particles at time t to a value N' that matches the chosen time step. The inflated population is subjected to a standard Monte Carlo operation that simulates the particulate processes of interest and produces a sample of the population at the end of the time step. Finally, the number of particles is deflated to N and the process is repeated. We formulate the mathematical basis of the algorithm and demonstrate the method in simulations of particle aggregation and breakup. We examine the accuracy as a function of the number of simulation particles and length of time step, and discuss the potential extension to 2- and 3-dimensional problems.

373g Validation of Fcmom for Bi-Variate Population Balance Equations

Matteo Strumendo and Hamid Arastoopour. Department of Chemical and Environmental Engineering, Illinois Institute of Technology, Chicago, IL 60616

FCMOM (Finite size domain Complete set of trial functions Method Of Moments) is a variation of the method of moments to solve PBE (population balance equations). FCMOM was presented recently [1] and afterwards validated for mono-variate PBE. In FCMOM, the solution of the PBE is sought, instead of the $[0,\infty]$ range, in the finite interval between the minimum and maximum particle size; the evolution of the minimum and maximum particle size is tracked imposing moving boundaries conditions. The main advantage of FCMOM is that it provides the solution of the PBE both in terms of the moments *and* of the reconstructed particle size distribution. In all the applications developed for mono-variate PBE (constant, linear, diffusion-controlled particle growth; particle dissolution; particle aggregation with constant, sum and product kernels), the solution obtained with FCMOM is in excellent agreement with the analytical solution, while the computational effort is low (comparable with QMOM).

In this work, FCMOM is applied to bi-variate PBE and is validated by comparison with analytical solutions of particle growth (constant, linear, diffusion-controlled), particle dissolution and particle aggregation (constant kernel).

In the bi-variate case, FCMOM performs as well as in the mono-variate case. In fact, the algorithm is still simple and efficient, and the computational time is low (typically, twice the computational time required for the corresponding problem in the mono-variate case). Moreover, the particle bi-variate (volume-surface area) distributions obtained by FCMOM are again in very good agreement with the analytical solutions.

Finally, applications to bi-variate particle aggregation with kernels derived by the kinetic theory of granular flows and to simultaneous particle aggregation and coalescence are discussed.

[1] Strumendo M., Arastoopour H., A new approach in solving PBE, Proceedings of the Fifth World Congress on Particle Technology, Orlando, Florida (USA), April 23-27, 2006.

399a Role of Counterion Replacement on the Deactivation of Co-Salen Catalysts in the Hydrolytic Kinetic Resolution of Epichlorohydrin

Surbhi Jain[1], Xiaolai Zheng[2], Christopher W. Jones[3], Marcus Weck[2], and Robert J. Davis[1]. (1) Chemical Engineering, University of Virginia, 102 Engineers Way, Charlottesville, VA 22904, (2) Georgia Institute of Technology, School of Chemical & Biomolecular Engineering, School of Chemistry and Biochemistry, Atlanta, GA 30332, (3) Chemical & Biomolecular Engineering, Georgia Institute of Technology, 311 Ferst Dr., Atlanta, GA 30332

The hydrolytic kinetic resolution (HKR) of terminal epoxides using Co-salen catalysts provides a convenient route to the synthesis of enantioenriched chiral compounds by selectively converting one enantiomer of the racemic mixture (with a maximum 50% yield and 100% ee). The utilization of water as the nucleophile makes this reaction straightforward to perform at a relatively low cost. The proposed mechanism for the HKR of terminal epoxides involves the cooperativity of two Co-salen complexes, where one Co metal center activates the electrophile (epoxide) and the other Co metal center activates the nucleophile (water). Hence, the observed rate depends on the square of the catalyst concentration. With this mechanism in mind, researchers have developed polymeric Co-salen catalysts that exhibit activities 1-2 orders of magnitude greater than that of the monomeric Co-salen catalyst. Recently, we have designed an oligo(cyclo-octene) supported Co-salen catalyst, with the idea that the presence of multiple Co-salen units as side chains on the oligomeric catalyst can enhance the reactivity and enantioselectivity in the HKR reaction. Most of these multimeric, homogeneous catalysts, as well as supported catalysts, are difficult to synthesize and need to be regenerated after the reaction. The general mode of catalyst deactivation has been reported in the literature to be the change of the active Co(III) salen complex to an inactive Co(II) salen complex. However, no detailed spectroscopic studies have been carried out to confirm this suggestion. Moreover, the mechanism of the proposed Co reduction is not clear.

In this work, we investigated the mode of catalyst deactivation during the HKR of epichlorohydrin by conducting recycling studies of homogeneous Co-salen catalysts without catalyst regeneration and following the chemical state of the catalyst by UV-Vis and X-ray spectroscopy. Although an active Co (III) salen catalyst deactivated substantially after multiple cycles without regeneration, the catalyst maintained its +3 oxidation state throughout the runs. Thus, deactivation of Co-salen during HKR was not the result of Co reduction. Results from various catalyst pretreatment tests as well as from catalysts containing various counterions (acetate, tosylate, chloride) indicated that gradual replacement of the Co-salen counterions with hydroxyl during the reaction deactivated the catalyst. The extent of counterion replacement by hydroxyl was influenced by the exposure time, the presence of epichlorohydrin and the nucleophilicity of the counterion. An oligo(cyclo-octene) supported Co-OAc salen catalyst, which was 25 times more active than the traditional Jacobsen Co-salen catalyst, was recycled multiple times with negligible deactivation.

399b Experimental and Dft Study of Base- and Ligand-Free Pd/c-Catalyzed Homocoupling Reactions of Arylboronic Acids

Jeng-Shiou Chen and Johannes Khinast. Chemical & Biochemical Engineering, Rutgers, the State University of New Jersey, 98 Brett Road, Piscataway, NJ 08854

Homocoupling of arylboronic acids (Fig. 1) was successfully carried out by Pd/C in water/2-propanol (9:1 in volume ratio) under air, to obtain symmetric biaryls in good yield. Symmetric biaryls are important precursors for advanced pharmaceuticals, such as glucocorticoid receptor antagonists [1]. The base-free, ligand-free system and the heterogeneous nature of the catalysts allow a practical and environmentally friendly operation. The influence of the co-solvent ratio was studied. Pd/C showed an increasing activity when the solvent fraction of water was increased from 10% to 90%. The influence of the reaction temperature was also investigated. It was shown that a higher reaction temperature leads to higher yields of desired biaryls. Furthermore, the effect of functional groups at para position was explored. DFT calculations indicate that the presence of sulfur in the function group hinders Pd catalysis and results in a slow reaction. DFT calculations also suggest that the first oxidative addition step in the proposed mechanism is not the rate-limiting step. Reference: [1] Bioorganic & Medicinal Chemistry Letters 16 (2006) 6086-6090.

399c Development of New Covalently Tethered Group 4 Metallocenes

Heidrun Woelfler, Georg Schitter, Rafael Eder, Claire Jeanquartier, and Johannes G. Khinast. Institute for Process Engineering, Graz University of Technology, Inffeldgasse 21/B, Graz, A-8010, Austria

Group 4 metallocenes bearing an ethylenebis(indenyl) (EBI) or an ethylenebis(tetra-hydroindenyl) (EBTHI) moiety are known to be very active and selective catalysts for reactions like hydrogenations, hydrosilylations and polymerizations. Although tremendous achievements have been made in the area of homogeneous chemical catalysis as well as in the implementation of these soluble complexes in industrial applications, such processes are not having as much impact on an industrial scale as may have been expected. The reasons may be that homogeneous catalysts lack the advantages of heterogeneous catalysts like easy separation and recycling of the catalysts, prevention of metal leaching, improvement of stability, etc. Within this project we present the development of heterogeneous Group 4 metallocenes which (i) are covalently bond to a solid support, such as H-terminated Si-surfaces or functional silica gel (ii) show high activity and selectivity for pharmaceutical GMP reactions (e.g., reduction of imines to chiral amines) and (iii) can be used to develop multifunctional catalysts, i.e., solid supports that feature two or more catalytic compounds. The development of these heterogeneous catalysts involves the preparation of functional ligands using a four step synthesis [1] in order to introduce tethers of various lengths and functionality to the metallocenes. The functional ligands and metallocenes are then immobilized on solid supports using different methods like UV-induced hydrosilylation [2] or chemical catalysis. State-of-the-art surface characterization methods are used in order to investigate the distribution and concentration of active sites on the surfaces. The development and optimization of the new metallocenes are accompanied with molecular modeling approaches using DFT methods (DFT = Density Functional Theory). Applying the B3LYP functional the relationship between function and structure of the catalysts as well as reaction profiles for the catalytic hydrosilylation of imines are studied [3]. The performance of the new heterogeneous metallocenes is compared to that of classic homogeneous catalysts, e.g., EBTHITibinol or EBTHITiCl2, which have proven to be potent catalysts for the hydrosilylation of imines leading to chiral amines with high yields and enantioselectivities.

[1] Panarello A. P., Vassylyev O., Khinast J. G., Synlett, 2005, 5, 797 [2] Langner A., Panarello A., Rivillon S., Vassylyev O., Khinast J. G., Chabal Y. J., J. Am. Chem Soc., 2005, 127, 12798 [3] Woelfler H., Flock M., Sassmannshausen J., Khinast J.G., manuscript in preparation

399d The Discovery of an Innovative and Very Green One-Step Synthesis of Tetronic Acid

Josef G. Schroer, Chemspeed Technologies, 7 Deer Park Drive, Suite L, Monmouth Junction, NJ 08852

Tetronic acid is an important key intermediate for different products (e.g. (+)-Biotin and Losigamon) of the fine chemical and pharmaceutical industry. The industrial production of tetronic acid can be achieved in a multi-step gas phase process, starting from 4-chloro aceto-acetate. It was found, that addition of 4-chloro acetoacetate to a cold, aqueous solution of sodium hydroxide yields a mixture of sodium chloroacetate and sodium acetate. In addition, the resulting yield of sodium tetronate is less than 30%. Based on these findings, an innovative one-step process for the synthesis of tetronic acid was targeted. Therefore, the pH and temperature dependency of this reaction was studied in a parallel, automated set of experiments.

This investigation was resulting in a innovative process, which saves several chemical steps over the pre-existing industrial synthesis of tetronic acid, and may result in major cost savings in the production of this important intermediate.

This and other relevant examples, presented herein, impressively support the enormous value of a parallel and automated High Output Experimentation approach.

399e Triflation Reaction Endpoint By A T R - F T I R, P L S - Based Chemometrics Model

James Marek, Shan Lin, Shekhar Viswanath, and Clifford Mitchell. GPRD Process R&D, Abbott Labs, 1401 Sheridan Road, R13-3, North Chicago, IL 60064

In-situ ATR-FTIR was used to monitor triflation of reactant R-OH with trifluoromethanesulfonic anhydride at -35 °C to -50 °C in a batch reactor. Data from three triflations were sufficient to evaluate models to correlate the spectral first derivatives with residual starting material levels derived from HPLC analyses. Approaches evaluated included simple linear regression, multiple regression, and full multivariate partial least squares (PLS). Spectral data collected in a fourth run was input to the PLS model, and the predicted residual R-OH levels compared favorably with the HPLC values. An updated PLS model was then generated by including data collected from all four batch reactions at -50 °C. The refined model was used to predict starting material levels in real-time in a subsequent triflation, predicting down to levels as low as 1 PA% ± 1 PA%.

399f Multi-Step Hydrogenations and Process Understanding for Api Manufacturing

Jason Mustakis and David J. Am Ende. Chemical R&D, Pfizer, Inc., Eastern Point Road Mail Stop 8156-70, Groton, CT 06340

Placeholder for 20 min presentation

399g Effect of Different Vessel Geometries on the Hydrodynamics of Partially Baffled Stirred Tank Reactor Systems Equipped with a Retreat Blade Impeller

Giuseppe Di Benedetto[1], Piero M. Armenante[1], and Billy R. Allen[2]. (1) Otto H. York Department of Chemical Engineering, New Jersey Institute of Technology, 323 King Blvd., Newark, NJ 07102-1982, (2) Lilly Corporate Center, Eli Lilly and Company, Indianapolis, IN 46285

Mixing and reaction/precipitation operations carried out in the pharmaceutical industry are frequently conducted in stirred tank reactors equipped with a single retreat-blade impeller and a single baffle. The walls of these vessels and their internal equipment, including the impeller and baffle, are typically glass lined in order to prevent contamination of the products and reduce the occurrence of corrosion.

Although partially baffled, retreat-blade impeller systems are critical in pharmaceutical production, there is very little information available to date on their hydrodynamics. Such knowledge is critical to understand processes, such as solid suspension, crystallization and chemical reaction, regularly conducted within them.

This research work is focused on the study of the effect of different vessel geometries on the velocity distribution inside retreat-blade impeller-vessel systems. The velocity distribution within these scaled-down systems was determined using Laser-Doppler Velocimetry (LDV) and Computational Fluid Dynamics (CFD). Each experimental apparatus consisted of a vessel equipped with a retreat-blade impeller and a single beaver-tail baffle. Differently styled vessel bottoms (flat-bottomed and hemispherical bottomed) were used during experimentation. All three velocity components and the turbulence intensity were measured at different radial positions on a number of horizontal planes inside the vessels. Numerical simulations of the velocity distribution and turbulence levels inside each vessel were conducted using a commercial mesh generator (Gambit) coupled with a computational fluid dynamic (CFD) package (Fluent). The full 360°-tank geometry was incorporated in the simulations.

The flow patterns in these partially baffled systems are generally complex. It was found that the flow in the lower section of each vessel is dominated by the tangential component of the velocity as in unbaffled systems, although the presence of the baffle produced a stronger axial flow in the upper region. This indicates that phenomena that are dominated by the hydrodynamics near the vessel bottom, such as solid suspension, may be more difficult to carry out in such vessels as compared to fully baffled systems.

409a Continuous Powder Blending For Pharmaceutical Applications

Michael Gentzler, Ecevit Bilgili, Brit L. Rudeen, Mano Ramasamy, and Steve J. Heidel. Merck & Co., Inc., PO Box 4, West Point, PA 19486

Apriori design of a continuous fine-powder blending module is currently quite difficult due to potentially complex granular flow behavior that impacts all performance aspects: 1) ingredient feeder fluctuations 2) maximum mixer inlet flow, or 'choke point' 3) temporal mixer efficiency, for damping of inlet concentration fluctuations 4) spatial mixer efficiency, for reduction of the length scale of segregation at the inlet to a level that is below the final scale of scrutiny.

The performance of two horizontal, tubular continuous mixers was experimentally assessed for a moderately cohesive pharmaceutical powder. Colorimetry was used to determine the axial Peclet number by the response to a step input for a wide range of impeller speeds [$0.2<Fr<6$] and flow rates. An intriguing correlation was obtained and found to be useful for evaluating mixer designs. For this class of mixer, axial dispersion is fast, as compared to rotating drums, and radial/spatial mixing is highly efficient. Overall module performance was found to be predictable with simple white-noise transfer function analysis, given experimentally determined feeder variances, mixer residence time and an axial dispersion coefficient.

409b Model Predictive Control on Roller Compaction for Pharmaceutical Manufacturing

Shuo-Huan Hsu[1], Girish Joglekar[1], Gintaras V. Reklaitis[2], and Venkat Venkatasubramanian[2]. (1) Discovery Park, Purdue University, West Lafayette, IN 47907, (2) School of Chemical Engineering, Purdue University, 480 Stadium Mall Drive, West Lafayette, IN 47907

When processing bulk powders with particle sizes smaller than 100 μm, problems such as poor flowability, segregation, and dust generation frequently arise. These problems can be prevented or reduced by agglomeration of powder particles, which is usually achieved by granulation processes (Shlieout et al. 2000). Granulation is the process in which primary powder particles are made to adhere to form larger, multiparticle entities called granules. It prevents segregation of the constituents of the powder mix, improves the flow properties and the compaction characteristics of the mix. The dry granulation processes such as roller compaction become attractive for pharmaceutical products, which are often moisture or heat sensitive. In addition, this technique is environmentally friendly since dust problems are minimized or avoided (Kleinebudde 2004).

Currently, the operating conditions are mostly found in experiments and only a basic regulatory control system on the gap width has been included in the roller compactor (Shlieout et al. 2002). This strategy is not suitable for continuous operation to which the pharmaceutical industry looks forward. Process control system, which is an important part of stable continuous operation, is usually designed based on a dynamic process model. However, there is no such a model existed for a roller compactor.

The possibility of model predictive control (MPC) scheme for roller compaction process is investigated in this paper. First a process dynamic model is proposed. It uses Johanson's rolling theory (Johanson 1965) to predict the stress and density profiles during the compaction and the continuity equation to describe the roll gap change. The proposed model, formulated as a differential-algebraic equation (DAE) system, considers the powder throughput as well as the roll force, so it becomes possible to design, optimize, and control the process based on the model. The simulation case studies showed the model can predict the ribbon density and gap width changes while varying roll pressure, feed speed and roll speed. The roll pressure influences the ribbon density much more than other two input variables, and the roll gap is affected by all three inputs. The interactions among the three input variables are also studied. If the ratio of feed speed to roll speed is kept constant, neither ribbon density nor gap width change, but the powder throughput.

Based on this dynamic model, a control scheme is then established. The control objective is to ensure that the product quality and the throughput follow the desired reference trajectories. The product quality is represented by the ribbon density, and the throughput is influenced by the gap width as well as the ribbon density. The roll pressure and feed speed are used for controlling these two variables. Since the linearization of the proposed model is difficult, the computation burden and stability of MPC are investigated, and model reduction based on process data is necessary to increase the computational speed. Simulation results provide the comparison of the control performance of the original model and the reduced order model.

Reference:

Johanson, J. R. (1965). "A rolling theory for granular solids." Trans. ASME: J. Appl. Mech. B., 32(4), 842-848.

Kleinebudde, P. (2004). "Roll compaction/dry granulation: pharmaceutical applications." Europ. J. Pharma. Biopharma., 58, 317-326.

Shlieout, G., Lammens, R. F., and Kleinebudde, P. (2000). "Dry granulation with a roller compactor. Part I: the functional units and operation modes." Pharma. Tech. Europ., 12(11), 24-35.

Shlieout, G., Lammens, R. F., Kleinebudde, P., and Bultmann, M. (2002). "Dry granulation with a roller compactor. Part II: Evaluating the operation modes." Pharma. Tech. Europ., 14(9), 32-39.

409c Experimental Study And Computational Modeling Of Continuous Powder Mixing Processes

Patricia M. Portillo, Marianthi G. Ierapetritou, and Fernando J. Muzzio. Rutgers University, Chemical and Biochemical Engineering, 98 Brett Road, Piscataway, NJ 08854

Powder mixing plays an important role in many processing stages within the pharmaceutical, catalysis, food, construction, and mineral industries. However due to stringent processing regulations, in the past, the focus of powder mixing research has been directed towards batch mixing. More recently, continuous processing has emerged as a significant process alternative and recent research efforts indicate that a well-controlled continuous mixing process can significantly enhance productivity (Marikh et. al [1]; Muerza et. al [2]). Our research focuses on investigating the design space of continuous mixing operations in order to optimize blending performance. We are targeting the development of a statistical and semi-empirical process models that allow for on-line control and optimization of powder mixing performance.

The continuous mixers considered are manufactured by GEA Buck Systems. The two mixers are cylindrical horizontal vessels with a series of adjustable blades they vary in size, speed, blade design, and motor power. In both units the powder is discharged through a weir in the form of a conical screen. The mixer has the flexibility of changing angles affecting the gravitational forces on the axial transport thereby affecting the powders residence time. Model blends have been formulated using the following materials: Milled Acetaminophen (30μm), Fast-Flo Lactose (75-250μm), and Lactose 125M. Powder samples are retrieved from the mixer outflow and characterized using Near Infrared Spectroscopy (NIR).

Different operating conditions, design parameters and pharmaceutical formulation are investigated. The results show that processing angle as well as rotation rates, affect mixing performance. Interestingly, faster rotation rates increase the intensity of electrostatics effect and reduce blend homogeneity (Portillo et al. [3]). Convective design has also shown to be an important factor in mixing performance as discussed by Laurent and Bridgwater [4]. Blade location, blade alignment, blade design, blade angle are all important factors within the convective design which affect the radial dispersion coefficient and drive powder flow. Another advantage towards utilizing continuous mixing is that segregation has not shown to play a role during the mixing process although electrostatic effects are clearly present.

After characterizing the mixer through a set of experiments, our focus has been to develop computationally feasible mixing process models that will enable the simulation, optimization, and control of mixing processes. Previously we examined a hybrid approach for convective mixing systems (Portillo et al [5]), which is less computationally expensive than an entirely discrete element approach. Nonetheless in order to further decrease computational time we examine alternative on-line modeling approaches. Berthiaux and coworkers [6] have modeled continuous mixing with Markov chain. However the difficulty of these methods is that without pre-existing particle dynamics the transitional matrices are difficult to obtain. In our work we have utilized the experimental data obtained from various process and operating conditions to develop and validate both statistical and empirical-based models using parameter estimation techniques.

We examine two modeling paradigms, one following an empirical modeling approach dependent on experimental data with a substantially low computational cost. The second approach is based on compartment modeling that has been shown to be very efficient for batch mixing (Portillo et al. [7]). Following compartment modeling our results demonstrate that CPU times are affected by the number of particles per compartment, number of compartments, and mixing rates. As an example a simulation with of 250,000 particles and 150 compartments requires only 45 CPU seconds. We can further reduce the CPU time by modifying the number of particles in the system without sacrificing modeling performance. Both modeling approaches are intended to describe blend homogeneity as a function of axial length within the mixer, thereby preventing or reducing undesirable outcomes by monitoring on-line operating conditions.

References:

[1] Marikh K., Berthiaux H., Mizonov V., Barantseva E., 2005, "Experimental study of the stirring conditions taking place in a pilot plant continuous mixer of particulate solids", Powder Technology, 157, 138-143.

[2] Muerza S., Berthiaux H., Massol-Chaudeur S., Thomas G., 2002, "A dynamic study of static mixing using on-line image analysis", Powder Technol., 128, 195-204.

[3] Portillo P.M., Ierapetritou M.G., Muzzio F.J., 2006, "Characterization of Continuous Convective Powder Mixing Processes", submitted for publication to Powder Technology in Feb 2007.

[4] Laurent B.F.C., Bridgwater J., 2002, "Performance of single and six-bladed powder mixers", Chemical Engineering Science, 57, 1695-1709.

[5] Portillo P.M., Muzzio F.J., Ierapetritou M.G., 2007, "Hybrid DEM-compartment modeling approach for granular mixing", AICHE, 53, 1, Pages 119-128.

[6] Berthiaux H., Marikh K., Mizonov V., Ponomarev D., Barantzeva E., 2004, "Modeling Continuous Powder Mixing by Means of the Theory of Markov Chains", Particulate Science and Technology, 22, 4, 379-389.

[7] Portillo P.M., Muzzio F.J., Ierapetritou M.G., 2006, "Characterizing powder mixing processes utilizing compartment models", International Journal of Pharmaceutics, 320, 14-22.

409d Determination Of Granulation Growth Rates: An Evaluation Between Fbrm And Sieve Analysis

Kevin Macias and M. Teresa Carvajal. Industrial and Physical Pharmacy, Purdue University, 575 Stadium Mall Drive, West Lafayette, IN 47907

Purpose: To develop an in-line method for the determination of particle size during a high-shear wet granulation process. The utility in such a method would be highly desirable for determining granulation growth rates during the process in order to determine and validate coalescence and breakage kernels used in population balance equations. Methods: Placebo wet granulations were performed in a Diosna six-liter high-shear granulator equipped with a bottom driven impeller and side mounted chopper. Granulations of lactose monohydrate (Vopak North America, TX) and pre-gelatinized maize starch (Cargill Food & Pharma Specialties, IA) were prepared using a 5% w/w aqueous solution of poly-vinylpyrrolidone (K~30-BASF, NJ). Operational conditions were held constant for each experiment with an impeller speed of 150 RPM, chopper speed of 1000 RPM and delivery rate of binder solution at 150 grams per minute. During the granulation process, monitoring with Lasentec focused beam reflectance method (FBRM) occurred. In addition, incremental granulations were completed to represent the progression of the granulation process. The latter granulations were fractionated using sieve analysis in order to monitor the change in particle size distribution for each subsequent granulation. Results: Focused beam reflectance method (FBRM) has been used to follow changes chord length distribution of granule growth during the wet granulation process. From the FBRM data, movement of material between size bins was calculated. Employing incremental granulations, changes in particle size distributions were measured using sieve analysis. Qualitatively, the results from each experimental procedure describe similar growth kinetics and mechanisms during the granulation process. Conclusions: The in-line method using FBRM appears to provide similar information compared to measuring agglomerate growth using off-line method. The methods put forth have been established to provide kinetic information of the granulation process. This provides methods that will contribute to experimental determination of appropriate coalescence kernels and subsequent verification of population balance equations.

Effect of stirrer design on the homogeneity of a pharmaceutical powder mixture

Henri Berthiaux, Cendrine Gatumel, Jean-Louis Dirion
Rapsodee, UMR CNRS 2392, Ecole des Mines d'Albi, campus Jarlard - route de Teillet, 81000 Albi, France

Introduction

Pharmaceutical industry, as well as food, plastic or cement industries produces tablets, capsules or packets that are currently mixtures of 5 to 15 ingredients, and may contain several active ingredients. Standards have been developed to estimate the quality of the mixtures with respect to each of the actives and authorize the release of the products on the market. As recycling is generally not allowed in such processes, products that are failing at this stage are destroyed, which of course increases the prices of the products and indirectly induces a social cost. This makes mixture homogeneity assessment a key factor, also enhanced by the recent "Process Analytical Technologies" (PAT) initiative from the FDA. Because of their great adaptability to process qualification and control, continuous processes may be considered to form part of this small revolution (Pernenkil and Cooney, 2006). In addition, when referring to process qualification, continuous mixers are offering a key advantage with respect to batch mixers during the scale-up procedure. In fact, a full-scale pharmaceutical process is qualified if its performance has been verified at one-tenth of the industrial scale. This is a tedious problem as there is no guarantee that what has been proved with a 100 l mixer will still hold for a 1 m^3 batch vessel. Conversely, continuous mixer qualification will just necessitate a full-scale validation of, for example, 1 hour if the industrial production time of a certain product is 10 hours. When considering a continuous mixing process, the factors to taken into account can be listed as: the physical and surface properties of the particulate solids, the geometrical and dynamical properties of the mixer, or the accuracy of the dosage system. Basically, a continuous mixer aims to depress dosage fluctuations (see Kehlenbeck and Sommer, 2003) and to put the different products into contact. In particular, the choice of the geometry and size of the stirring device has to take into account the properties and use of the powders to mix. For "fragile" powders, one has to employ a stirrer that may not provoke particle attrition, while for cohesive powders, the mobile may enhance shear forces to counterbalance interparticle forces. From another hand, mobile configuration is responsible for the enhancement of radial mixing and/or axial mixing … and even so, mixture segregation. Anyway, all this is still very intuitive as there is no universal method for mixer calculation, design and scale-up. In this work, we focus on quantifying the influence of stirrer design on the continuous mixing process for a pilot-scale apparatus. We experimentally compare the effect of the mobile type in terms of mixture quality. This is presented for a real OTC drug containing 3 actives for a total of 9 ingredients, one active being present at a low dosage (0,5 %).

Experimental set ups and methods

The pilot scale continuous mixer Gericke GCM 500® has been used in this study and is fully described in Marikh et al. (2005) and Marikh et al. (2006). The mixer itself is a hemi-cylindrical tank that can be classified in the convective mixer's family as the motion of the particles is due to the action of a mobile that rotates inside the bulk. The stirrer can be of two very different types (see figure 1): paddles mounted on a frame with internal screw (mobile A) or paddles mounted on a shaft (mobile B). Depending on the rotational speed, mobile B can induce a fluidized flow or a dense phase flow, while mobile A, probably most suitable for fragile products, only provokes dense flow. For instance, the two stirrers will be compared in the dense flow regime. The dosage system is made of three loss-in-weight feeders, which are in fact classical screw feeders, each being placed on a balance.

<u>Figure 1</u>: the two different stirrer designs considered in this study (mobile A - left ; mobile B - right).

The mixture studied corresponds to a real industrial OTC drug containing 3 actives for a total of 9 ingredients. Two of the actives, that will be further referred to as A_1 and A_2 are agglomerated with three other ingredients to form the Basic Mixture (BM). A_1 and A_2 are representing respectively 10 % and 4 % by weight of the final drug. The four other ingredients, three additives I_1, I_2, I_3 and the active A_3, are divided into two premixes P_1 and P_2, that are defined in table 1, as well as some of their physical characteristics. This finally defines a mixture made of three streams to be mixed:

BM, P_1, P_2. The flow rates attached to these streams have been calculated to cope with the industrial cadences of production, and of course, with the composition of the mixture. The overall mixture is also of low dosage in active A_3 (nearly 0.5 %), which can be considered as the "main key component". However, the mass of A_3 in a sample of unit dose is still detectable by conventional methods. About "physical" differences of the ingredients, the values do not really indicate an important risk of particle segregation by size or density. We may only note that flowability is quite worse for A_3 and I_3, which are to be considered as fully cohesive powders and may cause difficulties during dosage and mixing.

<u>Table 1:</u> *main characteristics of the powders used*

	Mean particle size (µm)	Carr index (%)	True specific gravity ($g.cm^{-3}$)***	Theoretical weight per unit dose (g)	Mixing configuration	Mass flows (kg/h)
BM containing A_1 and A_2	110**	15	1.48	4.275	BM	32.87
A_3	28*	21	1.22	0.025	$A_3 + I_2$	0.57
I2	59*	11	1.33	0.050		
I1	67*	15	1.31	0.200	$I_3 + I_1$	4.56
I3	53*	20	1.75	0.400		

Mixture quality assessment and standards

For continuous mixing processes, the assessment of mixture homogeneity must be performed at the outlet of the vessel either by considering the random sampling of n samples over the whole production time or by considering n consecutive samples for a given time period. In the second case, there is a chance to detect "local" defaults that cannot be detected by random sampling. However, there is also the risk to give a bad estimation because defaults can exist at a higher scale than that of the n consecutive samples. In this work, we consider the second strategy by placing a series of striated boxes on a conveyor belt at the outlet of the mixer, thus defining samples when the belt is operated. The conveyor also has adjustable speed, which in concordance to the flow rates used, can give rise to samples of desired sizes. The sample compositions are determined by HPLC for the through an industrially validated protocol.

In the pharmaceutical industry, standards on mixture homogeneity do exist and are actually used to release the products on the market. They are based on three criteria that must be fulfilled for each of the active ingredients after sampling of 10 samples:

- The estimated mean x_m must lay in a -7.5 % / + 7.5 % interval around the real mean content μ.
- Each individual value x_i must lay in a -15 % / + 15 % interval around μ (European pharmacopeia) or x_m (FDA).
- The coefficient of variation CV = s / x_m must be less than 6 %.

This procedure allows estimating mixture homogeneity and qualifying the sampling protocol at the same time. Also, it aims identifying possible deviance of individual content in the tablets.

Results and discussion

Because flow rates must be maintained constant to cope with the production constraints, mixers are compared for different rotational speeds only, and related to the above industrial standards applied to the lowest dosed active ingredient A_3.

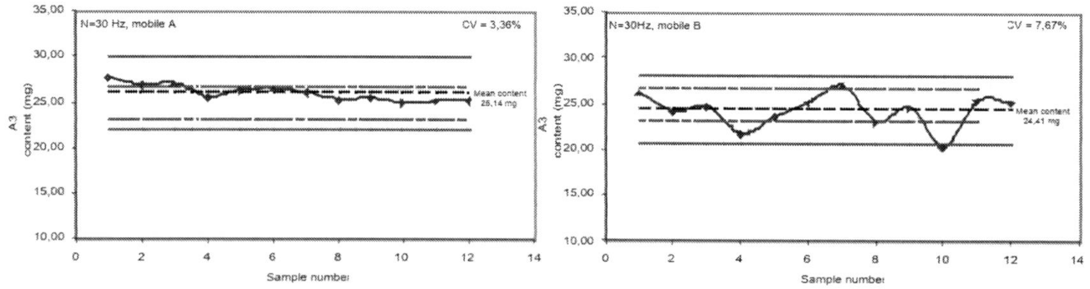

Figure 2: mixtures obtained with mobile A (left) and mobile B (right), other operating conditions being equal.

Figures 2 illustrate the effect of stirrer design in the dense phase flow regime (N=30 Hz). Individual content in A_3 is reported for each sample, as well as the mean of 12 consecutive samples. Limits of acceptability for x_i and x_m, as stated by standards, are also figured. Basically, the visual comparison of the two graphs is enough to state about effect of stirrer type. As for the model mixture, and as far as dense phase flow regime must be considered, the mobile A gives a better mixture than the mobile B for this rotational speed. In the present case, mobile B would have failed the conformity test and mobile A would have passed it successfully.

The action of the mobile B in the fluidised regime is illustrated by figures 3. At 60 Hz the mixture will have just passed the conformity test with respect to the mean, while at 80 Hz it will have failed it because of a too high CV value and a singularity on the individual contents. If one considers that lower hold-ups are attained with higher rotational speeds in a definite regime, this means that the dispersion aptitude of the mixer is getting worse with increasing N.

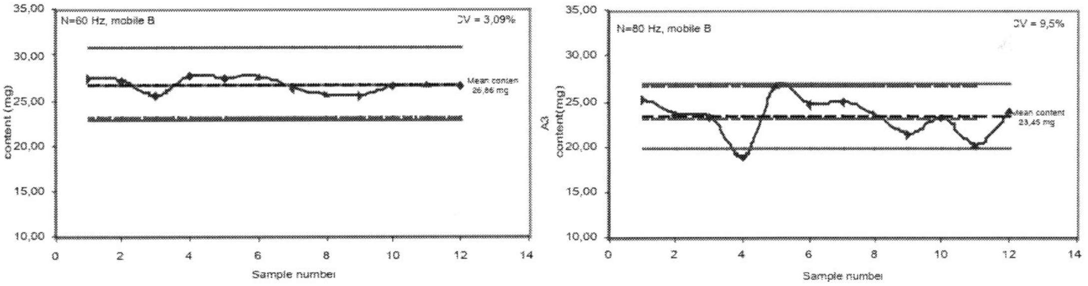

<u>Figure 3</u>: mixtures obtained with mobile B in the fluidised regime.

The results are synthesised in figure 4 for the coefficient of variation and the estimated mean respectively. In the dense phase flow regime, mobile A always gives CV values that are lower than mobile B. As the initial variances are all the same whatever the conditions studied, it can be concluded that mobile A is intrinsically better than mobile B for this "real" mixture, and as long as dense flow is to be maintained. In the fluidised regime, mobile B gives CV values of the same order as mobile A at 20-30 Hz, as long as the rotational speed of it is in the range 40-60 Hz. In the range above (70-90 Hz), the quality of the mixtures is worse, while it has been obtained for lower hold-ups. Again, we draw the attention to the fact that such a mobile looses its dispersion capability at too high rotational speeds. We may think that particles are hit too many times at those speeds to be able to transit efficiently in other parts of the vessel. From the viewpoint of acceptability of the mixtures, we have to focus on both graphs. From the CV criteria, five mixtures out of twelve meet the standard CV<6%. Two of them are obtained with mobile A, and the other three with mobile B. Focussing now on the mean values, one out of the three "a priori good" mixtures obtained with the mobile B fails the criteria and another one passes it for a difference of 0,01 mg ! The two good mixtures obtained with mobile A, are still acceptable, the one obtained at 30 Hz being the best mixture ever produced in the mixer.

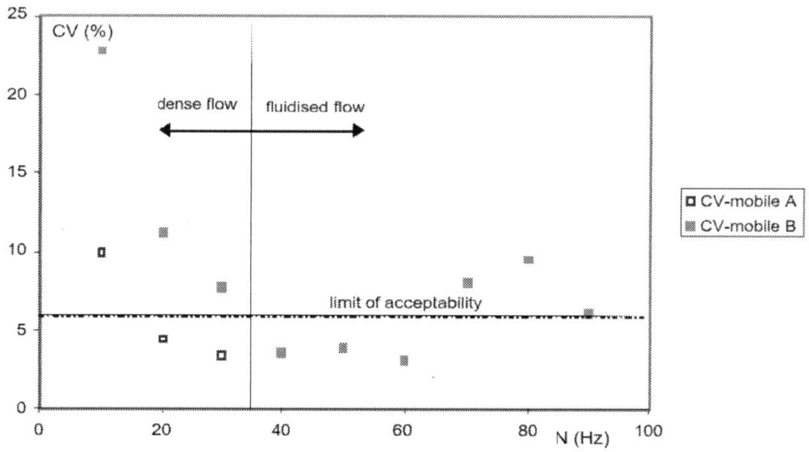

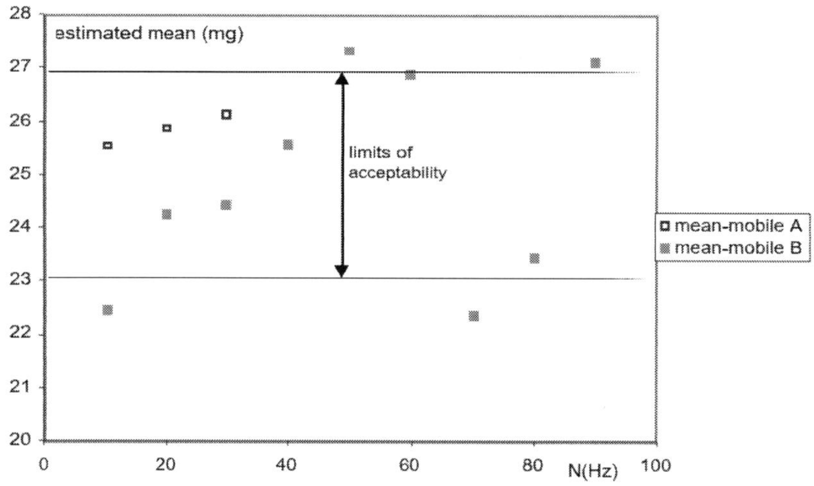

Figure 4: limits of acceptability of all the mixtures based on the CV (upper graph) and on the estimated mean content (lower graph).

Concluding remarks

The main conclusion is that under given experimental conditions, mobile A always gives better results in terms of mixture homogeneity than mobile B, as soon as the flow regimes are the same. The reason is that mobile B gives rise to higher hold-ups (and therefore higher mean residence time) with a dispersion capability that cannot counterbalance this effect. We suggest that stirrer design may be improved from the viewpoint of both dispersion and transport capacities, for example by Residence Time Distribution experiments and modeling.

All the mixtures investigated are far from passing this severe test, which is the rule in the industry. Indeed, continuous mixing process optimization through pharmaceutical constraints is not a trivial task, as it is for powder mixing in general. It is not a surprise to discover that some enquiries from the FDA (Hussain, 2004) are currently revealing rejection rates close to 50 % because of off-specification of batches.

Another important outcome that will be considered in future work of our team is concerned with continuous powder mixing process control. These recent years, a major effort has been employed to develop sensors for assessing homogeneity online or in-situ like NIR (Hailey et al., 1996), FT-Raman (Vergote et al., 200), or LIF (Pernenkil et al., 2006). The next step is to work out sensor models (chemiometrics) to convert the signal onto mixture homogeneity, mixing flow model to link homogeneity with agitation, sensibility to operating conditions to act on process variables in real time (see figure 5).

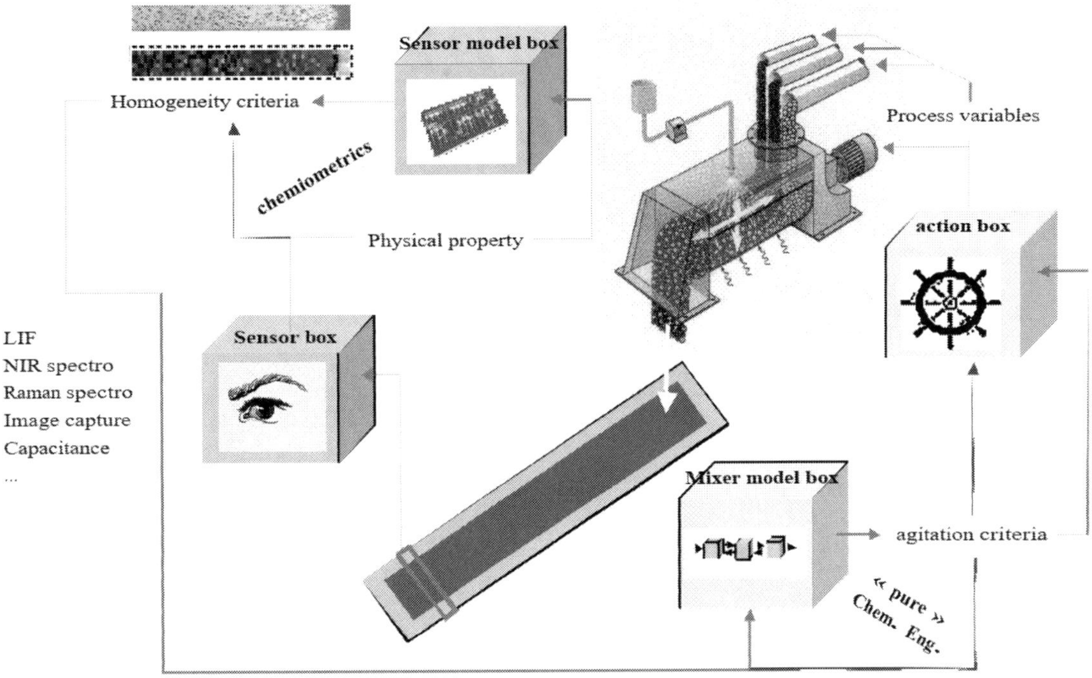

Figure 5: Process control - the real future of continuous powder mixing ?

References

Hailey, P.A., Doherty, P., Tapsell P., Oliver, T. and Aldridge, P.K., 1996, automated system for the on-line monitoring of powder blending processes using near-infra-red spectroscopy - 1. system development and control, Journal of Pharmaceutical and Biomedical Analysis, 14, 5, 551-559.

Hussain, A.S., The PAT initiative and the role of pharmaceutical engineering in the 21st century, Proceeedings of the Albi International Rencontres in Pharmaceutical Engineering, Albi, October 2004, 63-71.

Kehlenbeck, V. and Sommer, K., 2003, Possibilities to smooth the mass flow pulsations of volumetric feeders, CHOPS'4, Budapest, 14.13 - 14.19.

Marikh K., Berthiaux H., Mizonov V., Barantseva E., 2005. Experimental study of the stirring conditions taking place in a pilot plant continuous mixer of particulate solids, Powder Technology, 157, 138-143.

Marikh, K., Berthiaux, H., Mizonov, V., Barntseva, E. and Ponomarev, D., 2006. Flow analysis and Markov chain modeling to quantify the agitation effect in continuous powder mixer, Chemical Engineering Research and Design, 84 (A11), 1059-1074.

Pernenkil, L. and Cooney, C.L., 2006, A review on the continuous blending of powders, Chemical Engineering Science, 61, 720-742.

Pernenkil, L., Chirkot, T.P. and Cooney, C.L., 2006, Continuous operation of zigzag® blender for pharmaceutical applications, World Congress on Particle Technology 5, Orlando, Fl.

Vergote, G.J., DeBeer, T.R.M., Vervaet, C., Remon, J.P., Baeyens, W.R.G., Diericx, N. and F. Verpoort, F., 2004, in-line monitoring of a pharmaceutical blending process using FT-Raman spectroscopy, European Journal of Pharmaceutical Sciences, 21, 4, 479-485.

409f Application Of In-Line Transmission Nir For Improved Process Understanding And Control Of A Continuous Mixing Process

Gregory M. Troup[1], John P. Higgins[1], Craig A. Mckelvey[2], Michael Lowinger[2], Jeff S. Ko[3], Luke R. Schenck[3], Robert F. Meyer[3], and Brit L. Rudeen[3]. (1) Process Analytics, Merck & Co. Inc., 770 Sumneytown Pike, West Point, PA 19486, (2) Materials Characterization & Technology Assessment, Merck & Co. Inc., 770 Sumneytown Pike, West Point, PA 19486, (3) Pharmaceutical Commercialization Technology, Merck & Co. Inc., 770 Sumneytown Pike, West Point, PA 19486

This presentation discusses the application of an in-line spectroscopic technique during the process development stage of the production of a pharmaceutical intermediate in a continuous mixer. Specifically, we will show the development and implementation of a transmission NIR measurement system to monitor the composition of a multi-component process stream in real time. These real time data are used during process development to facilitate scale up, and to assess ability of the process to dampen input noises. Additionally during developmental processing this measurement system is used for real time process monitoring and detection of process upsets.

Continuous Blending of Powders: Effect of Process Variables and Material Properties

Lakshman Pernenkil*, Farzad Jalali-Yazdi*, Jason Whittaker*, Charles L. Cooney#

Introduction

Conventional batch blending processes in pharmaceutical manufacturing processes coupled with long quality analysis times increases the cycle time of secondary pharmaceutical manufacturing processes. Continuous dry powder blending offers the benefit of continuous monitoring that can be used in optimizing the process for the right quality of the product at the end of the process. In order to understand continuous dry powder blending processes, the microscopic particle level properties must be related to the macroscopic process variables.

This paper tries to characterize cohesive dry powder behavior in a continuous blending system. The blender performance is found to be related to the operational variables and material properties. Residence time distribution of the active ingredient in the blender allows us to characterize the blender performance with material properties.

Pernenkil and Cooney [1] reviewed the continuous blending literature and noted that cohesive powders were rarely studied in continuous blending literature. They also point out that medium to high shear mixers are needed for cohesive mixing. The work reported here uses a continuous double helical ribbon blender to mix cohesive caffeine and acetaminophen (as active ingredients) with lactose DCL 11, DC14 and Micro Crystalline Cellulose (MCC) Avicel and Celphere as excipients.

Axial Dispersion Model for Residence Time Distribution

Axially mixed systems can be represented by the Focker-Planck Equations for each component shown in equation 1 as discussed in Kehlenbeck and Sommer [2]. Following Fogler [3] we can derive the axial dispersion equation using the Fokker-Planck equations for one component non-dimensionalized using θ = tV/L, ξ=x/L, ψ=C/(M/V) and Bo=VL/D, where Bo is the Bodenstein number (Peclet number for powders), L is the length of the mixer and V is the average (mean) velocity of the powder in the mixer, we obtain equation (1).

$$\frac{\partial \psi}{\partial \theta} = -\frac{\partial \psi}{\partial \xi} + \frac{1}{Bo}\frac{\partial^2 \psi}{\partial \xi^2} \qquad (1)$$

* Department of Chemical Engineering, Massachusetts Institute of Technology, 77 Massachusetts Avenue, 56-454, Cambridge, MA 02139

Professor of Chemical and Biochemical Engineering, Massachusetts Institute of Technology, 77 Massachusetts Avenue, 56-469, Cambridge, MA 02139

This PDE can be solved using Danckwerts' Open-Open Boundary Conditions [4, 5]. Dispersion occurs both upstream and downstream of the mixing section in a blender (i.e. points upstream to tracer addition and downstream to tracer measurement), the boundary conditions are called open-open boundary conditions. Upon using these boundary conditions, and following Fogler [3] we can derive equation (2).

$$E(t)=\frac{1}{2\tau\sqrt{\frac{\pi t}{\tau Bo}}}\exp\left(-\frac{\left(1-\left(\frac{t}{\tau}\right)^{2}\right)}{\frac{4t}{\tau Bo}}\right) \qquad (2)$$

This axial dispersion model is applicable only if the length of the blender is considerably larger than the characteristic length of dispersion. In many real blending cases, this condition is satisfied. The most important parameter in this model is the Bodenstein number which can be estimated by performing non-linear regression of the observed response to a unit impulse in the continuous blender.

Description of Experiments

A continuous double helical ribbon blender (c.f. Figure 1) designed at MIT was fed using a Schenck Accurate (Schenck Accurate, Whitewater, WI) feeder for the excipient. The blender could take powder feeding rates up to 10 kg/hr. Powder was fed at the inlet hopper and the powder exit was kept completely open for all experiments. LIW feeders from Schenck Accurate Inc.(Whitewater, WI) were used to feed the excipients and active ingredients in this work. Light Induced Fluorescence and Near Infrared Spectroscopy were both used to determine the concentration of the active ingredient. Calibrations for these techniques using UV absorption spectroscopy indicated that the LIF sensor was more non-linear than the NIR sensor. PLS regression was used in calibration of the NIR sensor. The sensor was placed below the quartz glass window near the exit of the blender seen in Figure 1.

Two active pharmaceutical ingredients along with two different excipients were used in this work. Domike [6] and Ngai [7] studied the interparticle forces of adhesion for caffeine with micro-crystalline cellulose and lactose. Their work provided a basis to understand the effect of microscopic powder properties on macroscopic phenomena in continuous blending. USP grade caffeine powder was obtained from Sigma Aldrich (Sigma Inc., St. Louis, MO.) Caffeine is not a free flowing powder and usually agglomerates in the presence of atmospheric moisture. Therefore, caffeine was manually de-agglomerated for ten minutes using a crusher fitted with a 300 μm mesh before all experiments. Acetaminophen (N-(4-hydroxyphenyl)-Acetamide) was obtained from Sigma Aldrich (Sigma Inc, St. Louis, MO). Acetaminophen agglomerates in the presence of moisture and anhydrous acetaminophen was used in all the experiments.

Acetaminophen was manually de-clumped for ten minutes using a crusher fitted with a 300 μm mesh before all experiments.

Two grades of direct compressible lactose (DCL) were used in this work. Both were procured from DMV-International (DMV-Fonterra Excipients, Delhi, NY). Direct Compressible Lactose (DCL)-11 has a mean particle size of 100 μm ± 50 μm with a log-normal distribution. Direct Compressible Lactose (DCL)-14 has a mean particle size of 110 μm ± 70 μm with a log-normal distribution. In addition to these materials, DCL-14 was sieved into powders of two different particle size ranges 40 ≤ D50 ≤ 150 μm and 150 ≤ D50 ≤ 250 μm. Both these powders were evaluated for the effect of particle size on blending. Two grades of cellulose were used in this work. MCC is comparatively more cohesive than lactose. Avicel® was procured from FMC Biopolymer (FMC Biopolymer, Philadelphia, PA). Avicel® flow behavior is very erratic and is highly influenced by moisture. Celphere® microcrystalline cellulose CP-102 with a mean particle size of 150 μm is a spherical particle and was procured from Asahi-Kasei (Asahi-Kasei Corp., Tokyo, Japan). Celphere® flow behavior is less erratic than Avicel. However, Celphere® is less frequently used in the pharmaceutical formulations than Avicel®.

For the residence time distribution experiments, a single shot of the active ingredient with the weight equal to 5% by mass of powder that flows in the blender over 3 times the space time of the blender was used. Fill weight experiments to determine the powder behavior in the blender were also carried out at various flow rates, angle of inclines and shaft rotation rates.

Results and Discussion

Fill Weight Experiments

Increasing flow rate in the blender increased the fill weight of the blender linearly since the ribbon blender imparted moderate shear and no particle deformation or particle size reduction was seen. At +5º angle of incline, the slope of the line changed over at 4 kg/hr due to powder flowing out of the second outlet in addition to the outlet below shown in Figure 1. As the rotation rate increased, the fill weight in the blender decreased and could be explained because increasing rotation rate increased the rate of powder removal from the blender and caused a reduction in the fill weight of the powder in the blender. As the angle of incline decreased from +5º to -5º, the blender was being cleared of the powder at a greater rate due to gravity assisting the powder clearance in the blender.

Mean Residence Time

Increasing the shaft rotation rate increased the mean velocity of the powder in the blender and therefore caused a decrease in the mean residence time of the caffeine in the blender. The repeatability in mean residence time was very poor. RTD results from LIF data showed that excipient flow rate had a different effect at different angles of incline. At 0º as flow rate increased, the residence time decreased due to increased mean velocity of the material in the blender.

Increasing flow rate was supposed to reduce the mean residence time of the powder in the blender. However, the mean residence time was found to have cross interactions with angle of incline and shaft rotation rate. Increasing the angle of incline caused a maxima in the mean residence time as function of flow rate. In any given set of blender configuration and shaft rotation rate, the mean residence time was found to increase with the flow rate in the blender.

Increasing flow rate was expected to decrease the mean residence time of caffeine in the blender. However, this effect was observed to interact with the shaft rotation rate and the angle of incline. When angle of incline was increased, a maximum in mean residence time was observed near the critical flow rate where the second window was activated. When the angle of incline was decreased, the inner ribbon was not activated due to low fill weights and hence caused an increase in mean residence times as flow rates increased. It was found that within a given regime of operation, increasing flow rates decreased mean residence times due to increased mean velocities of particles in the blender.

Dispersion Coefficient

The dispersion coefficient was estimated from the Bodenstein number and the effect of operational variables on the dispersion coefficient was measured. Dispersion coefficients were computed using experimentally determined mean residence times and Bodenstein numbers.

The phenomenological behavior of powders depends on the microscopic properties of particles composing the bulk powder. The determination of this relationship is of critical importance in assuring that powder processing phenomena are well understood, characterized and controlled. Domike [6] and Ngai [7] related the microscopic properties of particles like particle size, cohesive and adhesive surface energies to macroscopically observed phenomena. The effect of particle size, cohesion and adhesion surface energies and particle shape were examined in this work and the results are described in subsequent sections.

By changing the particle size of the excipient (by sieving), it was found that an increase in size of the excipient particles that form the bulk of the powder mixture resulted in a decrease in the Bodenstein number as shown in Figure 2 at 0º incline. As the particle size of the excipient increased or the size ratio of excipient to active ingredient increased dispersive mixing. The larger particles are able to overcome the cohesive and adhesive forces due to gravity forces and would disperse more freely in the mixture.

Since micro-scale properties cannot be changed on the same particle (without changing the environmental conditions), materials were changed while maintaining other parameters like shape and particle size constant. In order to examine the effect of cohesion, acetaminophen and caffeine were chosen as APIs which have similar particle sizes but differing cohesion values. Similarly, in order to examine the effect of cohesion of excipient, lactose and MCC of similar particle size but strongly different cohesion values were used. In order to examine the effect of shape, MCC of different shape but similar particle size were used in the RTD experiments.

Principal component analysis of the dispersion coefficients, operational variables, types of excipients and APIs was carried out in SIMCA-P+. The results were plotted as a loadings plot between the first two components as shown in Figure 3 Micro Scale Properties and Effect on Dispersion Coefficient. Here excipient 1 was DCL-11, Excipient 2 was DCL-14, Excipient 3 was MCC Avicel PH102 and Excipient 4 was MCC Celphere CP-102. Active ingredient 1 was caffeine and 2 was acetaminophen. All correlations were found by taking the scalar product of the position vectors of the observed response in the loadings plot with the parameter of interest.

Changing the active ingredients, caffeine to acetaminophen, did not affect the mean residence time because the bulk of the powder flowing in the blender was the excipient. It was therefore expected that the excipient properties would dominate the dependence of mean residence time.

Higher cohesive forces were found to reduce the dispersion coefficient as seen by the negative correlation between MCC particles (3 and 4) and Dispersion Coefficient. As the particle shape was changed by changing the excipient from needle and flat plate MCC Avicel PH-102 particle to spherical Celphere CP-102 particles, the dispersion coefficient remained invariant as seen in Figure 3. However, the mean residence time was positively correlated with spherical Celphere MCC particles. A change in shape caused changes in density and increased the fill volume in the blender for the same fill weight and caused the activated the second window at the exit. Hence, the residence times of the particles were observed to be less for MCC Avicel PH-102 when compared with MCC Celphere CP-102.

Conclusion

Increasing cohesive forces were found to decrease the dispersion coefficient. Increasing particle sizes (of the excipient) gave gravity forces the upper hand in overcoming cohesive forces and increasing dispersive mixing. Residence time depends on the interaction between the particle shape, powder density, rotation rate and angle of incline and flow rate. Blending performance characterized by the variance reduction ratio will depend on both the particle dispersive mixing as well as the time spent by the particle in the blender. Hence particle properties play a critical role in determining the blender performance by determining both the intensity of dispersion and residence time.

Acknowledgments

The authors would like to thank Sohrab Singh Virk, Yiqun Bai, Louis Perna for their assistance in the experimental work, Dr. Bart Johnson of Axsun Technologies, Dr. Rajeev Garg and Dr. Michael Hoffman of Abbott Laboratories for their discussions on NIR analysis. The authors thank the Consortium for the Advancement of Manufacturing of Pharmaceuticals for financial support.

References

[1] Pernenkil L, Cooney CL. A review on the continuous blending of powders. Chemical Engineering Science. 2006;61(2):720-42.

[2] Kehlenbeck V, Sommer K. Modeling of the mixing process of very fine powders in a continuous dynamic mixer. Fourth International Conference on Conveying and Handling of Particulate Solids; 2003; Budapest, 2003; 2003. p. 9.26-9.31.

[3] Fogler HS, Brown LF. Analysis of Non-Ideal Reactors. In: Fogler HS, ed. *Elements of Chemical Reaction Engineering*. 2 ed. New York: Prentice Hall 1999:765-73.

[4] Danckwerts PV. Continuous Flow Systems. Chemical Engineering Science. 1953;2:1-13.

[5] Danckwerts PV, Sellers SM. The effect of hold up and mixing on a stream of fluctuating composition. Industrial Chemist. 1951;27:395.

[6] Domike RD. Pharmaceutical Powders in Experiment and Simulation. Cambridge, MA: Massachusetts Institute of Technology; 2003.

[7] Ngai SS. Multiscale Analysis and Simulation of Powder Blending in Pharmaceutical Manufacturing. Cambridge, MA: Massachusetts Institute of Technology; 2006.

Figures

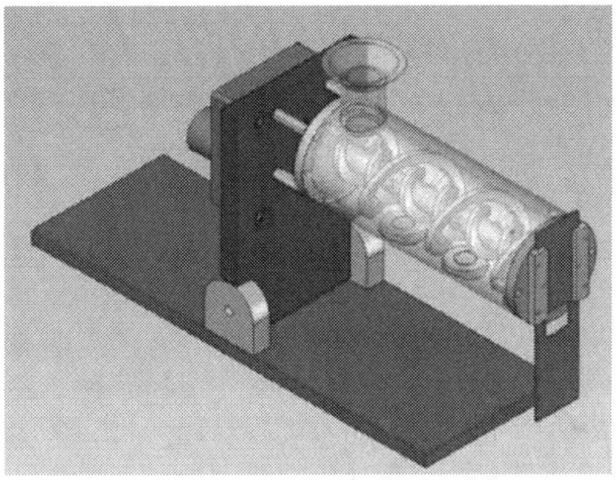

Figure 1 Double Helical Ribbon Blender

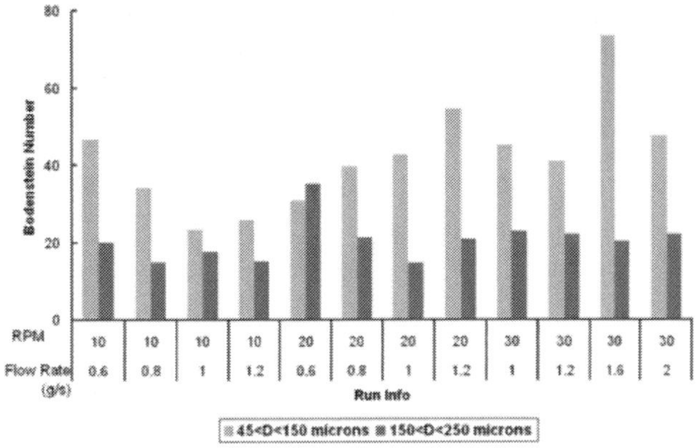

Figure 2 Effect of Excipient Particle Size on Bodenstein Number

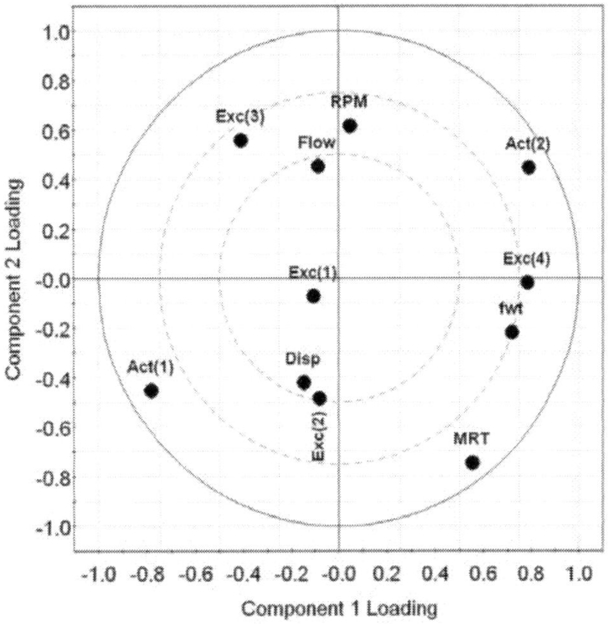

Figure 3 Micro Scale Properties and Effect on Dispersion Coefficient

410a Crystals And Crystallization Processes: A Crystallographic Perspective

Qi Gao, Analytical Research and Development, Bristol-Myers Squibb, 1 Squibb Drive, New Brunswick, NJ 08903 and William L. Parker, Process R&D, Bristol-Myers Squibb, 1 Squibb Drive, New Brunswick, NJ 08903.

Crystallization plays a critical role in pharmaceutical development not only because of its economical importance in large scale purification and great potential in optical resolution, but also because it is the only means of producing the desired crystalline forms of active pharmaceutical ingredients. Polymorph control is an essential process development activity, as polymorphism can alter physical properties that affect drug performance and stability. Although crystallization is one of the oldest technologies, establishing successful and robust conditions for producing specific polymorphs has been done by trial and error due to a lack of fundamental understanding of the thermodynamics and kinetics that govern the process. Knowledge of factors controlling the formation of crystals is thus demanded at both the bulk and molecular level. This paper attempts to link structural features of crystals, molecular conformations and intermolecular interactions, with nucleation where the atoms arrange in a defined and periodic manner that defines a crystal structure. Particular emphasis will be directed to observations of homochiral compounds that form crystals whose structures mimic the structures of their racemic forms. These pseudoracemic structures appear to have a significant impact on nucleation.

410b Relationship Between Self-Association Of Molecules In Supersaturated Solutions And Solid State Outcome

Deniz Erdemir[1], Soma Chattopadhyay[2], Liang Guo[2], Jan Ilavsky[3], Heinz Amenitsch[4], Carlo U. Segre[2], and Allan S. Myerson[1]. (1) Chemical Engineering, Illinois Institute of Technology, Chicago, IL 60616, (2) CSRRI and BCPS Department, Illinois Institute of Technology, Chicago, IL 60616, (3) UNICAT, Advanced Photon Source, Argonne National Laboratory, 9700 South Cass Avenue, Building 438 E, Sector 33, Argonne, IL 60439, (4) Institute of Biophysics and X-ray Structure Research, Austrian Academy of Sciences, Sincrotrone Trieste, SAXS Beamline, Strada Statale 14, Basovizza-Trieste, 34012, Italy

Understanding the link between solution and solid state is crucial in controlling the crystallization process in order to obtain products with desired properties, in particular the polymorphic form. In this work, small angle X-ray scattering was utilized to examine the early stages of crystallization in supersaturated solutions of glycine, in order to determine the role of initial molecular association in the polymorphic outcome. The majority of glycine molecules were found to exist as dimers in aqueous solutions, and monomers in acetic acid-water mixtures. Aqueous solutions rich in dimers crystallized into &alpha-glycine with a double layer structure composed of dimers, while the acidic solutions rich in monomers produced the &gamma-form with a polar chain structure built of monomeric units. These results suggest a direct correlation between the initial association of molecules in solution and the final crystal structure, which can be beneficial in selection of solvents favoring the nucleation of different polymorphic forms.

Selective Crystallization of the Metastable Anhydrate Form in the Enantiotropic Pseudo-Dimorph System of L-phenylalanine using Feedback Concentration Control

Nicholas C. S. Kee[†,§,‡], Paul D. Arendt[†], Reginald B. H. Tan[§,‡], and Richard D. Braatz[†]

[†] *University of Illinois at Urbana-Champaign, Illinois, USA;*
[§] *National University of Singapore, Singapore;*
[‡] *Institute of Chemical and Engineering Sciences, Singapore*

Introduction

Polymorphism is described as the ability of a compound to adopt more than one crystal structure.[1] Although chemically identical, each polymorph has its own unique combination of physical, thermal, and mechanical properties. These in turn exert an influence on the performance of the product; for example, the bioavailability and shelf-life of pharmaceutical compounds. It is therefore crucial to have a consistent and reliable production process for the specific polymorph to achieve feasible economic yield and for regulatory compliance. It is often possible to obtain the thermodynamically stable polymorph without much complication under suitable crystallization conditions such as temperature and agitation rate, for a long enough experiment. Producing large crystals of a metastable form is more challenging, the main challenge being to prevent cross nucleation of the stable form. A variation of polymorphism is the crystallization of hydrates or solvates in which solvent molecules are incorporated into the crystal structure at well-defined lattice positions; these crystalline forms are called pseudo-polymorphs.

L-phenylalanine (L-phe) in a mixed solvent system consisting of water and 2-propanol was chosen as the model enantiotropic pseudo-dimorph system. L-phe crystals exist in two different crystalline forms, the needle-like monohydrate form (Figure 1a) and the anhydrate form which appears as rhombic platelets (Figure 1b). The transition point was reported to be 37 °C in water.[2,3] The transformation rate of the monohydrate to the anhydrate form below the transition temperature has been studied previously[2] based on offline powder X-ray diffraction (PXRD) analysis of samples withdrawn at specific time intervals.

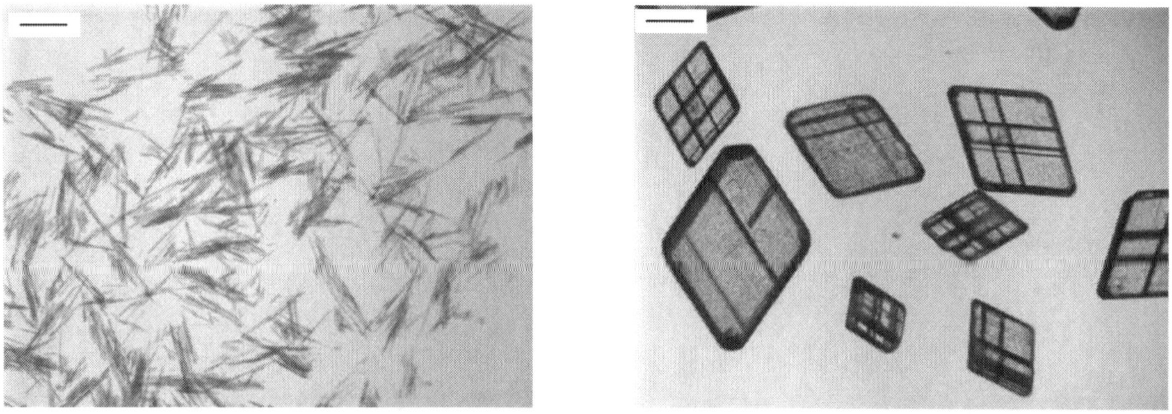

(a) (b)

Figure 1. Microscopy images of L-phe crystals:
(a) monohydrate form (scale bar 100 μm), (b) anhydrate form (scale bar 250 μm).

Recipes for industrial batch crystallization processes are typically based on optimal temperature or solvent addition profiles, which in turn are derived from accurate nucleation and crystal growth kinetics or through trial-and-error experimentation. The kinetic information can be obtained from a succession of crystallization experiments,[4,5] but this may prove laborious and time-consuming for complex systems. An alternative methodology, which does not require accurate kinetic data or repetitive trials, is to operate the crystallizer based on in-situ measurement of solution concentration so that it follows a supersaturation trajectory within the metastable zone to avoid unwanted nucleation.[6,7,8] An advantage of this concentration control strategy is its robustness against variation in growth or nucleation kinetics and practical disturbances and has been demonstrated for both cooling and antisolvent crystallizations in non-polymorphic systems.[9,10] More recently, this approach has been applied successfully toward the preferential crystallization of the metastable α-form crystals in the monotropic dimorph system of L-glutamic acid.[11]

This paper describes the application of this concentration control methodology for the pseudo-polymorph system of L-phe. The objective is to selective crystallize the anhydrate form in a seeded system to a temperature below the transition point in which this form is metastable. The lower and upper bounds of the operating regime are specified by the anhydrate form solubility curve and metastable limit, respectively. ATR-FTIR (attenuated total reflectance-Fourier transform infrared) spectroscopy coupled with a calibration model constructed using chemometrics techniques[12,13] was used to provide in-situ solution concentration measurement. FBRM (focused beam reflectance measurement), which measures in-situ the characteristics of crystal size distribution, was used to detect the metastable limit, as previously demonstrated.[9,14,15] In-situ video microscopy was used to assess the nucleated form. The seeded batch cooling crystallizations were implemented with the feedback concentration control approach using different supersaturation profiles to obtain the most appropriate recipe for selectively growing anhydrate L-phe crystals.

Experimental Procedures

Materials and Instruments

A Dipper-210 ATR immersion probe (Axiom Analytical) with ZnSe as the internal reflectance element attached to a Nicolet Protégé 460 FTIR spectrophotometer was used to obtain the spectra of the L-phe solution. The chord length distribution of the crystals in solution was measured using Lasentec FBRM and the in-situ images were collected using Lasentec Particle Vision and Measurement (PVM). The solution temperature was controlled by rationing hot and cold water to the jacket with a control valve as described previously.[9] The X-ray diffraction (XRD) patterns of L-phe crystals were collected offline based on powdered samples, using Bruker General Area Detector Diffraction System (GADDS, Bruker AXS, Inc.) with Cu $K_{\alpha1}$ and Cu $K_{\alpha2}$ (weighted sum) radiation and a step size of 0.02°. Figure 2 compares the PXRD patterns of both forms.

L-phe crystals obtained commercially (>98.5%, Sigma Aldrich) were verified using PXRD to be pure anhydrate form; the characteristic peaks of the monohydrate form were not observed. The monohydrate form was produced by rapidly cooling an initially undersaturated solution (4.5 g/100 g DI water) from 60°C to 25°C. The solid was filtered and dried following the procedures previously outlined,[2] in which the drying process did not result in dehydration.

Recrystallized anhydrate crystals were used in the metastable limit and batch crystallization experiments as seeds because these had a well defined rhombic shape as opposed to the commercially available crystals which were irregular platelets. The recrystallization procedures were as follows: an initially undersaturated solution (3.0 g/100 g mixed solvent) was cooled from 50°C to 42°C and monohydrate crystals were then added. The solvent mediated transformation of the monohydrate to the anhydrate form was complete after about 2.5 hours under isothermal conditions and constant stirring. The mixed solvent consisted of 75% DI water and 25% 2-propanol (ACS grade, Sigma Aldrich). The PXRD pattern of the recrystallized anhydrate form was identical to that obtained commercially; this shows than the recrystalllized form was simply of a different crystal habit and was not a different polymorph.

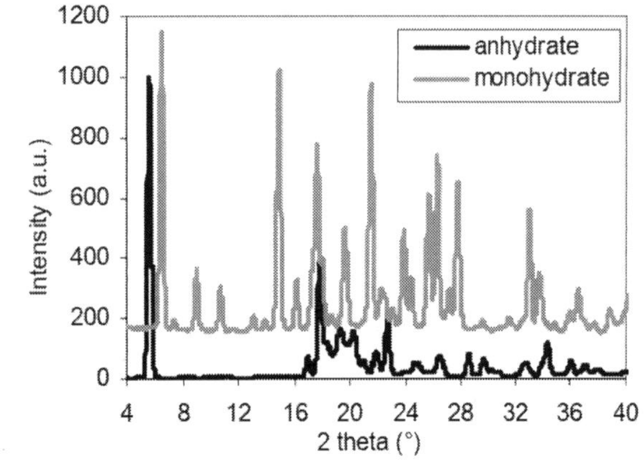

Figure 2. PXRD patterns for the monohydrate and anhydrate forms of L-phe.

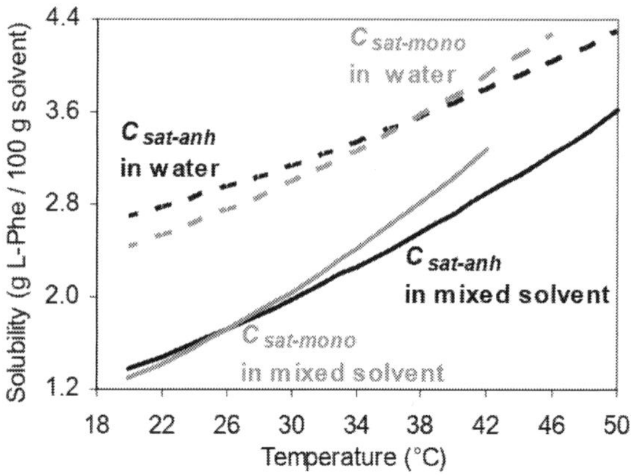

Figure 3. Anhydrate and monohydrate form solubility ($C_{sat\text{-}anh}$ and $C_{sat\text{-}mono}$ respectively) of L-phe in water[2] and in mixed solvent (75% water and 25% 2-propanol).

Calibration for Solution Concentration

Different solution concentrations of L-phe and the mixed solvent (300 g DI water and 100 g 2-propanol) were placed in a 500 ml jacketed round-bottom flask and heated until all the crystals dissolved. The solution was cooled at 0.5 °C/min, while the IR spectra were collected. The IR spectra in the range 1100–1650 cm^{-1} were used to construct the calibration model. The calibration model relating the IR spectra and temperature to solution concentration was determined as described previously using various chemometrics methods[16] such as principal component regression (PCR). The chemometrics forward selection PCR 2 (FPCR 2) method[17] was selected because it gave the smallest prediction interval; the prediction interval (0.06 g/100 g mixed solvent) was compatible with the deviation of the experimental data points from the fitted solubility curves.

Solubility and Metastable Limit Measurements

For each form, the IR spectra of L-phe slurries (saturated, and with excess crystals) were collected at different temperatures. The slurry was equilibrated for 30 minutes to an hour at a specified temperature while recording the IR spectra. The equilibrium solution concentration (solubility) was then calculated using the aforementioned calibration model. The metastable limit of L-phe solutions was determined at various solution concentrations, using the polythermal method.[18]

Each solution was heated above its saturation temperature to dissolve all crystals, before being cooled at 0.1 °C/min. As the main crystallization experiments utilized seeding, the metastable limit was determined for the seeded system; anhydrate crystals (0.5 g, 220-520 µm) were added at 0.6°C after the solution temperature crossed the anhydrate form saturation temperature. Upon seeding, PVM images were recorded to check for the appearance of the monohydrate needles. The chord length distribution of the crystals was monitored using FBRM. The start of sharp increase in the total counts of chord lengths / seconds after seeding was considered to be a nucleation event.

Seeded Batch Crystallization

Appropriate amounts of L-phe in 400 g of mixed solvent (300 g DI water and 100 g 2-propanol) was heated to above the anhydrate form saturation temperature in a 500-mL jacketed round bottom flask with an overhead mixer at 175 rpm, to create an undersaturated solution at a solution concentration of 2.3 g/100 g. The crystallizer was then cooled at 0.1 °C/min to 34°C (0.6°C below the anhydrate form saturation temperature), upon which anhydrate seed crystals (0.5 g, 220-520 µm) were added. Supersaturation setpoint profiles were followed during crystallization based on in-situ solution concentration measurement as described previously.[9,10] The control algorithm was started shortly after seeding. Supersaturation setpoint profiles were selected at different constant absolute supersaturation ($\Delta C = C - C_{sat}$) with respect to both the anhydrate and monohydrate form, where C is the solution concentration.

Results and Discussion

Solubility and Metastable Limit Measurements

Figure 3 compares the solubility curves of both forms of L-phe in water[2] and in the mixed solvent system. A decrease in the transition temperature from 37°C to 26°C was observed. This was also reported for the case of L-serine[19] (in water-methanol solutions), which was another amino acid with similar pseudo-polymorph behavior. The decrease in the transition point was attributed to a relatively larger increase in the heat of dissolution of the monohydrate form compared to that of the anhydrate form with the addition of methanol in the solvent. The increase in the FBRM total counts/second indicated a secondary nucleation event but did not directly identify the nucleated form. This was inferred depending on when the monohydrate needles were detected via PVM images relative to the rise in FBRM counts. Figure 4 plots the nucleation behavior for the seeded system. At temperatures below the transition point, needles were observed before the increase in FBMR counts; this meant that the latter was likely due to cross nucleation (or concomitant nucleation). The first limit determined via PVM images represents the onset zone of cross nucleation. In the interval between this and that measured by FBRM, cross nucleation was increasing and eventually substantial in registering a change in the FBRM measurements.

At temperatures above the transition point, needles were not detected prior to or at the onset of increase in FBRM counts but were only observed at a substantial later time; this proved that the nucleation event detected via FBRM was that of the seeded form.

The seeded metastable limit associated with both types of secondary nucleation provides an estimate of the maximum allowable supersaturation for the controlled growth of the anhydrate seed crystals, i.e. without cross nucleation which would impact the polymorph purity or nucleation of the seeded form which would widen the anhydrate product crystal distribution.

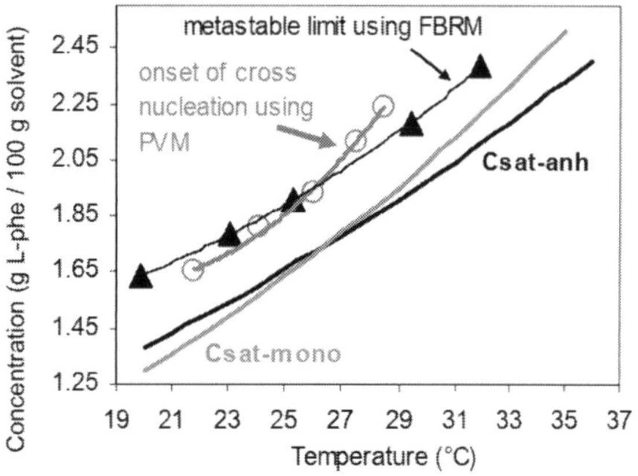

Figure 4. Metastable limits measured using FBRM and PVM.

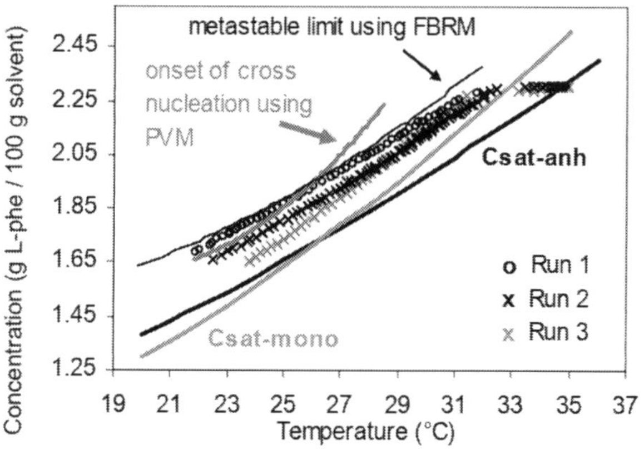

Figure 5. Supersaturation profiles implemented in the seeded batch crystallization runs.

Concentration-Controlled Batch Crystallization

Figure 5 shows the supersaturation profiles of the three seeded concentration-controlled crystallization carried out. Run 1 was implemented at constant ΔC at 0.2 g/100 g with respect to the anhydrate form; this was very close to the metastable limit (determined using FBRM) while exceeding the onset zone of cross nucleation (determined using PVM). Monohydrate needles were detected via PVM images (Figure 6) after 113 minutes (at about 28.5°C). However, the increase in the FBRM total counts / seconds (Figure 7) started earlier (after 45 minutes, at about 30.3°C); this was attributed to the nucleation of the anhydrate form. The appearance of the needles at 28.5°C even though the supersaturation profile did not cross the onset zone of cross nucleation at this point was likely due to the effect of increasing seed mass. Figures 8 and 9 compare the seed to the product crystal. Monohydrate needles were observed on anhydrate crystal surfaces while anhydrate crystals smaller than the seed size were also detected.* The polymorphic contamination was verified from PXRD analysis; the characteristic peaks of the monohydrate form were evident in the PXRD pattern of the product crystals.

Run 2 was implemented with a smaller driving force of constant ΔC at 0.15 g/100 g with respect to the anhydrate form. The supersaturation trajectory was within the metastable limits throughout the run (Figure 5). The FBRM total counts/second (Figure 7) showed a slight increase towards the end at 390 minutes (at 25.3°C). The supersaturation profile was near the

----

* Figure 9 shows seemingly lesser quantity of needles compared to Figure 6d. The needles agglomerated upon filtration (not shown in Figure 9), so that individual needles were not well dispersed in the dried sample compared to the solution.

onset zone which suggested that the cause of the increase was cross-nucleation; this was verified by the detection of small numbers of needles in the PVM images prior to this at 300 minutes (at about 27.6°C). The number of needles nucleated around 300 min was sufficiently small and the supersaturation with respect to this form was sufficiently small that it took ~90 min for the needles to be detected by the FBRM. Figure 10 shows the product crystals of this run which consisted of anhydrate crystals of more uniform size. Monohydrate needles were not readily observed in the microscopy images; the characteristic peaks of this form were also absent in the PXRD pattern of the product crystals. This showed that cross nucleation did not produce a significant amount of needles in this run.

One disadvantage of the earlier recipes was that by keeping the supersaturation with respect to the anhydrate form constant, the supersaturation with respect to the monohydrate was actually slowly increasing; this provided a rising tendency for cross nucleation as the run progressed. A better approach is to track and maintain a low supersaturation with respect to the monohydrate form, while growing the anhydrate form. Run 3 was implemented with constant ΔC at 0.1 g/100 g with respect to the monohydrate form; this translated into a varying driving force for the anhydrate form ranging from 0.19 g/100 g to 0.07 g/100 g. This recipe was selected such that the supersaturation was less than that of Run 1 from 30.3°C onwards to avoid secondary nucleation of the anhydrate form (otherwise observed in Run 1); similarly the supersaturation was below that of Run 2 from 27.6°C onwards to prevent cross nucleation (otherwise detected in Run 2). This recipe was successful in preventing any significant cross nucleation as shown by the constant FBRM total counts per seconds (Figure 7) throughout the run. Figure 11 shows the product crystals which consisted of anhydrate form crystals. Figure 12 compares the anhydrate product crystal size distribution for all three runs.

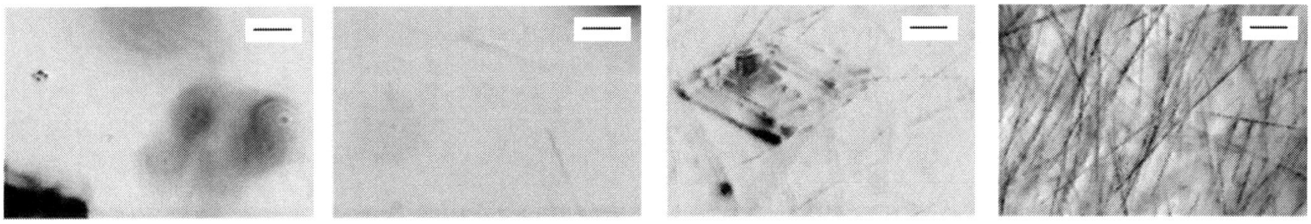

(a) (b) (c) (d)

Figure 6. PVM images from Run 1 at time (scale bar 100 μm): (a) 45 min (30.3°C), (b) 113 min (28.5°C), (c) 251.5 min (23.55°C), (d) 307.5 min (21.9°C).

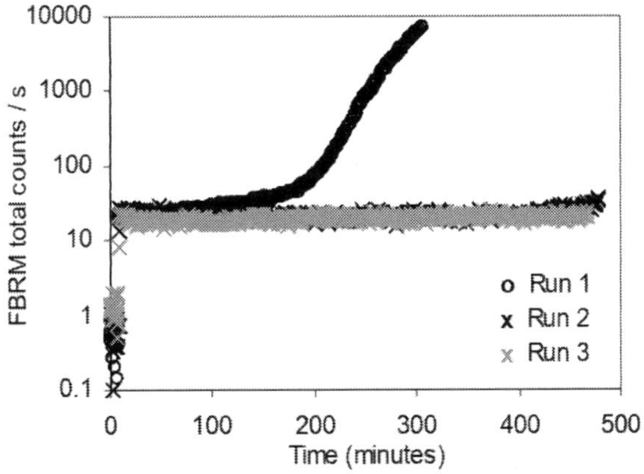

Figure 7. Total chord lengths counts / seconds using FBRM.

Figure 8. Microscopy images of anhydrate form seed crystals used in the crystallization experiments (scale bar 200 μm).

Figure 9. Microscopy images of product crystals from Run 1; monohydrate needles on anhydrate crystal surface (scale bar 200 μm).

Figure 10. Microscopy images of anhydrate form product crystals from Run 2 (scale bar 200 μm).

Figure 11. Microscopy images of anhydrate form product crystals from Run 3 (scale bar 200 μm).

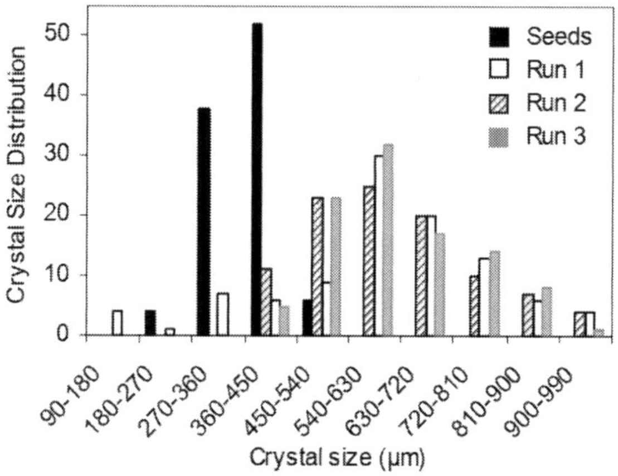

Figure 12. Seed and product crystal size distribution (based on largest diagonal length; sample size of 100 crystals).

Conclusions

A batch recipe for the selective crystallization of the metastable anhydrate form of an enantiotropic pseudo-polymorph system, L-phenylalanine in a mixed solvent system (75% water and 25% 2-propanol), was identified through direct design. The solubility curves of both pseudo-polymorphs were determined using ATR-FTIR spectroscopy, combined with chemometrics. Laser backscattering (FBRM) was used to determine the seeded metastable limit; the nucleated forms were identified using in-situ video microscopy (PVM). The concentration control strategy, based on in-situ solution concentration measurement and feedback control of the temperature, enabled batch operations at various supersaturation profiles. Undesired secondary nucleation of both forms at the metastable limit provided the operating constraints in terms of the maximum allowable supersaturation. Batch crystallization implemented at low constant absolute supersaturation with respect to the monohydrate form prevented cross nucleation and was successful in selectively growing large anhydrate crystals with relatively more uniform sizes to temperatures below the transition point where the anhydrate form is metastable. Similar operations stopping above the transition temperature (where the anhydrate form is stable) would obviously face lesser complications in preferentially growing crystals of this form but would have limited product yield. This methodology for the selective growth of metastable crystals extends the useful range of the phase diagram to increase product yield. It is also a practical method applicable to other enantiotropic systems, because of its direct design approach and flexibility in using different seed loadings or sizes.

Acknowledgements

Institute of Chemical and Engineering Sciences and Agency for Science, Technology and Research, Singapore are acknowledged for their support of this project. M. Fujiwara is acknowledged for providing technical advice on this project. The authors also wish to thank S. Wilson and M. Ramesh from the George L. Clark X-ray facility for the PXRD data collection.

References

1. Davey, R. J.; Garside, J. *From Molecules to Crystallizers*; Oxford: Oxford University Press, **2000**.
2. Mohan, R.; Koo, K. K.; Strege, C.; Myerson, A. S. *Ind. Eng. Chem. Res.* **2001**, *40*, 6111-6117.
3. Sato, T.; Sano, C. *European Patent #0703214A2*, **1995**.
4. Togkalidou, T; Tung, H. H.; Sun, Y.; Andrews, A. T.; Braatz, R. D. *Ind. Eng. Chem. Res.* **2004**, *43*, 6168-6181.
5. Worlitschek, J.; Mazzotti, M. *Crystal Growth & Design* **2004**, *4*, 891-903.
6. Feng, L. L.; Berglund, K. A. *Crystal Growth & Design* **2002**, *2*, 449-452.
7. Liotta, V.; Sabesan, V. *Org. Process Res. Dev.* **2004**, *8*, 488-494.
8. Gron, H.; Borissova, A.; Roberts, K. J. *Ind. Eng. Chem. Res.* **2003**, *42*, 198-206.
9. Fujiwara, M.; Chow, P.; Ma, D.; Braatz, R. D. *Crystal Growth & Design* **2002**, *2*, 363-370.
10. Zhou, G. X.; Fujiwara, M.; Woo, X. Y.; Rusli, E.; Tung, H. H.; Starbuck, C.; Davidson, O.; Ge, Z.; Braatz, R. D. *Crystal Growth & Design* **2006**, *6*, 892-898.
11. Kee, N. C. S.; Tan, R. B. H.; Braatz, R. D. *Application of Engineering Fundamentals to API Process Development - PAT Applications*, AIChE Annual Meeting, Paper 684b, **2006**.
12. Workman, J. J.; Mobley, P.; Kowalski, B.; Bro, R. *Appl. Spec. Rev.* **1996**, *31*, 73-124.
13. Mobley, P. R.; Kowalski, B. R.; Workman, J. J.; Bro, R. *Appl. Spec. Rev.* **1996**, *31*, 347-368.
14. Tahti, T.; Louhi-Kultanen, M.; Palosaari, S. In Proceedings of the 14th *International Symposium Industrial Crystallization*; Institution of Chemical Engineers: Rugby, UK, 1999; pp 1-9.
15. Barrett, P.; Glennon, B. *Trans. Inst. Chem. Eng.* **2002**, *80*, 799-805.
16. Togkalidou, T; Fujiwara, M; Patel, S.; Braatz, R. D. *J. Crystal Growth* **2001**, *231*, 534-543.
17. Xie, Y.; Kalivas, J. *Anal. Chim. Acta* **1997**, *348*, 19.
18. Nyvlt, J.; Sohnel. O.; Matuchova, M.; Broul, M. *The Kinetics of Industrial Crystallization*; Elsevier: New York, 1985.
19. Luk, J. C.; Rousseau, R. W. *Crystal Growth & Design* **2006**, *6*, 1808-1812.

410d Growth Prediction For Molecular Crystals Of Api-Complexity

Ryan C. Snyder, Jacob P. Sizemore, and Michael F. Doherty. University of California Santa Barbara, Department of Chemical Engineering, Santa Barbara, CA 93106

Modeling crystal growth of organic compounds is of paramount interest to the pharmaceutical, dyes and food products industries. Crystal growth plays an important role in both crystal morphology and polymorph transformations. Steady state and dynamic crystal morphology affect down stream processes such as filtering, washing and drying as well as particle flowability and agglomeration. Furthermore, the quality and efficacy of these materials are affected by crystal shape and polymorph. The shape affects which surface structures are present and their relative sizes. The polymorph can affect the solubility, color, bioavailability, stability and patentability of a crystalline product. Polymorphic phase transformations often take place by a solution mediated mechanism, which is characterized by the growth of a more stable polymorph and the simultaneous dissolution of a less stable polymorph. Thus, predicting growth of the stable polymorph facilitates the calculation of a polymorphic phase transformation rate. The abilities to predict crystal shapes and polymorphic phase transformations enable both process and product improvments.

Models of crystal growth shape have evolved from those that depend purely on crystal structure and thermodynamics to those that incorporate mechanistic phenomena such as the incorporation of solute at kinks on steps flowing across surfaces. We have focused on a class of these kinetic models and the development of steps on crystal faces through the spiral growth mechanism that was pioneered by Burton, Cabrera and Frank (1951) and further developed by Chernov (1984). Winn and Doherty (1998) further enabled the models to account for the effect of solvent. However, these models have primarily focused on systems where the molecules and crystal interactions are highly symmetric. On the other hand, pharmaceutical level molecular organic and biological crystals have complex bonding structures, molecular arrangements and energetic interactions.

We have developed a spiral growth model to incorporate these important quantities, which result in numerous complicated microscopic features such as non-isotropic spiral shapes, non-uniform velocity distributions and spirals of alternating step height. Some of these features (for example non-isotropic spiral shapes) have previously been measured by experiment; however, existing models do not account for their influence. In addition, for these complex systems, classical kinks (i.e. half crystal positions) do not exist since there are many locations where the interactions at a site cannot be divided in half. Thus, the velocity for each side of a spiral cannot be determined as being proportional to the probability of finding a kink site. Rather, the probability of a molecule attaching to each of the different sites must be enumerated. In this talk, key modeling details will be presented alongside examples of complex crystal systems where each of the new modeling concepts is needed. Finally, the results for growth shape predictions are compared to experiment.

Burton, W.K., Cabrera, N., and Frank, F.C., "The Growth of Crystals and the Equilibrium Structure of their Surfaces," Phil. Trans. R. Soc., 243, 299 (1951).

Chernov, A.A., Modern Crystallography III: Crystal Growth, Berlin: Springer-Verlag (1984).

Winn, D. and M.F. Doherty, "A New Technique for Predicting the Shape of Solutions-Grown Organic Systems," AIChE J., 44 (11), 2501 (1998).

410e Solution Speciation Of Impurities: A Force Field Simulation Of Its Impact On α-Glycine Crystal Habit

Sendhil Poornachary[1], Pui Shan Chow[2], and Reginald B.H. Tan[2]. (1) Department of Chemical & Biomolecular Engineering, National University of Singapore, 4 Engineering Drive 4, Singapore, Singapore, (2) Crystallization and Particle Science, Institute of Chemical & Engineering Sciences, 1 Pesek Road, Jurong Island, Singapore, Singapore

Solution crystallization of small organic molecules is of central importance to industrial manufacturing of food, pharmaceuticals and specialty chemicals. It is requisite that the particulate products obtained from the crystallization process have desired solid state properties such as shape (crystal habit), form (crystal structure) and size distribution. These properties, in turn, impact the downstream processing and formulation of the final product. In industrial systems, it is not uncommon that impurities, whose molecular structures are similar to the primary solute, come from the upstream processes. From previous studies, it is known that this class of impurities interacts with a developing crystal resulting in habit modification and crystallization of unexpected polymorphs. Isomorphic impurities also form solid solution in the crystalline state and affect crystal purity and solubility (Koolman and Rousseau 1996). Alternatively,"tailor-made" additives can be used to alter crystal growth kinetics and therefore achieve the desired crystalline properties. Recent studies are focused on theoretical simulation of impurity effects on the crystal growth process. In this work, we have used glycine (a simple amino acid) as the primary solute, and selected higher homologous α-amino acids as the impurities. Glycine is usually crystallized from pure aqueous solution as the metastable α-form. On the basis of this fact, initially, we have used atom-atom potential energy calculations to predict the theoretical crystal habit of α-glycine from its crystal structure (Materials Studio Modeling,Accelrys Software Inc.). The morphologically significant crystal facets were identified by comparing the experimental crystal habit with the one simulated. Subsequently, specific crystal surfaces were cleaved from the optimized α-glycine crystal structure and then built as a superstructure. The adsorption of impurity molecules that exist as zwitterions and anions was modeled by making a stereospecific docking on these crystal surfaces. Preferential interaction of impurity molecules with the glycine crystal, and the subsequent inhibition of crystal growth were quantitatively determined through "layer" and "attachment energy" calculations. The sensitivity of these calculations on the force field parameters was also investigated. The results obtained are compared with the experimentally observed habit modification (Poornachary et al. 2007).
References Koolman, H. C. and Rousseau, R. W. "Effects of isomorphic compounds on the purity and morphology of L-Isoleucine crystals" *AIChE J.*, 42 (1), 147 (1996) Poornachary, S. K., Chow, P. S., Tan, R. B. H. and Davey R. J. "Molecular speciation controlling stereoselectivity of additives: Impact on the habit modification in α-glycine crystals" *Crystal Growth & Design*,7 (2), 254 (2007)

410f Quantitative Analysis Of H-Bonding Interactions: Relation To Crystal Shape And Size

Charles Acquah[1], Luke E. K. Achenie[2], Arunprakash T. Karunanithi[3], Shanthakumar Sithambaram[4], Steven Suib[5], and Jose Gascon[4]. (1) Chemical Engineering, University of Connecticut, Storrs, CT 06269, (2) Department of Chemical Engineering, Virginia Polytechnic Institute and State University, Randolph Hall 133, Blacksburg, VA 24061, (3) U.S. Environmental Protection Agency, 26 W. Martin Luther King Drive, Cincinnati, OH 45268, (4) Department of Chemistry, University of Connecticut, Storrs, CT 06269, (5) Department of Chemistry, Chemical Engineering and Institute of Material Science, University of Connecticut, Storrs, CT 06269

This study was aimed at exploring the role of hydrogen bonding interaction between crystal and solvent molecules in the determination of crystal morphology. Two different approaches were used. The first approach involved using Fourier Transform Infra-Red (FTIR) spectroscopy analytical experiments to study crystal-solvent interactions. The FTIR spectrum of different solvents ranging from non-polar to highly polar are examined for differences in spectra shift on addition of a small but fixed amount of solute. Solutes used in this study were succinic acid, adipic acid and ibuprofen. The differences in spectra shift was used as a measure of the hydrogen bonding interaction. The second approach employs molecular dynamics (MD) to investigate crystal-solvent interactions which are mainly controlled by interfacial tension between the two phases. Preliminary results show that hydrogen bonding (H-bonding) interaction in some solvents (e.g. methanol, ethanol) was intermolecular. However, for some other group of solvents (e.g. ethylene glycol, propylene glycol and glycerol), which are characteristically viscous, bonding tended to be intramolecular and this greatly limits their intermolecular interaction. These solvents therefore behave like "non-polar" solvents. This observation was consistent for both the FTIR and MD approaches. In the context of crystal morphology, we expect these group of solvents to exhibit morphology somewhat similar to that of the morphology of crystals grown from non-polar solvents. Insights gained from this work provides useful information for designing and selecting solvents for use in pharmaceutical crystallization processes.

NANOPARTICLE FORMATION OF HIGHLY WATER-SOLUBLE PHARMACEUTICALS COATED WITH POORLY WATER-SOLUBLE MATERIAL BY USING THE RAPID EXPANSION OF WATER-IN-SUPERCRITICAL CO2 MICROEMULSION

Worawut Temphuwapat, The University of Tokyo, Tokyo, JAPAN
Masahiro Ikeda, Gifu University, Gifu, JAPAN,
Hideki Ichikawa, Kobe Gakuin University, Kobe, JAPAN
Yoshinobu Fukumori, Kobe Gakuin University, Kobe, JAPAN
Atsushi Tsutsumi, The University of Tokyo, Tokyo, JAPAN*

Abstract

A novel technique for nanoparticle production and nanoparticle encapsulation are investigated using the rapid expansion of water-in-supercritical CO_2 microemulsion. The aqueous cores of micellar nanodroplets filled up with Gd-DPTA, highly water-soluble anticancer drug, are formed and dispersed in the supercritical CO_2 to form the inverse microemulsion system. The rapid expansion of the supercritical microemulsion into the atmosphere is then carried out for the production of Gd-DPTA nanoparticle. For Gd-DPTA particle encapsulation, paraffin, encapsulating material, is additionally dissolved into the supercritical CO_2 phase. The morphologies of particle samples are examined by the SEM and TEM observation with EDS analysis. The composite nanoparticles consisted of AOT and Gd-DPTA were success fully produced at the average size of 350 nm and relatively narrow size distribution. During the rapid expansion process, the supersaturation of AOT dissolved at the interface of polar micelle filled up with Gd-DPTA and supercritical CO_2 take place. The precipitation of AOT capturing Gd-DPTA micelles leads to the formation of the composite particle. In addition, the composite particle of AOT and Gd-DPTA was successfully encapsulated with paraffin with a coating layer in nanoscale. During the rapid expansion, the supersaturation of AOT, dissolved at the interface of polar micelle filled up with Gd-DPTA and supercritical CO_2, and the supersaturation of paraffin dissolved in the supercritical CO_2 take place. Then, the precipitated paraffin nuclei are likely to deposit on the surface of the Gd-AOT composite particle precipitated simultaneously, leading to the encapsulation with paraffin. The sizes of the micelles seemed to play a significant role in mechanism of particle production and particle encapsulation. The results obtained have demonstrated that the rapid expansion of water-in-supercritical CO_2 microemulsion is a promising method for continual combination of nanoparticle production and their encapsulation, especially for highly water-soluble pharmaceuticals.

Introduction

Nanoparticle has been widely used in pharmaceutical field. The pharmarceutical nano-sized allows crossing of biological barriers, ameliorates tissue tolerance and improves cellular uptake and transport, so enabling efficient delivery to the target sites like liver, brain and solid tumor [1-2]. Also, by modulating the surface properties, composition and milieu, the desired release pattern of the therapeutic agent and biodistribution can be achieved. The drug nanoparticle have revealed a great potential in tackling problems like low oral bioavailability, high intersubject variability and poor or suboptimal therapeutic response associated with delivery of poorly water-soluble drugs. The submicronic nature of these systems helps in improving therapeutic and pharmacokinetic performance of such drugs, thus enabling effective drug delivery by various routes, especially parenteral and peroral [3-4]. However, The nanosized of highly water-soluble pharmaceutical has been widely used also, especially for highly water-soluble pharmaceutical. For example, the microcapsules containing highly water-soluble Gd-DPTA were prepared and studied as a promising anti-cancer drug used in neutron capture therapy by Fukumori and Ichikawa [5]. Because of the highly water solubility of Gd-DPTA, intravasular injected Gd-DPTA use rapidly excreted from the systemic circulation and even intra tumor injected Gd-DPTA cannot maintain adequate levels of Gd that are required for an efficient therapeutic index.

Up to now, many techniques have been developing for processing of pharmaceutical nanoparticle so far; including microemulsion template, high-gravity reactive precipitation, solvent displacement, and supercritical fluids. Most of them require the use of large quantity of organic solvent in the processes but supercritical fluids-based process. Regarding this characteristic, supercritical fluids-based process is attractive and has advantages over the others due to its solventless requirement or smaller usage of organic solvent in the process.

Supercritical fluid (SCFs) is a state of a substance when temperature and pressure are above its critical point. Supercritical fluids exhibit significant variation in thermodynamic and transport properties as a function of pressure and temperature. The unique properties of SCFs, especially the pressure-dependent solubility, have led to their widespread industrial applications as solvents and reaction media. Supercritical carbon dioxide has been the most widely used because of its mild critical pressure, chemical stability, nonflammability and non-toxicity. Hence, it is an environmentally friendly substitute for other organic solvents. Moreover, the low critical temperature of carbon dioxide allows mild operating temperatures, which is attractive for processing heat sensitive pharmaceuticals.

Several supercritical CO_2-based technologies for nanoparticle processing have been developing for twenty years, for example, RESS, SAS, PGSS, etc. These processes use different nucleation and growth mechanisms to precipitate very fine particles depend on the role of supercritical carbon dioxide. The solubility of the core material in supercritical carbon dioxide is the key to the production of very fine particle containing the core material with supercritical-based technology, especially for RESS process [6-8]. Then, major limitation with supercritical carbon dioxide is the low solvent strength, incapable of dissolving most polar compound. In the past decade, several approaches have been explored to enhance the solubility of polar substance in supercritical carbon dioxide. For example, the addition of polar cosolvent such as alcohols or acetone can increase the polarity, and therefore the solvent power. However, most of water soluble pharmaceutical materials are still insufficiently soluble in those supercritical carbon dioxide medium.

The objective of this research is to develop the novel supercritical CO_2-based nanoparticle processing that is able to produce the nanoparticle of highly water-soluble pharmaceuticals and their encapsulation with controlled-release coating material. The process so-called the rapid expansion of water-in-supercritical carbon dioxide microemulsion developed here is proposed to process the highly water-soluble Gadolinium-diethylenetrianminepenaacetic acid (Gd-DPTA), which is now being developed as an anti-cancer drug used in Neutron Capture Therapy. Production of pure Gd-DPTA nanoparticle and encapsulation of Gd-DPTA nanoparticle are investigated by using the rapid expansion of Gd-DPTA polar micelles in pure supercritical carbon dioxide, and the rapid expansion of Gd-DPTA polar micelles in supercritical solution of paraffin/CO_2, respectively, to the atmosphere. Supercritical carbon dioxide functions as a solvent and a dispersing phase. For the sample characterisation, SEM (Hitachi, S-900) and SEM with EDS analysis (JEOL, JSM-7000F) are used to analyze the morphology and element of the collected particles. In addition, TEM observation with EDS analysis, TEM (JEOL, JEM-2010F), is used to analyze the structure and the element of the collected particles.

Experimental

Materials

Sodium bis (2-ethylhexyl) sulfosuccinate (AOT, minimum 99%, Sigma Ultra, MW 444.56) was purchased from Sigma Chemical Co., Ltd. (USA) and used as surfactant. 2,2,3,3,4,4,5,5-octafluoro-1-pentanol (F-pentanol) was purchased from Tokyo Kasei Kogyo Co., Ltd. and used as

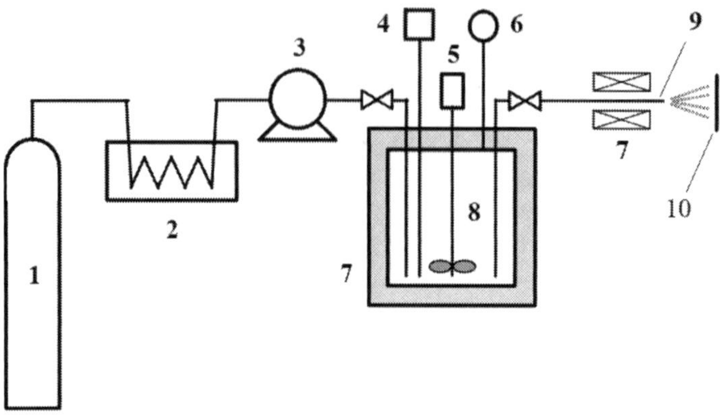

1. CO_2 Cylinder
2. Liquefier
3. High Pressure Pump
4. Thermocouple
5. Mixer
6. Pressure gauge
7. Heater
8 High pressure vessel
9. Nozzle
10. Target plate

Figure 1. Experimental Apparatus

co-surfactant. Lysozyme (from egg white) was purchased from Merck KGaA (GER). Gadolinium-diethylenetrianminepenaacetic acid (Gd-DPTA) was donated by Prof. Fukumori and Prof. Ichikawa from Kobe gakuin University. Carbon dioxide (critical temperature Tc = 304 K, critical pressure Pc = 7.4 MPa) purchased from SusukiShokan Co., Ltd was used as continuous phase and solvent. The coating material, paraffin (melting point 321-323K), and the ultrapure water were purchased from Kanto Kagaku Co., Ltd. All of chemicals are used without any treatment.

Apparatus

The experimental apparatus is schematically illustrated in Fig. 1. The whole system mainly consists of a high pressure pump, a high pressure vessel with a mixing paddle, a nozzle and a targeted plate. The high pressure vessel has a volume of 1500 cm^3 and was consisted of four sapphire windows for visual observation of the phase behavior. The stainless steel capillary nozzle with an inner diameter of 0.1 mm and a length of 10 mm was used for the rapid expansion process.

Procedure

First, a measurable amount of AOT, F-pentanol, aqueous solution of Gd-DPTA, aqueous solution of lysozyme, and paraffin were added into the high pressure autoclave as shown in Table 1. After the autoclave was pre-heated to the operating temperature, then the autoclave was sealed and carbon dioxide liquefied at 268–273 K by the liquefier was fed to the autoclave by the high pressure pump. After having reached the desired pressure and temperature, an optically transparent phase, which occurs when the CO_2 density was raised to a cloud point at which the microemulsion system becomes stable, was confirmed by the visual observation via four sapphire windows equipped on the high pressure autoclave. The microemulsion shown in Table 1 was stirred for 30 minutes, at the mixer's rotation speed of 300 rpm, to achieve the equilibrium. The microemulsion was assured the existence of the optically transparent phase by visual observation again, and then was expanded through the capillary nozzle, which was heated and maintained at the temperature of 453 K, into the atmospheric pressure where a target plate covered with aluminum paper was placed at a distance of 20 cm from nozzle.

Table 1. List of the amounts of chemicals used and experimental parameters

Condition	AOT	F-pentanol	Filled solution	Paraffin	Expansion condition
A	0.02 mol/l	0.3 mol/l	Lysozyme, Wo = 5	1 g	26 MPa, 305 K
B	-	-	-	1 g	26 MPa, 305 K
C	0.02 mol/l	0.3 mol/l	Lysozyme, Wo = 8	1 g	26 MPa, 305 K
D	0.02 mol/l	0.3 mol/l	Gd-DPTA, Wo = 5	-	26 MPa, 305 K
E	0.02 mol/l	0.3 mol/l	Gd-DPTA, Wo = 5	0.8 g	26 MPa, 305 K

Figure 2. Comparison of particles produced by different process
(a) The rapid expansion of water-in-supercritical CO_2 microemulsion (Condition A)
(b) Rapid Expansion of Supercritical Solution (Condition B)

SEM and TEM observations with EDS analysis

At each run of experiments, particles attached on the aluminum paper were used for the further characterization. SEM (JEOL, JSM-7000F) was used to analyze the morphology and element of the collected particles. As to the sample preparation for SEM observation, a small piece of aluminum paper was cut and mounted on the specimen stub with the conductive carbon tape, and then was sputter coated with platinum. For TEM (JEOL, JEM-2010F) observation, samples were prepared by directly spraying the microemulsion to the copper grid with collodion film.

Results and Discussions

Effect of the existence of micelles

Figure 2(a) shows the SEM image of the particles produced under condition A. There are two types of particles on the target plate, conic particles and spherical particles. Figure 2(b) shows the SEM image of the precipitated paraffin samples obtained under condition B, without the existence of polar micelles, under the same pre-expansion pressure. We can say that the experimental in condition B represents the particle formation by using the rapid expansion of supercritical solution (RESS) technique. Only the conic particles were collected. Comparison between these two figures suggests that an existence of polar micelles in the supercritical microemulsion leads to the formation of spherical particles.

Effect of Wo Value on Particle Encapsulation with Paraffin

The TEM images were also taken in order to affirm the existence of the encapsulation. Figure 3 show the TEM images of particles obtained from the experiments at Wo of 5 and 8. Figure 3(a) shows that the shade of coating layer is darker than that of center area. Moreover, it is likely that the fracture of paraffin layer was observed. On the other hand, Figure 3(b) shows the shade of center area is darker than that of coating layer. These imply that there was an open space and a lysozyme solution encapsulated inside the paraffin particles shown in Fig. 3(a) and Fig. 3(b), respectively.

In addition, comparison between these two figures suggested that a smaller encapsulated particle is able to produce in the microemulsion system which has lower Wo value. According to correlation between Wo value and size of micelle in reported in the small-angle neutron scattering (SANS) investigation of the water-in-CO_2 microemulsion [9], it also suggested that the higher the Wo value, the larger the sizes of micelles. Thus, that implied the size of the micelles might play a significant role in mechanism of encapsulation process, causing a difference of particle size.

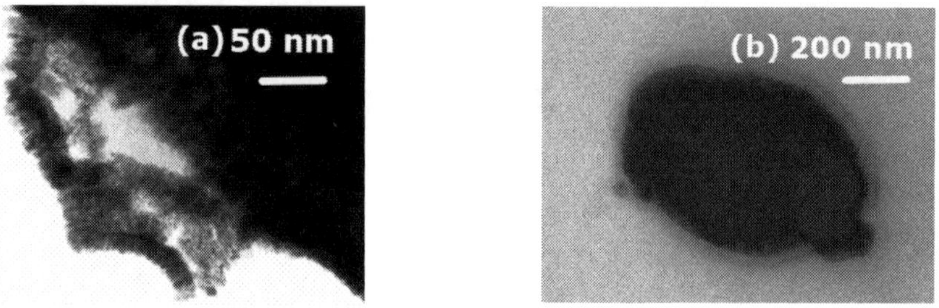

Figure 3. TEM images show the effect of Wo value
(a) $Wo = 5$, $P = 26$ MPa, $T = 305$ K; Condition A
(b) $Wo = 8$, $P = 26$ MPa, $T = 305$ K; Condition C

Production of Gd-DPTA Particles

Figure 4(a) shows the SEM image of the particles produced by the rapid expansion of water-in-supercritical CO_2 microemulsion, in which Gd-DPTA was dissolved in the micellar phase, which molar ratio of water to surfactant (Wo Value) of 5 at pre-expansion pressure of 26 MPa. The spherical particles with rough morphology were found on the target plate. Figure 4(b) shows the EDS profiles of the particle produced by the rapid expansion of water-in-supercritical CO_2 microemulsion, in which Gd-DPTA was dissolved in the polar micelles, from batch A also. The elements of C, O, Na, S, Al, and Gd detected affirms the existence of AOT, identified by Na and S, and Gd-DPTA, identified by Gd, inside the particle. Detection of Al element is because of the effect from aluminium paper, which was used as a target plate. Figure 4(c) shows the TEM image of particle taken from condition A. It shows that there are no different in shade of colour in particle. It is suggested that this particle has a composite structure of AOT and Gd-DPTA. The high dispersion and relatively narrow size distribution of particle size shown in Fig. 5, suggest that the mechanism for the formation of Gd-DPTA particle depended on the sizes of the micelles. Because the optically transparent phase of pre-expansion water-in-supercritical CO_2 microemulsion indicated that pre-expansion micelles have narrow size distribution and size less than 100 nm, it is likely that the micelles grew their sizes in the capillary nozzle during the rapid expansion process, increasing the sizes of precipitated particles.

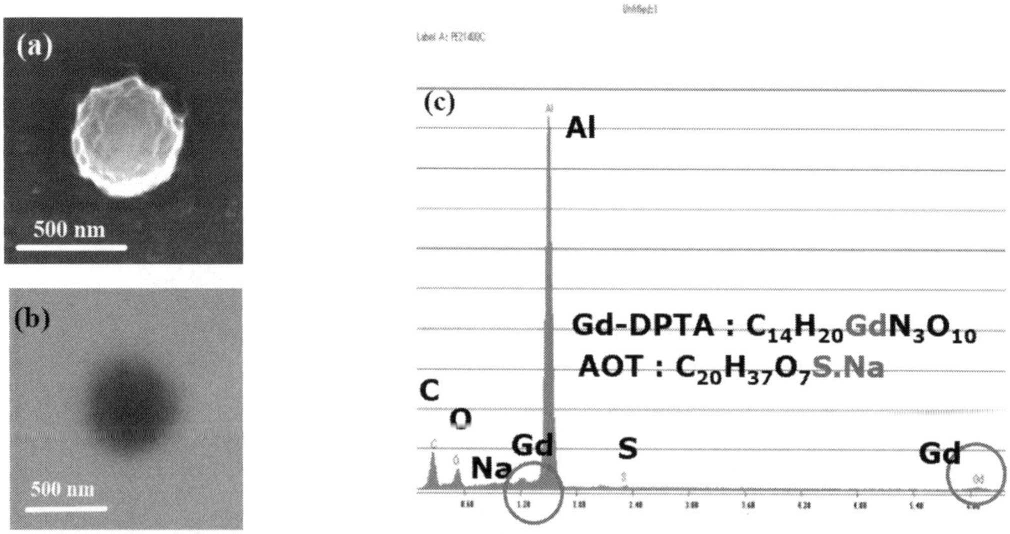

Figure 4. Physical properties of Gd-DPTA particles collected from condition D
(a) SEM picture (b) TEM picture (c) EDS analysis

Figure 5. Relatively narrow particle size distribution of Gd-DPTA particles (Left) Unprocess, (Right) Produced from condition E

Encapsulation of Gd-DPTA Particles with Paraffin

For the SEM observation, when the electron acceleration voltage is constant, the higher the magnification is, the higher the intensity of the electronic beam on the scanning area is. Part of the kinetic energy of the electronic beam is absorbed by the surface of particle sample, which makes the temperature of the scanning area increase. Because of low melting point of paraffin (only 321–323 K), the electronic beam of the SEM could melt the thin film of paraffin on the surface of coated particles [7]. By focusing the electronic beam on a coated particle for a relatively long time, the changes of the coated surface during the melting process could be observed. Thus, the coating of the core particles could be characterized qualitatively by the SEM observation.

When comparing with the surface morphology of particle produced under condition D, no addition of paraffin, in Fig. 4(a), the outline of particle produced under condition E shown in Fig 6(a) was seen to become spread. This indicates that paraffin covered the surface of this particle sample and the paraffin layer began to melt with increasing temperatures. Note that the expansion of coating layer due to paraffin melting is very uniform. This suggests that the surface of core particle was evenly coated by paraffin film. With respect to the coated particle shown in Fig. 4(a), the diameter of core particle was about 400 nm and thickness of the paraffin layer was around 30–40 nm. The core particle is suggested to be the composite particle of AOT and Gd-DPTA, like particle produced under condition A, because of the detection of Gd, Na, and S elements shown in Fig. 4(b). Comparison between EDS profiles shown in Fig. 4(c) and Fig. 6(b), the higher ratio of C to other elements in Fig 6(b) indicates the existence of paraffin coating layer on the particle produced under condition E, success in encapsulation.

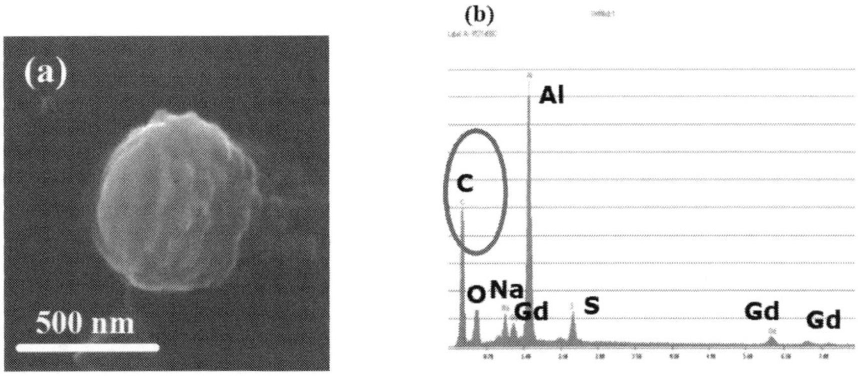

Figure 6. Gd-DPTA particle coated with paraffin layer (a) SEM picture (b) EDS analysis

Mechanism of Particle Production and Particle Encapsulation

(a) Particle Production

According to a structure model of water-in-supercritical CO_2 with AOT as surfactant and F-pentanol as co-surfactant proposed by J. Liu et al. [10], it suggest that the highly water soluble Gd-DPTA or lysozyme is dissolved in the nano-sized droplets of polar micelles disperse in the supercritical CO_2 because of similar polarity. At the interface of the polar micelle and supercritical CO_2, AOT and F-pentanol act as a surfactant stabilizing the formation of pre-expansion microemulsion. Because the solubility of AOT in the CO_2 depends on the pressure, during the rapid expansion to the atmospheric pressure, the solubility of AOT in CO_2 decreases dramatically. A very high supersaturation ratio leads to the formation of AOT nuclei capturing the Gd-DPTA solution. Then, the composite particle of AOT and Gd-DPTA is formed. Mechanism of composite particle formation by using the rapid expansion of supercritical CO_2 microemulsion is presented in Fig. 7 (a). In this process, supercritical CO_2, first, acts as continuous phase of the microemulsion system to be a dispersing phase of nano-sized polar micelles. In addition, supercritical CO_2 also acts as a solvent to dissolve the AOT, the Gd-DPTA capturing material, before expansion.

When comparing with the theoretical micelle size (< 100nm) of the microemulsion system, the size of the composite particle of Gd-DPTA with AOT is about 3 times larger. Because the pre-expansion microemulsion was optically transparent, micelles size would be lower than 100 nm, it is likely that micelles increased their sizes inside the nozzle. During the rapid expansion that can consider as an adiabatic process [11], the dramatically decreasing of temperature and pressure affect the stability of the water-in-supercritical carbon dioxide microemulsion system. It is considered that the aggregation of lower-100nm micelles take place, increasing their sizes to about the particle size at the outlet of the nozzle (Fig. 7 (b)).

(b)Particle Encapsulation

According to a structure model of water-in-supercritical CO_2 with AOT as surfactant and F-pentanol as co-surfactant by J. Liu et al. [10], it suggest that the highly water soluble Gd-DPTA or lysozyme is dissolved in the nano-sized droplets of polar micelles, in the other hand, paraffin is dissolved in the supercritical CO_2 because of similar polarity. At the interface of the polar micelle filled up with Gd-DPTA and supercritical CO_2 filled up with paraffin, AOT and F-pentanol act as a surfactant stabilizing the formation of pre-expansion microemulsion. Because the solubility of AOT and paraffin in the CO_2 depend on the pressure, during the rapid expansion to the atmospheric pressure, the solubility of AOT and paraffin in CO_2 decreases dramatically. A very high

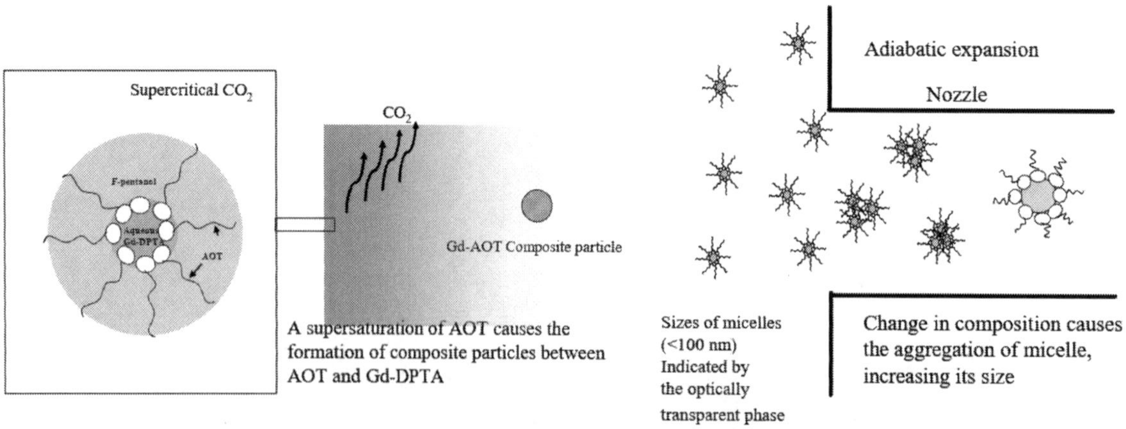

Figure 7. Mechanism of Gd-DPTA particle formation by the rapid expansion of water-in-supercritical CO_2 microemulsion (a) Overall Process (b) Aggregation of micelles at nozzle

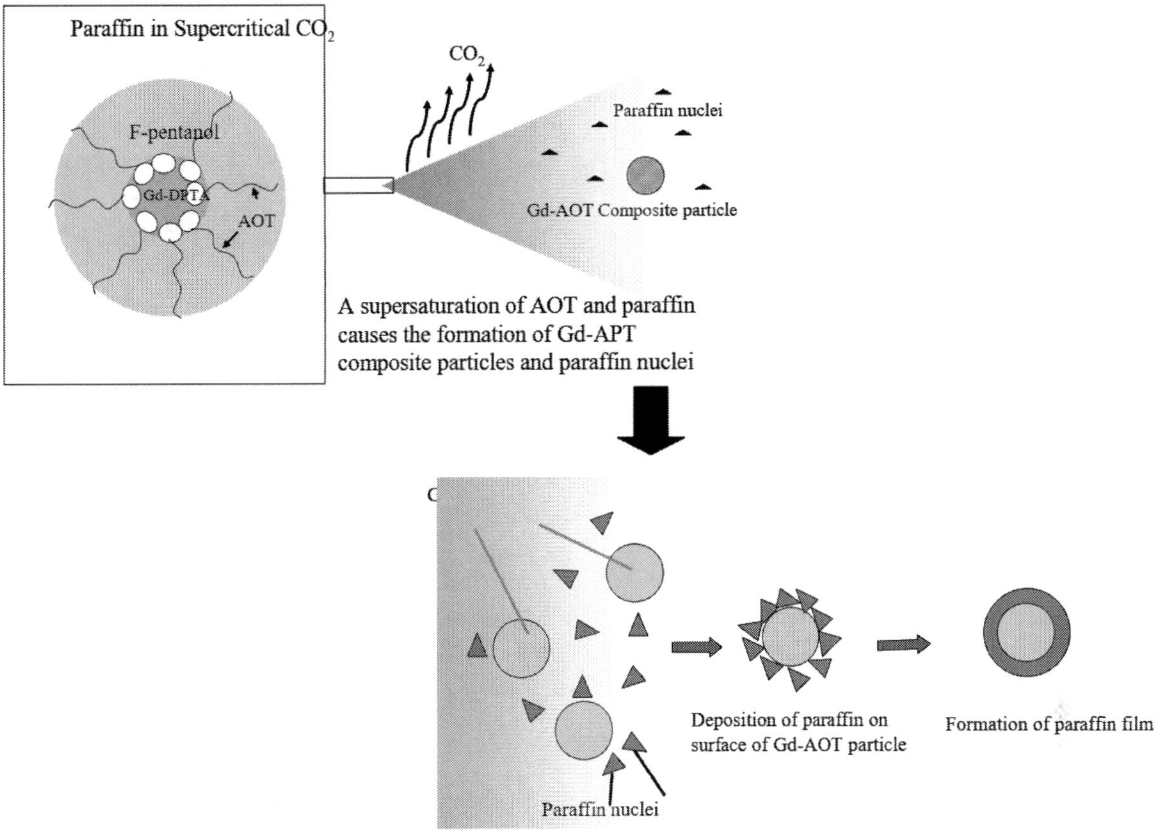

Figure 8. Mechanism of Gd-DPTA particle encapsulation with paraffin by the rapid expansion of water-in-supercritical CO_2 microemulsion

supersaturation ratio leads to the formation of Gd-AOT composite particle, as mentioned in previous part, and the formation of paraffin nuclei. Subsequently, the paraffin nuclei deposit on the surface of composite particles, leading to the encapsulation. The existence of conic particles, shown in Fig. 2a), is considered to be the paraffin nuclei that did not attach and deposit on the surface of Gd-AOT composite spherical particle. Mechanism of particle encapsulation by using the rapid expansion of supercritical CO_2 microemulsion is presented in Fig. 8. In this encapsulation process, supercritical CO_2, first, acts as continuous phase of the microemulsion system to be a dispersing phase of nano-sized polar micelles. In addition, supercritical CO_2 also acts as a solvent to dissolve the AOT, the Gd-DPTA capturing material, and paraffin, coating material of the Gd-AOT composite particle, before expansion.

Conclusions

We have proposed a novel nanoparticle processing by using rapid expansion of water-in-supercritical CO_2 microemulsion. The composite nanoparticle consisted of AOT and Gd-DPTA was success fully produced at the average size of 350 nm and relatively narrow size distribution. In addition, the composite particle of AOT and Gd-DPTA was successfully encapsulated with paraffin with a coating layer in nanoscale. During the rapid expansion, the supersaturation of AOT, dissolved at the interface of polar micelle filled up with Gd-DPTA and supercritical CO_2, and the supersaturation of paraffin dissolved in the supercritical CO_2 take place. Then, the precipitated paraffin nuclei are likely to deposit on the surface of the Gd-AOT composite particle precipitated simultaneously, leading to the encapsulation with paraffin. The size of the micelles seemed to play a significant role in mechanism of particle production and particle encapsulation. The results obtained have demonstrated that the rapid expansion of water-in-supercritical CO_2 microemulsion is a promising method for continual combination of nanoparticle production and their encapsulation, especially for highly water-soluble pharmaceuticals.

References

1. J. Kreuter, Nanoparticulate systems for brain delivery of drugs, *Advance Drug Delivery Reviews* **47**, 65 (2001).
2. Y. Nishioka, H. Yoshino, Lymphatic targeting with nanoparticulate system, *Advance Drug Delivery Reviews* **47**, 55 (2001).
3. S.M. Moghimi, A.C. Hunter, J.C. Murray, Long-circulating and target-specific nanoparticles: theory to practice, *Pharmacological Reviews* **53**, 283 (2001).
4. K. Mader, W. Mehnert, Solid lipid nanoparticles: production, characterization and applications, *Advance Drug Delivery Reviews* **47**, 165 (2001).
5. Y. Fukumori and H. Ichikawa, Nanoparticles for Cancer Therapy and Diagnosis, *Advanced Powder Technology* **17**, 1-28 (2006).
6. A. Tsutsumi, S. Nakamoto, T. Mineo, K. Yoshida, A novel fluidized bed coating of fine particles by rapid expansion of supercritical fluid solutions, *Powder Technology* **85**, 275 (1995).
7. A. Tsutsumi, M. Ikeda, W. Chen, J. Iwatsuki, A nano-coating process by the rapid expansion of supercritical suspensions in impinging-stream reactors, *Powder Technology* **138**, 211 (2003).
8. K. Moribe, S.-I. Tsutsumi, S. Morishita, H. Shinozaki, Y. Tozuka, T. Oguchi, K. Yamamoto, Micronization of phenylbutazone by rapid expansion of supercritical CO_2 solution, *Chemical and Pharmaceutical Bulletin* **53**, 1025 (2005).
9. C.T Lee Jr., K.P Johnston, H.J Dai, H.D. Cochran, Y.B. Melnichenko, G.D. Wignal, Droplet interactions in water-in-carbon dioxide microemulsions near the critical point: a small-angle neutron scattering study, *Journal of Physical Chemistry B* **105**, 3540 (2001).
10. J. Liu, Z. Shervania, P. Raveendran, Y. Ikushima, Micellization of sodium bis(2-ethylhexyl)sulfosuccinate in supercritical CO_2 with fluorinated co-surfactant and its solubilization of hydrophilic species, *Journal of Supercritical Fluids* **33**, 121 (2005).
11. T.-J. Wang, A. Tsutsumi, H. Hasegawa, T. Mineo, Mechanism of particle coating granulation with RESS process in a fluidized bed, *Powder Technology* **118**, 229 (2001).

440b Processing of Superparamagnetic Iron Oxide Nanoparticles by the Dense Gas Technology

Un Teng Lam[1], Raffaella Mammucari[2], and Neil R. Foster[2]. (1) School of Chemical Sciences and Engineering, University of New South Wales, Level 2, Applied Science Building, NSW 2052, Sydney, Australia, (2) School of Chemical Sceinces and Engineering, University of New South Wales, Level 2, Applied Science Building, NSW 2052, Sydney, Australia

Superparamagnetic iron oxide nanoparticles (SPIONs) have gained an important role in recent developments in drug targeting, cancer therapy and biomedical imaging. Intense research efforts have been directed toward the application of SPIONs as drug carriers in magnetic drug targeting, as contrast agents in magnetic resonance imaging (MRI) and as heating mediators in hyperthermia treatments. Surface modification of SPION is required to improve their stability, biocompatibility and therapeutic efficacy. Surface modification of SPION is usually achieved by applying a biocompatible polymeric coating. Current coating processes for SPION involve in-situ polymerization of deposited monomers. In situ-polymerization presents the drawbacks of presence of residual solvents in the final product and need of intensive post-processing. The dense gas technology can be an attractive option to conventional coating processes for SPION. The technology of dense gases is able to generate solvent-free products by applying single-step processes. Moreover, it has a proven ability to produce submicron particles with narrow particle size distribution, which is a critical aspect in different biomedical applications of SPIONs. Magnetite-polymer composite below 1 μm has been produced by Aerosol Solvent Extraction System (ASES) using carbon dioxide as the antisolvent. Biocompatible pH-responsive polymer Eudragit® S100 was used in order to prepare multi-responsive composite with potential applications as oral contrast agents for gastrointestinal MRI and as drug carriers for colon specific delivery. Effects of different processing parameters on magnetite content, product morphology, particle size, encapsulation efficiency and magnetic properties are presented.

440c Application Of Supercritical Anti-Solvent To Synthesize Delivery Systems For A Sustained Release Of Food Antimicrobials

Minfeng Jin, Food Science and Technology, The University of Tennessee, 2509 River Drive, Knoxville, TN 37996 and Qixin Zhong, Department of Food Science and Technology, University of Tennessee, 2509 River Drive, 115B McLeod Building, Knoxville, TN 37996-4539.

Supercritical carbon dioxide is an ideal solvent to synthesize particulate delivery systems suitable for food applications. However, regulations require solvents and ingredients are either generally-recognized-as-safe or approved by FDA. We utilized unique features of the supercritical anti-solvent process to synthesize systems suitable for delivering antimicrobials in foods. The polymer we used was zein, prolamines, that is a category of alcohol-soluble corn proteins, and lysozyme was used as a model antimicrobial. Zein is insoluble in aqueous solutions and is ideal for a sustained delivery of antimicrobials into food products that have a long shelf-life. Our results showed that lysozyme was gradually released from particles for more than 30 days. A sustained release is required to inhibit microorganisms over a long period of time because free antimicrobials lose their biological activity in a few days. Our work therefore exemplified a system suitable to enhance the microbial safety of food products.

440d Visualization And Conception Of The Atomized Rapid Injection For Solvent Extraction (Arise) Process For The Production Of Highly Respirable Powders

Roderick Sih and Neil R. Foster. School of Chemical Sciences and Engineering, The University of New South Wales, Applied Science Building, Kensington, NSW 2052, Australia

Supercritical fluid (SCF) processing is a viable alternative to conventional methods during the downstream formulation of active pharmaceutical ingredients (API). By harnessing the highly tunable, versatile and unique properties of SCFs, innovative processes have been used to generate API products with characteristics that surpass traditional benchmarks. The characteristics of APIs obtained from SCF processing are often distinctive and cannot be replicated with conventional processing techniques.

Experience with established SCF processes such as the Aerosol Solvent Extraction process (ASES) and the Gas Antisolvent System (GAS) at research and pilot scales have revealed certain opportunities for operational improvement. Processing improvements and the implementation of a novel method of solution coalescence have inspired the conception of the Atomized Rapid Injection for Solvent Extraction (ARISE) process.

In the ARISE process, organic solutions containing dissolved APIs are delivered into supercritical carbon dioxide as a single bolus injection under quasi-isothermal and isochoric conditions to effect homogenization and solvent extraction. With the energized rapid release of organic solutions into the anti-solvent environment, the ARISE process eliminates the use of capillary nozzles and low solution flowrates for atomization, thereby reducing processing times significantly. By effecting precipitation over a larger volume, the ARISE process is also capable of processing with increased nucleation homogeneity, decreased nucleation densities and hence, decreased bulk densities.

The feasibility and tunability of the ARISE process was successfully demonstrated with the generation of dry powders of pharmaceutical actives. At different operating conditions of the ARISE process, dry powders generated was either of very narrow particle size distributions or of extremely low bulk densities. In-vitro analyses of low bulk density products indicated excellent aerodynamic properties and dry dispersibility.

440e Encapsulated Quercetin Particles Using a Supercritical Antisolvent Process

Gunilla B. Jacobson-Andrews[1], Elizabeth M. Szilágyi[1], Can Quan[2], Charlotta Turner[2], and Richard N. Zare[1]. (1) Department of Chemistry, Stanford University, 333 Campus Drive, Stanford, CA 94305, (2) Department of Physical and Analytical Chemistry, Uppsala University, Uppsala, Sweden

Many drugs exhibit poor solubility in water and lead to an extremely low oral bioavailability in human. In the pharmaceutical industry, micronization techniques are used to improve dissolution rates of drugs into the biological environment. Several conventional techniques have been utilized for particle size reduction. These include mechanical processes (crushing, grinding, and milling), spray-drying, freeze-drying, and recrystallization of the solute particles from solutions using liquid anti-solvents. A novel technique, the supercritical anti-solvent (SAS), was developed for micronization of particles. In the SAS process, an active substance and a carrier dissolved or suspended in an organic solvent are sprayed together or separately in an anti-solvent. The anti-solvent expands the solvent(s) t]which leads to microparticle formation by instant precipitation. One advantage of the SAS process is the ability to produce solvent-free product without the need for additional solvents or surfactants to induce precipitation. Low critical temperature solvents such as CO2 (Tc=31.1 °C) can be used for the precipitation of thermally labile materials without the risk of degradation.

Quercetin belongs to the chemical class of flavonoid and is widely distributed in vegetables and plants. It has been demonstrated to possess a wide array of biological effects that are considered beneficial to health, including antioxidative, free radical scavenging and anticancer. However, quercetin is sparingly soluble in water, which has limited its absorption upon oral administration.

The effects of various process parameters including the co-solvent, temperature, pressure, and flow rates on the particle size and morphology of pure quercetin was studied. In most cases long needles of quercetin were formed, ranging from 1 to several microns long. With very low concentrations of quercetin rod like particles in the 500 nm range were formed. The anti-oxidative activity of quercetin is maintained following processing with the SAS system as shown using 2,2-diphenyl-1-picrylhydrazyl (DPPH) as a radical scavenger.

The particles were also encapsulated with a biodegradable polymer, poly(lactic acid), for controlled-release of drug and to maintain antioxidative activity during storage and delivery. The effect on encapsulation yield and release of quercetin was shown to vary with polymer concentration, choice of co-solvents, and flow rates. The SAS method proved to be a suitable process for encapsulating quercetin and these particles are now further studied in animal models to follow their fate in vivo.

467a Estimation Of Nucleation Kinetics From The Kinetic Limit Of Metastable Zone Width

Venkateswarlu Bhamidi, Sameer Talreja, Guangwen He, Paul J. A. Kenis, and Charles F. Zukoski. Department of Chemical and Biomolecular Engineering, University of Illinois at Urbana-Champaign, 600 S. Mathews Ave., Urbana, IL 61801

Crystallization of pharmaceutical compounds from solution is usually limited by an inadequate understanding of crystal nucleation and growth processes. Significant interest exists in industry and academia towards innovative methods and techniques that facilitate rapid determination of useful kinetic data. In this work, we suggest a method to extract this kinetic information from the limiting cases of metastable zone width (MZW).

Our approach relies on the existence of a boundary for the MZW in nucleation from solutions. We measure this limit of a given compound by introducing small droplets of solution at various initial conditions into a micro-device, in which crystallization is driven by different rates of evaporation of the solvent. The droplets are observed periodically using an automated imaging system and the nucleation time for any crystals formed is recorded. From this information, we determine the limit of MZW at which the solution crystallizes. We measured this limit for wide range of model compounds in our earlier studies [1].

In this work, we propose a way of estimating nucleation kinetics from this limit of MZW using the classical nucleation theory (CNT). Since the rate of evaporation of solvent (and thus the rate of generation of supersaturation) is very slow, we consider the droplet to be at an equilibrium steady state at all points of time and examine the evolution of the 'quasi-equilibrium cluster size distribution' in the droplet. The limit of the metastable zone can then be interpreted as the minimum driving force required for the formation of a 'critical nucleus' in the droplet. CNT expresses the nucleation kinetics in terms of a kinetic prefactor and the surface energy between the solution and the nucleus. We estimate the pre-factor by considering how rapidly the cluster size distribution evolves to populate a single nucleus and we calculate the surface energy from the supersaturation at which the droplet nucleates. We compare our estimates of kinetics with the information available in literature for various systems and discuss the role of rate of supersaturation in the methods used for the measurement of nucleation kinetics.

References:

[1] Guangwen He, Venkateswarlu Bhamidi, Reginald B. H. Tan, Paul J. A. Kenis and Charles F. Zukoski, Cryst. Growth & Design, 6 (2006), 1175

Correlating Molecular Interactions with Solubility of Alpha-amylase from *Bacillus Licheniformis*.

Chirag M. Mehta[1], Edward T. White[1], James D. Litster[1], Matthew T. Hardin[1] and Abraham Lenhoff[2], (1) Chemical Engineering Department, The University of Queensland, 74 College Road, Brisbane, 4072, Australia, (2) Department of Chemical Engineering, University of Delaware, 226 Colburn Lab, Newark, DE 19716-3110.

Abstract

There is a strong link between solubility, and thus crystallisation, and the molecular interactions of proteins in dilute salt solutions. Such molecular interactions are governed by the weak interaction forces (electrostatic, hydration and hydrophobic). In this work molecular interactions were determined and correlated with solubility for the enzyme, alpha-amylase from *Bacillus licheniformis* (BLA) which finds an important application in starch hydrolysis.

Molecular interactions were determined using self interaction chromatography (SIC) which evaluates the second virial coefficient (B_{22}). The second virial coefficient was found to be highly dependent on the pH, salt type and salt concentration. Strong molecular interactions were observed at high (>1.2 M) salt concentration in ammonium sulphate solutions. Correspondingly salting-out behaviour was observed during solubility experiments, strongly supporting the usefulness of the second virial coefficient in predicting the crystallisation behaviour of enzymes in salt solutions.

Introduction

Protein-protein interactions impact throughout the range of bio-separation processing environments, for instance aggregation, precipitation and gelation. The forces responsible for such interactions are believed to be short range (force occurring in the vicinity of less than the diameter of the globular protein molecule) such as electrostatic forces, hydrogen bonds, van der Waals forces and hydrophobic effects.

The potential of the mean force can be used to quantitatively determine such forces for solutions in thermodynamic equilibrium. The second virial coefficient (B_{22}) can be correlated to the potential of the mean force (w) which accounts for all types of forces between two protein molecules in dilute solution [1].

$$B_{22} = 2\pi\int_0^\infty \left(1 - e^{-w/kT}\right) r^2 dr \qquad (1)$$

where B_{22} is the average virial coefficient, r is the intermolecular separation, k is the Boltzmann constant, and T is the absolute temperature. George and Wilson [2] were the first to correlate the experimentally obtained second virial coefficients with the appearance of successful and unsuccessful conditions for crystallisation in dilute protein solutions. They postulated that crystallisation is favoured in the region of B_{22} from -1 to -8 x 10^{-4} (mol.ml.g^{-2}). Since then interest has developed to correlate the second virial coefficient with various physical and thermodynamic quantities such as solubility [3], nucleation [4] and enzyme stability [5].

The solubility of a chemical species (solute) refers to the equilibrium state between the amount of solute dissolved in solution and the excess of un-dissolved solid. It is determined under defined solution conditions (i.e. temperature, precipitant concentration, pH and pressure). The degree to which a solute dissolves in solution depends upon polar interactions with the aqueous solvent, ionic interactions with any salt present and repulsive electrostatic forces between like-charged molecules [6].

In this work, both second virial coefficients and solubilities were determined in salt solutions at various conditions. There are a few empirical and mathematical models to correlate solubility with the second virial coefficient [3, 7-9]. Such correlation can successfully predict the solubility behaviour of enzymes as an alternative to doing expensive, tedious and time consuming solubility experiments. The Hass, Drenth and Wilson (HDL) [7] model is used here to correlate solubility with the second virial coefficient. The HDL model is based on the interaction potential between protein molecules derived using the square well potential and it accounts for a restricted range of interactions. The present work also shows an avenue for the rapid development of scalable, reproducible, and robust crystallization processes for BLA recovery and formulation.

Experimental

Self Interaction Chromatography

For this study, BLA was obtained from Sigma Aldrich (St Louis, MO) as a lyophilized powder (A4551). It is understood Sigma purifies a commercial enzyme obtained from Novozymes A/S (Denmark), known as "Termamyl 300L, Type DX". The enzyme was dissolved in MES buffer at pH 6 and dialyzed using a Slide-A-Lyzer dialysis cassette from Pierce (Rockford, IL) with a 10 kDa molecular wt cut off. SDS-PAGE showed some higher molecular weight impurities but at less than 5%. These impurities were not further removed. Individual enzyme solutions of concentration 5 mg/ml were prepared by dissolving the dialysed BLA in the required buffer solution.

The procedure used follows that given in the literature [10]. EDC and NHS chemistry was used to covalently attach the BLA onto the chromatography particles [10]. For immobilization, dialysed BLA solution was diluted in a pH 6 buffer (50mM MES, 10mM $CaCl_2$) to get a 15 mg/ml final solution. AF-Amino-650 M chromatography particles (Pierce) (1.0 ml of settled volume) were washed with 100 ml of pH 6 buffer and then combined with 3 ml of the BLA solution (15 mg/ml). The particle slurry was mixed well on a rotating disk mixer for half an hour. During that period two different solutions were prepared, one containing 150 mg of EDC in 1 ml of solution and the other 15 mg of NHS. 100 µl of each solution was added to the particle–enzyme mixture. NHS was added first to the solution and than EDC, as EDC was very reactive. The resultant mixture was stirred on the rotating disk mixer overnight. The mixture was centrifuged the next day and a 1 ml aliquot of filtered enzyme solution was collected for later analysis for enzyme concentration (by BCA and UV). The particles were washed with 500 ml of 10 mM MES at pH 6. The BLA-modified particles were placed back in the rotary mixer overnight with 3HP (0.7 ml, 0.2 M), EDC (2.0 mg), and NHS (0.5 mg) for the purpose of capping any unreacted amine groups on the particles. These particles were washed with the original MES buffer and stored at 4°C until used. In addition, enzyme free AF-Amino particles capped with 3HP were also prepared. As determined by a Bicinchonic acid (BCA) assay, approximately 15 mg of BLA per ml of settled particles was immobilized by the above procedure.

The BLA immobilized particles were then loaded in a 3 mm bore chromatography columns of 100 mm length (dead volume 0.35 ml). The column integrity was determined by passing a pulse of acetone through the bed and confirming that the un-retained peak was symmetrical. We used the peak maximum for the retention volume measurement. The chromatography experiments and analysis were conducted using an Amersham Biosciences (Piscataway, NJ) AKTA Explorer unit. The second virial coefficient was determined using the theory proposed by Tessier et al. [10] for self-interaction chromatography. All values of the second virial coefficient are quoted in mol ml g^{-2}.

Solubility

A commercial BLA (Termamyl 300 L, Type DX) from Novozymes A/S (Denmark) was used for solubility measurement. The BLA was dialyzed three times by volume using an Amicon Ultrafiltration unit (10 kDa molecular weight cut-off membrane) with ultrapure water of 18.2 $M\Omega.cm^{-1}$ conductivity (Millipore, Sydney, Australia). The dialysed solution was concentrated in the same unit to attain a concentration of 50-60 mg/ml. The concentrated solution was then transferred to a glass beaker and pH was adjusted to 7.0 using 2M sodium hydroxide solution. The solution was then crystallised by dissolving crystals of ammonium sulphate slowly which dissolved with constant high speed stirring until the solution turn murky. Hexagonal bipyramidal crystals of BLA were formed with sizes in the range of 1-5 microns. The crystals were recovered by centrifugation (7500 rpm, 20°C, 10 min) and were re-dissolved in water. The re-crystallisation process was repeated twice more by the same method again using ammonium sulphate. The triply crystallised enzymes were redissolved in water and were dialysed extensively using the Amicon Ultrafilteration unit (10 kDa molecular weight cut-off membrane). The dialysed solution was then freeze dried at -30°C at a pressure less than 1 Torr. The freeze dried enzymes was stored at 4°C until further used.

For solubility experiments solutions were prepared by mixing the required amounts of 20 mM MES buffer and 3.6 M ammonium sulphate in 20 mM MES buffer. The pH of the solution was adjusted using strong base or acid in the pH range 5.5 to 7.5. The solution was prepared in 600 microlitre vials and excess of the triply crystallized freeze dried BLA was added. Solutions were left agitated on the rotating mixer in a temperature control room at 20°C. The crystals were observed for polymorphs after 24hrs. The vials were then centrifuged at 6000 rpm for 2 mins and the supernatant solution was recovered and analysed for BLA concentration using the UV spectrophotometer. An absorbance coefficient at 280 nm for freeze-dried BLA was experimentally determined at pH 7.0 ($A_{1cm}^{1\%}$= 24.03). The supernatant concentration did not change significantly (i.e., < 1 %) on holding for 48 hrs, indicating that the equilibrium concentration has been achieved within the experimental time employed in the study.

Purity analysis

The purity of the freeze dried crystals was determined using SDS-PAGE and IEF gels. SDS-PAGE was carried out using 4-20% Tris-HCl ready gels (Biorad, Hercules CA) in a Mini-Protean II cell (Biorad) with Precision protein standards molecular weight markers (Biorad). Isoelectric focusing was performed using 10-well Novex gels (Invitrogen, Australia) with a pH range of 3 to 7 and the gel was calibrated using a commercially obtained protein mixture (SERVA liquid mix IEF markers pH 3 to 7, Invitrogen). The IEF run was conducted in an XCell Surelock Mini-Cell using Novex IEF anode and cathode buffers of pH 3 to 7 (Invitrogen) according to the manufacturer's instructions. All gels from SDS-PAGE and IEF runs were stained with Coomassie Brilliant Blue.

Theoretical Model

The model proposed by Haas, Drenth, and Wilson [7] was used in this work to correlate the BLA solubility values (S) with the second virial coefficient data (B_{22}). The molecular based theoretical correlation expression (equation 2) depends upon the interaction parameter (A) and the coordination number (z) of the molecules in the crystal.

$$B_{22} = \left(\frac{4}{M\rho}\right)\left[1 - A\left\{\left(\phi_s/m\right)^{(-2/z)} - 1\right\}\right] \qquad (2)$$

$$m = \frac{M}{18\rho}; \qquad \phi_s = \frac{S}{\rho 10^3}$$

where M is the molecular weight of the protein and ρ is the density of the protein crystal (~1.36 $g{\cdot}cm^{-3}$). The fitting parameter, A ($A = \mathbf{p}\,(\lambda^3 - 1)$), depends on a combination of the anisotropy, **p** (**p** = 1 for isotropic interactions and **p** < 1 for anisotropic interactions), and the range of interactions (λ) between protein molecules. Anisotropy (**p**) and the range of the interaction (λ) were found to be nearly the same for all solvents for one protein and also independent of the temperature and the pH [7]. The coordination number (z) depends upon the crystal-polymorph. Here z=3 was used as the space group of BLA crystals was found to be $P\,6_1$ [11].

Results and Discussion

Purity analysis

We have purified the BLA by triple crystallising using ammonium sulphate. The purity of the crystallised BLA was investigated using SDS-PAGE and IEF gels. The SDS-PAGE stained with Coomassie blue showed one major band at 55 kDa as shown in Figure 1a. The SDS-PAGE also showed the effect of re-crystallisation on the purity of the material by comparing band A with B in Figure 1a.

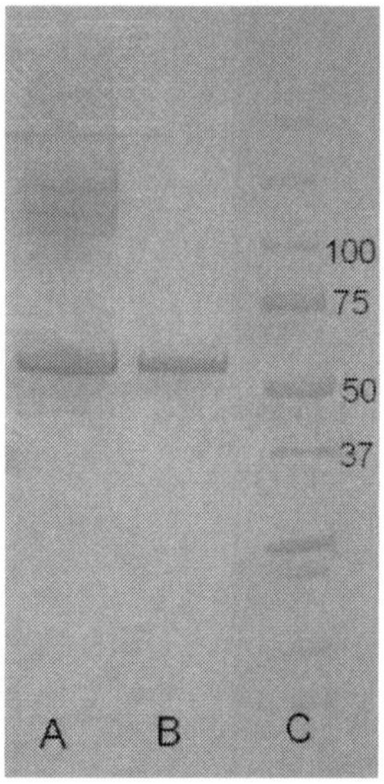

(1a)

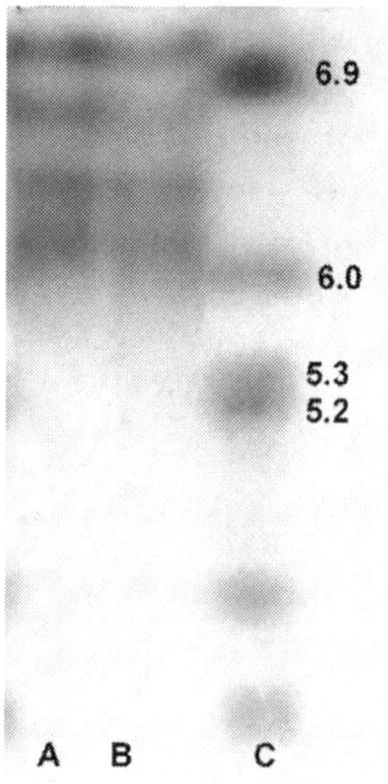

(1b)

Figure 1: Purity analysis of BLA: (1a) SDS-PAGE; A: Original solution, B: Triply crystallised BLA, C: Molecular markers. (1b) IEF results A and B: Triply crystallised BLA C: IEF markers from pH 3-7.

The isoelectric focussing results (Figure 1b) showed four major bands in the pH range 6 to 7. However all these band corresponded to 55 kDa on SDS-PAGE. Similar observation was made by Faber et. al. [12], where purified BLA showed five major bands in the pH range 6 to 7.5. They suspected that it could be the presence of isoforms of the same enzyme, as all isoforms exhibited BLA activity [12]. The exact isoelectric point of our enzyme could not be predicted because of the presence of these isoforms. Interestingly all four isoforms appeared to be present in approximately the same amounts. A further study is required to re-analyse the isoelectric focusing data using the 3 to 10 pH range. The fractions of each isoforms also need to be determined. The BLA literature gives estimates for the isoelectric point of near 5.2 or 5.4 [13] and 6.9 [14].

Solubility results

Solubility is the thermodynamic equilibrium between the species in the solution and the species present as crystal. The solubility of a protein can be measured from both initially under-saturated and super-saturated solutions. In this work solubility was measured from under-saturated solutions where an excess amount of protein crystals is added to the solution which is kept until the supernatant solution reaches equilibrium. This is also known as the dissolution technique and it takes lesser time and is more reliable then the super-saturated method.

Crystals of BLA obtained from freeze-drying were dissolved in ammonium sulphate solution (0.8 to 1.6 M) in the pH range 5.5 to 7. Solubilities were not measured at lower pH (pH < 5.5) as alpha-amylase showed loss in activity due to denaturation at acidic pH [15, 16]. Bipyramidal hexagonal crystals were formed after 24 hrs for all condition used in this work, as similarly reported by Machius et. al. [11] working on BLA obtained from Novozymes. The solubility results are shown in Figure 2.

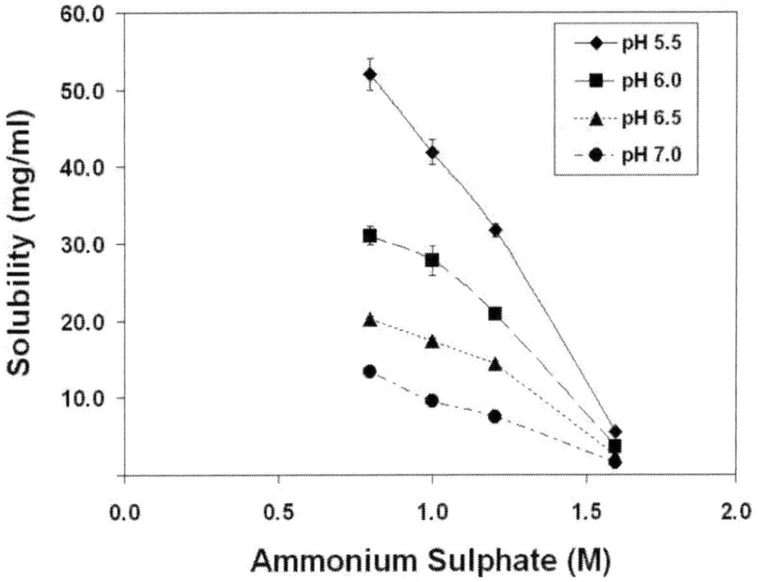

Figure 2: Solubility of BLA in ammonium sulphate at 20°C showing salting-out effect.

A salting-out behaviour was observed with increases in ammonium sulphate concentration. This suggests that BLA, being acidic in nature followed the lyotropic series proposed by Hofmeister [17]. BLA has a high number of hydrophobic amino acids, similar to ovalbumin, and they both show similar salting-out behaviour in ammonium sulphate. Veesler [18] reported a similar effect for the solubility of the orthorhombic polymorph A of alpha-amylase. Monoclinic crystal of alpha-amylase can also be produced by the addition of salts such as ammonium sulphate [19, 20] and sodium sulphate [21]. Ammonium sulphate appears to be a good precipitant choice for the crystallisation of BLA.

The solubility of BLA in ammonium sulphate solutions also decreases as the pH is increased (Figure 2). The lowest solubilities were found around the isoelectric point of BLA (6.1 to 6.9). Such behaviour was expected because the overall charge on enzyme molecules is neutral and intermolecular electrostatic repulsion is a minimum at the isoelectric point (pI). With a decrease in pH, excess negative charges accumulate on the surface of enzymes which then intensifies electrostatic intermolecular repulsion.

Self interaction results

Figure 3 shows the second virial coefficient (B_{22}) results for BLA as a function of ammonium sulphate concentration (0.2 to 1.4M) in the pH range 5.5 – 7.0. At low salt concentration (< 1M) the response is substantially constant i.e. values of second virial coefficient were independent of pH and salt concentration. At high salt concentration (> 1M), values of B_{22} fall quite steeply to large negative values. Such behaviour has been observed for other proteins such as lysozyme [22, 23], equine serum albumin [24], ovalbumin [24], bovine serum albumin [10, 23] and myoglobin [10] in ammonium sulphate solutions.

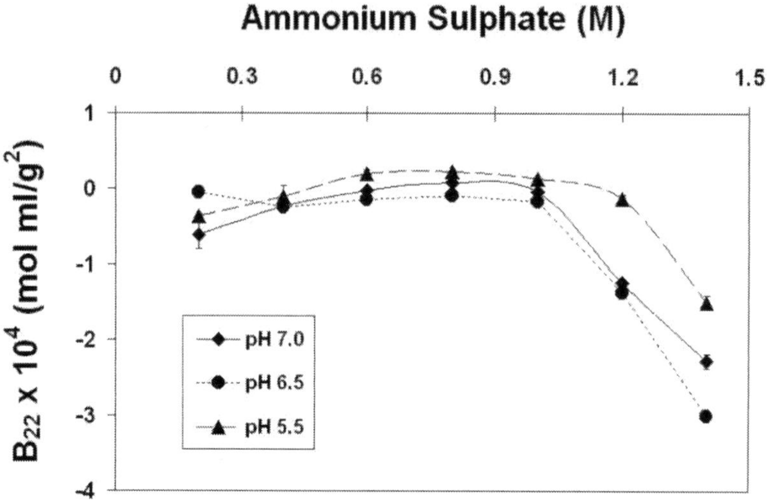

Figure 3: Second virial coefficient for BLA in ammonium sulphate solutions indicating increased interaction between molecules at high salt concentration.

Interactions are stronger near the isoelectric point (6 to 6.9) compared to pH 5.5. Correspondingly lower solubilities were observed near the isoelectric point (Figure 2). Similar pH dependent virial coefficient curves were obtained for *Pseudomonas* alpha-amylase which has isoelectric point 4.2 [5]. Strong pH dependence has been reported previously for amylase [18].

Comparing experimental results with the HDL model

Figure 4 shows the experimental BLA solubility data plotted versus the second virial coefficient. A correlation is indicated. The data was fitted with the HDL model (equation 2) and this is shown as the curve.

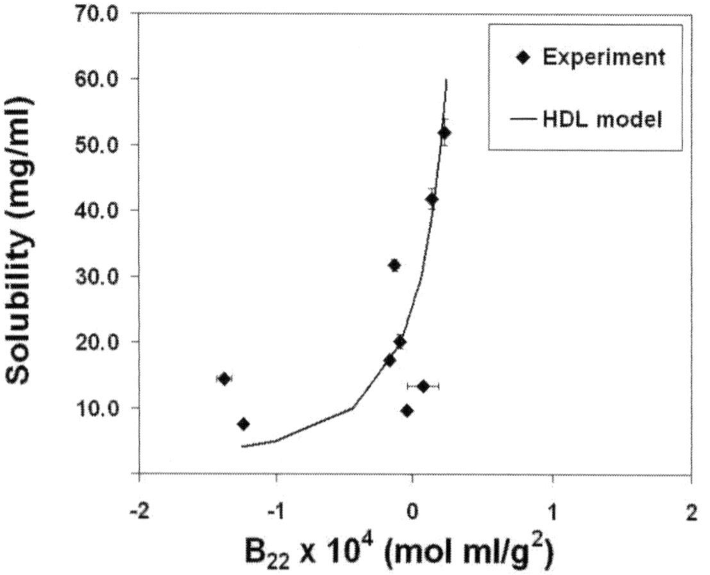

Figure 4: Correlation of second virial coefficient (B_{22}) with solubility for BLA showing HDL model result.

For the BLA used in this work with molecular weight M = 55 kDa, coordination number is z=3 and the interaction parameter A was found to be 0.00040. The solubility data fitted was in the range of 7.5 to 52.0 mg/ml and the corresponding second viral coefficient data ranged from -1.37 to 0.23 x 10^{-4} mol ml g^{-2}. The HDL model has previously been successfully used for lysozyme in sodium chloride and ovalbumin in ammonium sulphate solutions [7].

Conclusions

In this work pure BLA was successfully separated as crystals from commercially available Termamyl 300L, Type DX (Novozymes A/S, Denmark). Experimentally estimated second virial coefficients showed negative values at high (>1.2 M) salt concentration in ammonium sulphate solutions. Corresponding salting-out behaviour was observed during solubility with increase in ammonium sulphate concentrations. The HDL model gave a good fit between the solubility and the second virial coefficient data. This works confirms the usefulness of the second virial coefficient for considering crystallisation conditions.

Acknowledgments

We thank Alex Chan from the University of Queensland for his assistance on Isoelectric focusing. This research was funded by an ARC linkage grant and from Dupont, Australia.

References

1. Chandler, D., Introduction to Modern Statistical Mechanics. 1987, Oxford: Oxford University Press.
2. George, A. and W.W. Wilson, Predicting protein crystallization from a dilute solution property. Acta Crystallographica Section D: Biological Crystallography, 1994. D50: p. 361-365.
3. Guo, B., S. Kao, H. McDonald, A. Asanov, L.L. Combs, and W. W. Wilson, Correlation of second virial coefficients and solubilities useful in protein crystal growth. Journal of Crystal Growth, 1999. 196(2-4): p. 424-433.
4. Bhamidi, V., S. Varanasi, and C.A. Schall, Protein Crystal Nucleation: Is the Pair Interaction Potential the Primary Determinant of Kinetics? Langmuir, 2005. 21: p. 9044-9050.
5. Valente, J.J., B.G. Fryksdale, D.A. Dale, A.L. Gaertner, and C.S. Henry, Screening for physical stability of a Pseudomonas amylase using self-interaction chromatography. Analytical Biochemistry, 2006. 357(11): p. 35-42.
6. Scopes, R.K., Protein Purification: Principles and Practice. 1994, New York: Springer Verlag. p. 71-101.
7. Haas, C., J. Drenth, and W.W. Wilson, Relation between the solubility of proteins in aqueous solutions and the second virial coefficient of the solution. Journal of Physical Chemistry, 1999. 103: p. 2808-2811.
8. Agena, S.M., D.L. Bogle, and F.L.P. Pessoas, An activity coefficient model for proteins. Biotechnology and Bioengineering, 1997. 55: p. 65-71.

9. Ruppert, S., S.I. Sandler, and A.M. Lenhoff, Correlation between the osmotic second virial coefficient and the solubility of proteins. Biotechnology Progress, 2001. 17: p. 182-187.
10. Tessier, P., S. Vandrey, B.W. Berger, R. Pazhianur, S.I. Sandler, and A.M. Lenhoff, Self-interaction chromatography: a novel screening method for rational protein crystallization. Acta Crystallization, 2002. D58: p. 1531-1535.
11. Machius, M., N. Declerck, R. Huber, and G. Wiegand, Activation of Bacillus licheniformis a-amylase through a disorder → order transition of the substrate-binding site mediated by a calcium–sodium–calcium metal triad. Structure, 1998. 6(3): p. 281-292.
12. Faber, C., T.J. Hobley, J. Mollerup, O.R.T. Thomas, and S.G. Kaasgaard, Study of the solubility of a modified Bacillus licheniformis alpha-amylase around the isoelectric point. Journal of Chemical and Engineering Data, 2007. 52: p. 707-713.
13. Vihinen, M. and P. Mantsala, Microbial amylolytic enzymes. Critical Reviews in Biochemistry and Molecular Biology, 1989. 24: p. 329-418.
14. Ivanova, V.N., E.P. Dobreva, and E.I. Emanuilova, Purification and characterization of a thermostable alpha-amylase from Bacillus licheniformis. Journal of Biotechnology, 1993. 28: p. 277-289.
15. Shaw, A., R. Bott, and A. Day, Protein engineering of alpha-amylase for low pH performance. Current Opinion in Biotechnology, 1999. 10: p. 349-352.
16. Nielsen, J.E., T.V. Borchert, and G. Vriend, The determinants of a-amylase pH-activity profiles. Protein Engineering, 2001. 114(7): p. 505-512.
17. Hofmeister, F., Zur lehre von der wirkung der salze. Arch Exp Pathol Pharmacol, 1888. 24: p. 247–260.
18. Veesler, S., S. Lafont, S. Marcq, J.P. Astier, and R. Boistelle, Prenucleation, crystal growth and polymorphism of some proteins. Journal of Crystal Growth, 1996. 168: p. 124-129.
19. Lee, S.Y., S. Kim, R.M. Sweet, and S.W. Sub, Crystallization and a preliminary x-ray crystallographic study of alpha-amylase from Bacillus licheniformis. Archives of Biochemistry and Biophysics, 1991. 291(2): p. 255-257.
20. Ninomiya, K., T. Yamamotob, T. Ohedab, K. Satob, G. Sazakic, and Y. Matsuuraa, Morphology and solubility of multiple crystal forms of Taka-amylase A. Journal of Crystal Growth, 2001. 222: p. 311-316.
21. Suzuki, A., T. Yamane, Y. Ito, T. Nishio, H. Fujiwara, and T. Ashida, Crystallization and preliminary crystallographic study of bacterial a-amylases. Journal of Biochemistry, 1990. 108: p. 379-381.
22. Rosenbaum, D.F., A. Kulkarni, S. Ramakrishnan, and C.F. Zukoskia, Protein interactions and phase behavior: Sensitivity to the form of the pair potential. Journal of Chemical Physics, 1999. 111(21): p. 9882-9890.
23. Moon, Y.U., C.O. Anderson, H.W. Blanch, and J.M. Prausnitz, Osmotic pressures and second virial coefficients for aqueous saline solutions of lysozyme. Fluid Phase Equilibria, 2000. 168: p. 229–239.
24. Demoruelle, K., B. Guo, S. Kao, H.M. McDonald, D.B. Nikic, S.C. Holman, and W.W. Wilson, Correlation between the osmotic second virial coefficient and solubility for equine serum albumin and ovalbumin. Acta Crystallographica, 2002. D58: p. 1544-1548.

467b Molecular Dynamics Simulation of the Hydration and Interaction between Bovine Insulin Hexamers

Vishal Koparde, Chemical Engineering, Penn State University, 174 Fenske Lab, University Park, PA 16802 and Kristen A. Fichthorn, Dept. of Chemical Engineering, The Pennsylvania State University, 164 Fenske Laboratory, University Park, PA 16802.

Water is known to play a vital role in the structure and dynamics of proteins, thereby affecting their biological functionality. Similarly, it is believed that protein-protein interactions that lead to self-assembly are governed to a large extent by hydration forces. We are particularly interested in studying the interactions between two insulin hexamers. Insulin is a small protein that is comprised of 51 amino acid residues per monomer. Its hexamer complex includes two zinc ions. In this work, we use molecular dynamics to analyze hexamer-hexamer interactions in water and to evaluate the role of hydration forces in these interactions. We use NAMD, along with the CHARMM force field to model the protein and the TIP3P potential to model the water molecules. Simulations of a single insulin hexamer are carried out to determine the variation of water residence times and water coordination number with protein residues. Simulation results can be directly compared with experiments. Simulations to determine the potential of mean force between two insulin hexamers are performed with hydrophobic and hydrophilic surfaces of the hexamers facing each other. The structure and density of water in the region between the two hexamers is analyzed in detail and linked with the hexamer force profiles. Our simulations show that hydrophobic interactions can be an important driving force to direct protein assembly.

467d Microfluidic Platforms For In-Meso Membrane Protein Crystallization

Griffin W. Roberts, S.L. Perry, J.D. Tice, and P.J.A. Kenis. Department of Chemical & Biomolecular Engineering, University of Illinois at Urbana-Champaign, 800 S. Mathews, Urbana, IL 61801

Crystallization of membrane proteins continues to be a field of study that proves to be more of an art than science. Membrane proteins are a fundamental component of many biological processes such as transport and ion regulation through a cell membrane. Membrane proteins also serve as a channel for signal transduction from one side of the cell membrane to the other and establish contact with the proteins of neighboring cells to form tissues such as muscles and organs. Proteomics has shown how the functionality of membrane proteins can be determined by their structure. Structural determinations are performed using various methods such as X-ray diffraction, electron microscopy, and solid state NMR. All methods are well developed and understood however; they all require a sample with some degree of crystallinity. Crystallization is the limiting step behind the proteomics of membrane proteins. Factors such as quantity, amphiphilicity, and non-rigid conformations increase the difficulty of determining crystallization conditions. The 1988 Nobel Prize in Chemistry was given in recognition for the 3D structural determination of the membrane protein bacteriorhodopsin from a crystal grown using detergent solublized protein. This is now referred to as the in-surfo crystallization method. However, the in-surfo method has serious limitations such as the choice of detergent for solubilization, denaturation due to the loss of the lateral membrane pressure, and applicability to various proteins. The in-meso crystallization method alleviates these problems by allowing the proteins to be crystallized from an artificial membrane-like environment. In this method it has been proposed that the bicontinuous region supplies protein for crystallization as nucleation and growth occur in a locally lamellar region formed from a precipitant-induced concentration change to form stacked 2D sheets. The rate of protein diffusion has been shown to be a function of the curvature of the lipid membrane. The formation of this artificial mesophase involves the mixing of non-Newtonian fluids having viscosities differing by ~30x. Traditionally this is done using a manual syringe or centrifuge-style mixer, requiring volumes of 20 μL of protein or more. Once the mesophase is formed high throughput matrix screening can be used to determine crystallization conditions in well-established well plate technology. These methods involve the use of large expensive fluid handling equipment, which can be on a scale larger than the amount of protein available. We report on the design and testing of a microfluidic platform designed for mixing viscous liquid to achieve in-meso membrane protein crystallization on-chip at volumes of >20 nL per sample. This platform allows for the preparation of multiple mesophase samples at once, while also facilitating high throughput screening of multiple precipitant solutions on a single chip. In addition to screening for appropriate crystallization conditions, independent control over precipitant injection has the potential to allow for kinetic control of crystal growth done in-meso through the use of additives to control the rate of protein diffusion by affecting the curvature of the protein-containing mesophases. The membrane-bound proton pump bacteriorhodopsin is used as a model system.

Analysis of Concentration Field within an Evaporating Hanging Droplet

Li May Goh and Richard D. Braatz*

Department of Chemical and Biomolecular Engineering, University of Illinois at Urbana-Champaign, 600 South Matthews Avenue, Urbana, Illinois 61801
*Corresponding author: Phone: 217-333-5073, Fax: 217-333-5052, E-mail: braatz@uiuc.edu.

I. Introduction

In recent years it has become common to estimate nucleation and growth kinetics for the crystallization of pharmaceuticals, amino acids, and proteins from induction times measured in droplets, typically within microfluidic devices (e.g., Galkin & Vekilov, 1999; Talreja et al, 2005; 2007). Most of the methods developed to determine nucleation kinetics assumes the droplets to be spatial uniform. In this case there is equal chance for nucleation to take place at any position within the system. This paper evaluates the accuracy of this assumption. Although the analysis and simulation carried out in this study are applicable to any crystallization within droplets, for specificity the system under consideration is an evaporation-based crystallization involving a hanging drop (e.g., Talreja et al, 2005; He et al, 2006).

Several researchers have argued that the concentration gradients within drops are negligible due to mixing caused by buoyancy and/or surface-tension-driven convection, i.e., Marangoni convection. For example, Grant and Saville (1991) performed a quasi-steady state analysis which indicated that, with natural convection due to buoyancy, the solution concentration at the crystal surface is essentially equal to the bulk concentration, so that the growth rate is completely controlled by surface kinetics and not limited by mass transport. On the other hand, a later study by the same authors that considered the growth of lysozyme crystals indicated that mass transport limitations were important (Grant & Saville, 1995). A computational simulation by Savino and Monti (1996) within droplets of different configurations showed that concentration gradient within the hanging droplet is small due to buoyancy and Marangoni convection. However, Kawaji (2003) reported that no natural convection was detected. Yet, in another report, the existence of natural convection was proven physically for the ethanol-water system (Kang et al, 2003). Section II of this paper discusses an order-of-magnitude analysis for glycine to assess whether perfect mixing can be assumed. It is found that convective time scale computed using the Grashof number is several orders smaller than all other time scales including the process time of forming one crystal (the induction time) which indicates that the solution is well-mixed throughout the induction period. Both buoyancy and Marangoni convection (Deen 1998; Ostrach 1982) are considered.

In Section III, the three-dimensional solution concentration field within evaporating droplets is simulated for aqueous glycine solution. The purpose of this study is to assess the accuracy of order-of-magnitude scaling analysis and to validate the assumption of spatial uniformity used in modeling nucleation kinetics in evaporation-based microfluidic crystallization (e.g., Goh & Braatz, 2007, and citations therein). The governing equations are closely related to those of Savino and Monti (1996), which assumes the Boussinesq approximation and the quasi-stationary approximation, i.e., that the effects of the moving liquid-gas interface due to evaporation are negligible. Axial symmetry is assumed with respect to drop axis. The coupled nonlinear distributed parameter equations for the velocity and species concentration fields are solved. The order-of-magnitude analysis on the characteristic velocity of the system agrees well with the simulation results. In addition, the simulation results show that the maximum concentration is located at the surface at the bottom of the drop and the maximum difference between concentration and the average concentration increases with time. This maximum difference is small enough to suggest that spatial uniformity is a reasonable assumption for some systems.

II. Order-of Magnitude Analysis

The system that was analyzed is aqueous glycine at 0.1, 0.3, and 0.5 wt% (glycine + water only) of 5 μL volume. This is one of the compounds studied by He et al (2006) and with complete information available for analysis. An order-of-magnitude analysis is useful to quickly assess whether the mixing effect by natural convection is significant within the droplet; here an analysis is performed analogous to that of Grant (1991), Deen (1998), and Ostrach (1982). The characteristic velocity, U, depends on the dominant forces with the droplet. For an evaporating droplet, several transport processes take place simultaneously. When water evaporates from the droplet, two phenomena occur at the surface: (1) the glycine concentration increases at the surface and (2) the temperature drops due to evaporation (Figure 1) (Savino et al, 2002). For glycine solution, the solution density increases with glycine concentration and decreases with temperature, hence, at the surface, solution density increases (Shen et al, 2000).

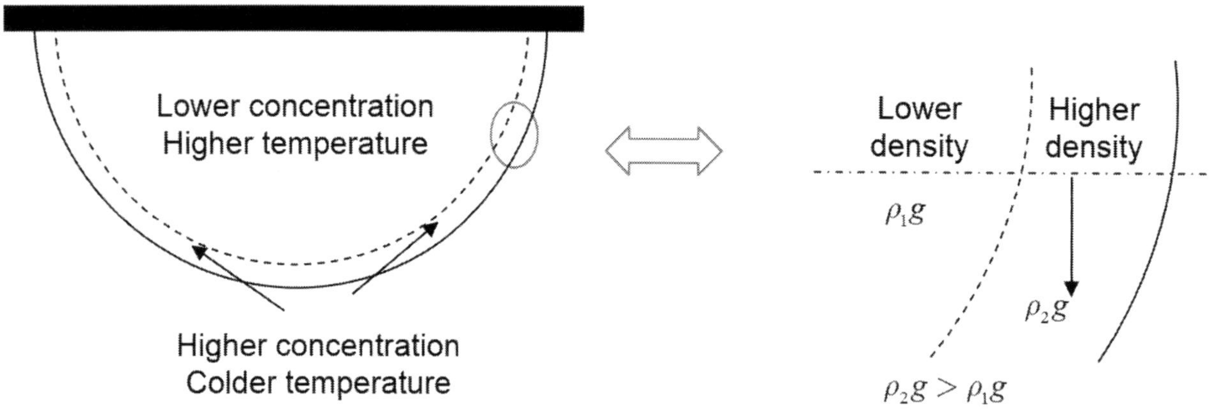

Figure 1. Flow due to buoyancy.

The strength of motion due to buoyancy convection is characterized by the Grashof number (Gr), the ratio of buoyancy force driving the motion to the viscous resistance (Deen, 1998). The Grashof number is Gr = $L^3\beta_c g\Delta c/\nu^2$ or $L^3\beta_T g\Delta T/\nu^2$, where g = gravitational acceleration, Δc, ΔT = concentration/temperature difference between two reference points, β_i = expansion coefficient (i = T or c), ν = kinematic viscosity, and L = characteristic length (equal the radius for a hanging droplet). In order to compute the Grashof number, the concentration/temperature difference between two reference points within the droplet has to be approximated. The concentration gradient at the surface can be approximated by (Savino & Monti, 1996): $\partial c/\partial r \approx \Delta c/\Delta r = (J_w/\rho D)(c_{gly}/c_{water})$, where c = concentration (wt%) J_w =

evaporation flux (kg/m^2-s), ρ = density of solution, and D is the diffusivity of glycine in water. According to penetration analysis, for short times $t << L^2/D$, assuming convection has not set in, the penetration depth due to diffusion can be approximated using $\delta \approx (Dt)^{1/2}$. For a 5 μL hanging drop of aqueous glycine, the radius of the hanging drop, taken to be the characteristic length scale, is $L = 1.337\times10^{-3}$ m. The diffusivity $D = 1\times10^{-9}$ m^2/s is determined from Chang et al (1986). Hence, for short time after forming the drop at t = 10 s, $\delta \approx 1\times10^{-4}$ m, $\Delta c \approx 2.2\times10^{-7}$, and $Gr_c \approx 2.5\times10^{-3}$ (Badarayani et al, 2004; Shen et al, 2000). The characteristic velocity due to solutal buoyancy-driven convection, U_b, which can be computed from $Gr^{1/2}(\nu/L)$ is 3.3×10^{-5} m/s (Deen, 1998). This is very close to the simulated value of 2.5×10^{-5} m/s which will be further discussed in Section III.

When buoyancy sets in, the higher density region of the drop flows along the surface which results in a concentration/temperature difference along the interface. Consequently, another mode of flow is generated by the surface tension gradient ($\partial\sigma/\partial\theta$). This is known as the Marangoni solutal or thermal convection (see Figure 2). Such flow results in a drag force or shear stress, τ, along the surface as described by (Leal, 2007): $\tau = \mu(\partial v_\theta/\partial r) = -(1/R)(\partial\sigma/\partial\theta) = -(1/R)(\partial\sigma/\partial c)(\partial c/\partial\theta) = -(\gamma/R)(\partial c/\partial\theta)$. Surface tension usually decreases with temperature but may increase or decrease with concentration depending on the solutions (Molenkamp, 1998). For example, Matubayasi et al (2002) reported that glycine and L-alanine demonstrated an increase in surface tension with respect to concentration whereas L-leucine and L-valine demonstrated a decrease in surface tension with respect to concentration. Since glycine has a positive surface tension gradient, Marangoni convection takes place along the surface towards the higher concentration region which is in the same direction as the buoyancy flow.

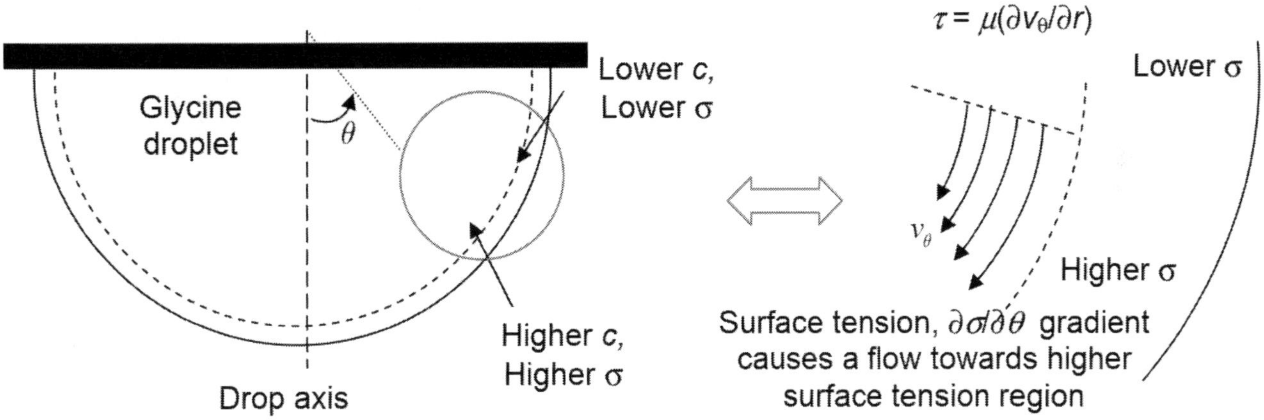

Figure 2. Flow due to surface tension gradient (Marangoni effect)

Using the same values for computing buoyancy, $\Delta c = 2.2\times10^7$ and $\delta \approx 1\times10^{-4}$, characteristic velocity due to Marangoni convection, U_m, during the early stages of forming the drop can also be approximated using order-of-magnitude analysis (Ostrach, 1982). It is found that $U_m = |\gamma|\Delta c\delta/(\mu L) = 2.1\times10^{-7}$ m/s which is two orders of magnitude smaller than U_b. The characteristic diffusive velocity, $U_d = D/R$, is computed to be 7.5×10^{-7} m/s, is also two orders of magnitude smaller than U_b. Hence, solutal buoyancy convection is the dominant force and U_b can be used to represent the characteristic velocity of the flow within the hanging drop.

III. Simulation of Concentration Profile within Hanging Drop

The velocity and concentration fields for 0.1, 0.3, and 0.5 wt% aqueous glycine solutions were simulated. The model describes the 3D solutal buoyancy and Marangoni convection behavior within the glycine hanging drop. The dynamic momentum equation was coupled with a diffusive-convection material balance equation. The Marangoni convection induced at the interface was coupled as a weak boundary PDE. The drop was built as a quadrant using axisymmetric coordinates and Femlab's built-in 2nd Bezier Curve (Figure 3a). Using the Boussinesq approximation, density differences are assumed to be negligible except in terms multiplied by the gravitational acceleration which results in a significant forcing term in the momentum equation in the vertical downwards direction. Hence, the vertical force due to gravity is $F_z = g\rho_0\beta_c(\rho - \rho_0)$ while $F_r = 0$. The pressure considered here is the dynamic pressure where $\nabla P \equiv \nabla \tilde{P} - \rho_0 g$ (Deen, 1998).

Momentum Equation: $\rho \frac{\partial \underline{u}}{\partial t} - \mu \nabla^2 \underline{u} + \rho(\underline{u} \cdot \nabla)\underline{u} + \nabla P = \underline{F}$

Continuity Equation: $\nabla \cdot \underline{u} = 0$

Material Balance: $\frac{\partial c}{\partial t} = \nabla \cdot (D \nabla c)$

Stress Balance at interface: $\tau_{rz} = \mu \left(\frac{\partial v_r}{\partial z} + \frac{\partial v_z}{\partial r} \right) = -\gamma \left(\frac{\partial c}{\partial z} + \frac{\partial c}{\partial r} \right)$ (axisymmetric coordinates)

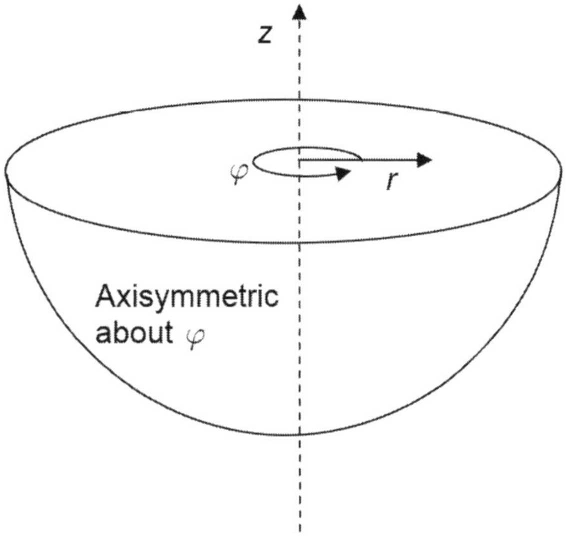

Figure 3a. Hanging drop built in Femlab

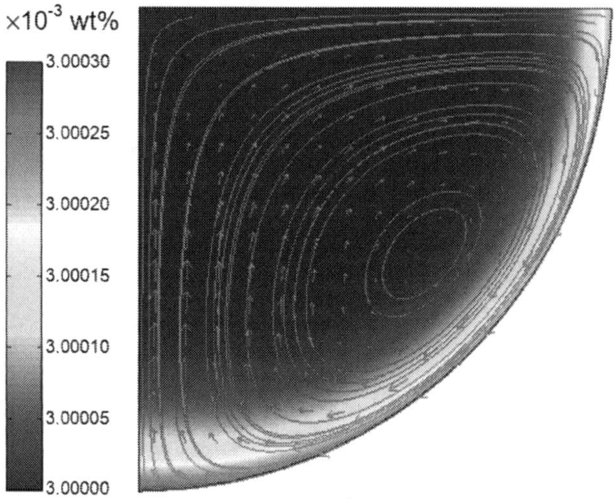

Figure 3b. Streamlines and concentration profile for 0.3 wt% at 10 s plot (radius = 1.337 mm)

Figure 3bc show the streamlines and concentration profiles. From the simulation, it is found that for 0.3 wt% aqueous glycine solution at 10 s, the maximum Euclidean norm of the velocity vectors is 2.5×10^{-5} m/s. The order-of-magnitude analysis in Section II approximated the characteristic velocity to be 3.3×10^{-5} m/s, which agrees well with the simulated values. Using the same order-of-magnitude analysis for 0.5 wt% aqueous glycine solution at 10 s, the characteristic velocity is approximated to be 4.4×10^{-5} m/s, which also agrees well with the simulated velocity of 4.3×10^{-5} m/s (maximum Euclidean norm).

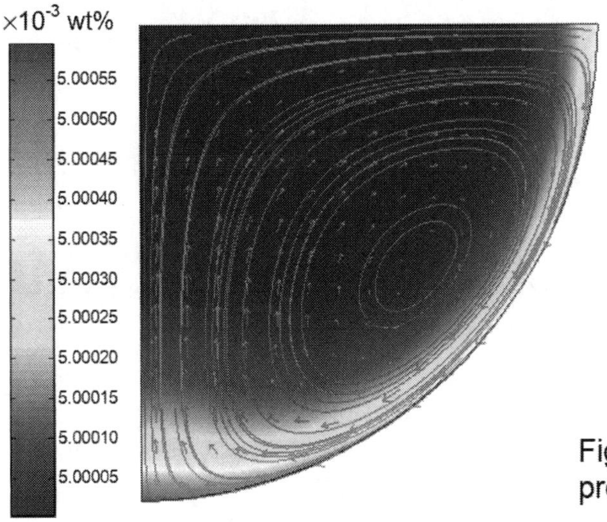

Figure 3c. Streamlines and concentration profile for 0.5 wt% at 10 s plot (radius = 1.337 mm)

The simulations verified the accuracy of the order-of-magnitude analysis, which can be used to predict the average velocity within a hanging drop as well as to show that velocity increases with initial concentration. Of course, the simulations also provide the detailed velocity and concentration fields for the entire induction time, which are not given by order-of-magnitude analysis. Simulation was also carried for 0.1 wt% aqueous glycine solution from 10 s to the average induction time of 50 hr at room temperature, where the average induction time was obtained from experimental data collected over a wide range of evaporation rates and initial concentrations as reported by He et al (2006). The plot of maximum Euclidean norm of the

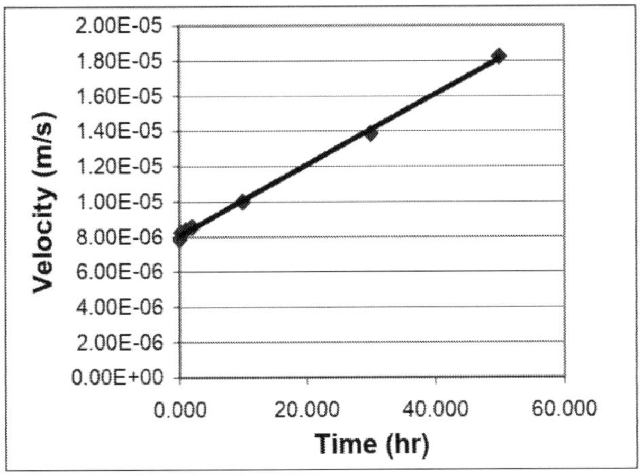

Figure 4a. Simulated velocities vs. time for 0.1 wt% glycine

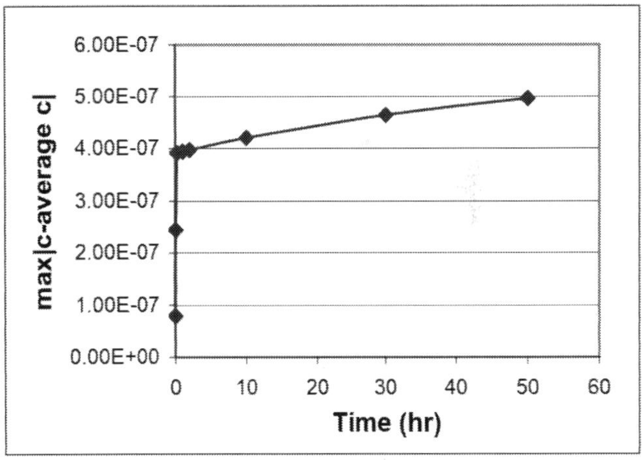

Figure 4b. Simulated max|*c* – average *c*| vs. time for 0.1 wt% glycine

velocity versus time shows that the velocity increases gradually with time (Figure 4a). The characteristic velocity determined from the order-of-magnitude analysis for short time serves not only as a good approximation for the flow velocity for short time and but also serves as an approximate lower bound on the velocity at long time. That is, the order-of-magnitude analysis provided a conservative assessment of the extent of mixing throughout the induction time. Given such a high extent of mixing due to natural convection, it is reasonable to assume well-mixed conditions within the drop. Figure 4b shows that the maximum absolute difference between concentration and average concentration is very small throughout the induction period which supports the assumption that the drop is well-mixed. It is shown by Goh & Braatz,

2007 that induction time data for paracetamol were well fit to the classical primary nucleation expression with the assumption of spatial uniformity.

IV. Discussion and Thermal Analysis

The order-of-magnitude analysis was carried out for short times after creating the hanging droplet assuming convection had not yet set in. Based on the penetration approximation, if $t << L/D$, the penetration depth due to diffusion can be approximated using $\delta \approx (Dt)^{1/2}$. It was found that the convection flow due to buoyancy was dominant compared to diffusion and Marangoni convection, hence U_b, where subscript b refers to buoyancy, is used to represent the characteristic velocity. The characteristic velocities are found to be of the same order of magnitude as the simulated results for a range of glycine initial concentrations. The characteristic time scale, L/U_b is about 12 s, is much less than the diffusive time scale of about 1800 s and Marangoni convection time scale of about 6400 s. Hence, the convection flow sufficiently mixes the concentration gradient that would be generated by diffusion, i.e., approximately 150 cycles for each diffusive time scale.

It was also found that the approximated characteristic and simulated velocity increases with initial concentration i.e., for 0.3 wt%, U_b = 3.3×10^{-5} m/s (order-of-magnitude approximation) and 2.5×10^{-5} m/s (simulation); for 0.5 wt%, U_b = 4.4×10^{-5} m/s (order-of-magnitude approximation) and 4.3×10^{-5} m/s (simulation). In the actual experiment for the evaporation-crystallization platform, the actual initial concentration was about 10 times higher than that used here which implies that its convection flow is greater. The plot of maximum Euclidean norm of velocity versus time for 0.1 wt% aqueous glycine shows that the velocity increases gradually within the induction period. The characteristic velocity determined from the order-of-magnitude analysis for short time provides a conservative assessment of the extent of mixing throughout the induction time. The high extent of mixing due to natural convection makes it reasonable to assume well-mixed conditions within the drop. Simulation also shows that the maximum absolute difference between concentration and average concentration is very small throughout the induction period which supports the assumption that the drop is well-mixed. We can conclude here that the hanging drop is well-mixed due to solutal buoyancy convection. It is shown by Goh & Braatz, 2007 that induction time data for paracetamol were well fit to the classical primary nucleation expression with the assumption of spatial uniformity.

To gain more insights on the flow within the drop, a similar analysis can be performed for flow due to thermal buoyancy and thermal Marangoni convection. According to Molenkamp (1998), solutal Marangoni convection in protein solution is important only for short times after creating the hanging drop because surface tension ceases to change after the protein concentration reaches a critical micelle concentration. Using the same order-of-magnitude analysis, assuming evaporation is the only heat flux at the interface, it can be determined that $\delta_T \approx 2\times10^{-3}$, $Gr_T \approx 0.02$, $U_b \approx 1\times10^{-4}$ m/s for thermal buoyancy at 0.5 wt% aqueous glycine concentration (Savino, 2002). The ratio of solutal and thermal buoyancy, $N = \beta_c\Delta c/\beta_T\Delta T = 4\times10^{-3}$. Hence thermal buoyancy is dominant and the characteristic velocity should be higher than that approximated earlier. Since thermal convection is postulated to be stronger than solutal convection, we can safely justify the assumption of Goh and Braatz (2007) that the hanging drop is well-mixed throughout the induction period based on the solutal convection analysis alone.

References

- Badarayani, R., Kumar, A. (2004). Viscometric study of glycine, L-alanine, glycylglycine in tetrat-n-alkylammonium bromide solutions at 298.15 K. *J. Chem. Thermodynamics*, **36**, 983.
- Chang, Y.C., Myerson, A.S. (1986) Diffusivity of glycine in concentrated saturated and supersaturated aqueous solutions. *AIChE J.*, **32**, 1567.
- Deen, W.M. (1998). *Analysis of Transport Phenomenon*, Oxford University Press.
- Galkin, O., Vekilov, P.G. (1999). Direct determination of the nucleation rates of protein crystals. *J. Phys. Chem. B*, **103**, 10965.
- Goh, L.M. (2007). M.S. Thesis. *University of Illinois at Urbana-Champaign.*
- Goh, L.M., Chen, K.J., He, G.H., Bhamidi, V., Kenis, P.J.A., Zukoski, C.F., Braatz, R.D. (2007). Nucleation kinetics determination in high-throughput microfluidic systems. *AIChE Annual Meeting*, Paper 353d.
- Grant, M.L., Saville, D.A. (1991). The role of transport phenomena in protein crystal-growth. *J. Crystal Growth*, **108**, 8.
- Grant, M.L., Saville, D.A. (1995). Long-term studies on tetragonal lysozyme crystals grown in quiescent and forced-convection environments. *J. Crystal Growth*, **153**, 42.
- He, G., Bhamidi, V., Tan, R.B.H., Kenis, P.J.A., Zukoski, C.F. (2006). Determination of critical supersaturation from microdroplet evaporation experiments. *Crystal Growth & Design*, **6**, 1175.
- Kang, K.H., Lee, S.L., Lee, C.M. (2003). Visualization of flow inside a small evaporating droplet. *5th International Symposium on Particle Image Velocimetry*, Korea, Paper 3242.
- Kawaji, M., Gamache, O., Hwang, D.H., Ichikawa, N., Viola, J.P., Sygusch, J. (2003). Investigaion of Marangoni and natural convection during protein crystal growth. J. Crystal Growth, **258**, 420.
- Leal, L.G. (2007) *Advanced Transport Phenomenon: Fluid Mechanics and Convective Transport Processes*. Cambridge University Press.
- Matubayasi, N., Miyamoto, H., Namihira, J., Yano, K., Tanaka, T. (2002). Thermodynamic quantities of surface formation of aqueous electrolyte solutions: V. Aqueous solutions of aliphatic amino acids. *J. Colloid and Interface Science*, **250**, 431
- Molenkamp, T. (1998). *Marangoni convection, mass transfer and microgravity*. Thesis Chapter 5. Online Publication: http://www.desc.med.vu.nl/Frames.htm
- Ostrach, S. (1982). Low-gravity fluid flows. *Ann. Rev. Fluid Mech.*, **14**, 313.
- Savino, R., Monti, R. (1996). Bouyancy and surface-tension-driven convection in hanging-drop proten crystallizer. *J. Crystal Growth*, **165**, pp. 308-318.
- Savino, R., Paterna, D., Favaloro, N. (2002). Bouyancy and marangoni effects in an evaporating drop. *J. Thermophysics and Heat Transfer*, **16**, 562.
- Shen, J.L., Li, Z.F., Wang, B.H., Zhang, Y.M. (2000). Partial molar volumes of some amino acids and a peptide in water, DMSO, NaCl, and DMSO/NaCl aqueous solutions. *J. Chem. Thermodynamics*, **32**, 805.
- Talreja, S., Kenis, P.J.A., Zukoski, C.F. (2007). A kinetic model to simulate protein crystal growth in an evaporation-based crystallization platform. *Langmuir*, **23**, 4516-4522.
- Talreja, S., Kim, D.Y., Mirarefi, A.Y., Zukoski, C.F., Kenis, P.J.A. (2005). Determination of critical supersaturation from microdroplet evaporation experiments. *J. Appl. Cryst.*, **38**, 988.

467f Crystallization Of Spherical Agglomerates Of A Peptide

Shailendra Bordawekar, Abbott Laboratories, 1401 N. Sheridan Rd., Bldg. R8-3, Dept. R452, North Chicago, IL 60064, Samrat Mukherjee, Process R&D, Abbott Laboratories, Dept R452, Bldg R8-3, 1401 Sheridan Rd., North Chicago, IL 60064, and John Tolle, Process R&D, Abbott Labs, 1401 Sheridan Road, R13-3, North Chicago, IL 60064.

During the development of a crystallization process for a peptide, two distinct crystal forms were observed; an anhydrous form and a hydrate form. The hydrate was more stable of the two forms under crystallization conditions. However, the anhydrous form provided a processing advantage due to better filtration characteristics, thereby reducing the downstream processing required for an acceptable quality peptide. The processing advantage was further enhanced by making spherical agglomerates of the anhydrous form. This was achieved by controlling the rate of desupersaturation during crystallization. Since the incoming peptide solution had inherently high supersaturation, controlling the rate of desupersaturation was critical to prevent precipitation of amorphous material. The effects of various crystallization parameters on desupersaturation rate were studied to design a robust process for making spherical agglomerates of the anhydrous form. The results of these studies and scale-up data for the process will be discussed.

Acceleration of the Formation Rate of Silver Nano Particles with Sonochemical Process by the Addition of Microbubbles

Satoshi Takeda[1], Akihiro Yamasaki[2], Yukio Yanagisawa[1]

[1]*School of Frontier Sciences, The University y of Tokyo, 5-1-5, Kashiwanoha, Kashiwa, Chiba 277-8563, Japan*

[2]National Institute of Advanced Industrial Science and Technology, 1-16 Onogawa, Tsukuba, Ibaraki 305-8569, Japan

Abstract

The addition of microbubbles with about 25 μm average diameter significantly accelerated the formation rate of nano particles of silver in a sonochemical process. The size distribution of the produced silver nano particles was almost unaffected by the addition of microbubbles. These results suggested that the microbubbles will provide the similar reaction sites as the cavitation bubbles in the ultrasonic reaction field.

1. Introduction

Sonochemical processes have been applied to a wide range of applications [1]. In the sonochemical processes, cavitation bubbles are formed by applying an ultrasonic field. The cavitation bubbles will undergo a cycle of adiabatic compression and expansion by the ultrasonic radiation. The cycle resulted in a high temperature (~ several thousands K) and high pressure (several hundreds atm) conditions near the cavitation bubbles [2]. Water molecules will be decomposed under such a severe condition to generate reactive radicals, which may initiate a series of chemical reactions in the aqueous or gas phases. Thus, the cavitation bubbles would serve cites for the chemical reactions in the sonochemical processes. The size of the cavitation bubbles are known to be in the range of several to several tens microns [3], which corresponds to that of "microbubbles". The microbubbles are bubbles of which the size is smaller than several tens microns [4]. Many kinds of applications have been proposed with microbubbles in the used of their very high specific surface area or some other special properties. Because of the size ranges of the microbubbles are similar to those of the cavitation bubbles, it can be anticipated that the microbubbles added in the sonochemical system will undergo the same process as the cavitation bubbles. The addition of microbubbles will increase the reaction cites for the sonochemical process, and consequently, the reaction rate will be accelerated. In addition, the properties of the product can be controlled by the microbubbles addition with a controlled size distribution. In this study, effect of the addition of microbubbles into the particle formation process of silver by the ultrasonic radiation, which is known to produce nano-sized silver particles [5], was experimentally investigated.

2. Generation of micro bubbles

A variety of methods have been proposed for the generation of microbubbles [6]. By considering the application to the continuous chemical reaction system, the porous membrane method [7] was employed in this study. The outline of the microbubble formation system is the following. A porous membrane was mounted in a stainless steel cell. One side of the membrane was exposed to the gas phase, while the other side was exposed to the liquid (aqueous) phase. With increasing the gas phase pressure, the gas can be penetrated into the liquid phase through the membrane pores when the gas phase pressure is higher than the bubble point pressure. The bubble point pressure, ΔP_{bp} (=P_G-P_L, where P_G is the gas phase pressure, P_L is the liquid phase pressure), is given by

$\Delta P_{bp} = 4\gamma\cos\theta/d_p$, where d_p is the pore diameter, γ is the surface tension, and θ is the contact angle [8]. The microbubbles in the liquid phase can be taken out from the cell in a water flow through two L-shaped nozzles, and the size distribution can be measured with a laser scattering method (Shimadzu SALD-2100).

Figure 1 shows the area-based size distribution of microbubbles generated with nitrocellulose membranes NC0.025 provided by Millipore Ltd., of which the pore size is 0.025 μm. The median diameters of the generated bubbles are 26.0 μm and 28.3 μm, respectively, for the trans-membrane pressure at 1350 kPa and 1600 kPa. The size distribution was sharper for the case with the higher trans-membrane pressure. The diameter of the generated bubbles was several hundreds times larger than the average pore diameter of the membrane, probably due to the coalescence of the generated bubbles in the liquid phase. The scattering intensity by the microbubbles, which reflects the volume density of the bubbles in the liquid phase, was increased with increasing the trans-membrane pressure. The volume ratio of the bubbles can be estimated by the scattering experiments with glass beads of which the volume fraction is known. The bubble density was 0.18 cm^3/L for 1350 kPa and 0.74 cm^3/L for 1600 kPa.

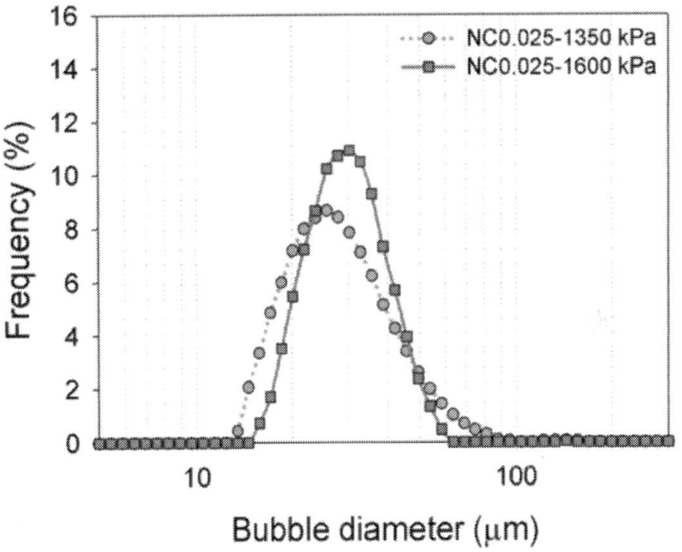

Figure 1. Size Distribution of the Micro bubbles generated with the Polymeric Porous Membrane.

3. Formation of metal nano particles

The generated micro bubbles were then introduced to the sonochemical reaction system for the production of silver particles. The liquid phase was an aqueous solution of silver nitrate (10 mM) and sodium dodecyl sulfate (SDS, 32 mM). SDS was added as a reduction reagent for silver ion and dispersion reagent for the produced silver particles. The gas phase for the microbubble formation was argon. The reaction system was a circulated flow system driven by a liquid pump, where the bubble generation part (stainless steel cell) was connected to the reaction part made of a stainless steel pipe. The reaction part was located on an ultrasonic oscillator of which the highest power was 480 W with 180 kHz frequency (Nippon Alex Co.). A small amount of the aqueous phase was sampled at a given time interval, and the UV-vis absorbance was measured.

Figure 2 shows the UV-vis absorbance for the aqueous solutions after 40 min of irradiation of ultrasonic. The peak appearing at about 400 nm wavelength can be assigned as small particles of silver [9]. Without microbubbles, no detectable peak was observed at about 400 nm wave length. On the other hand, a clear peak was observed for the each case with the microbubble addition,

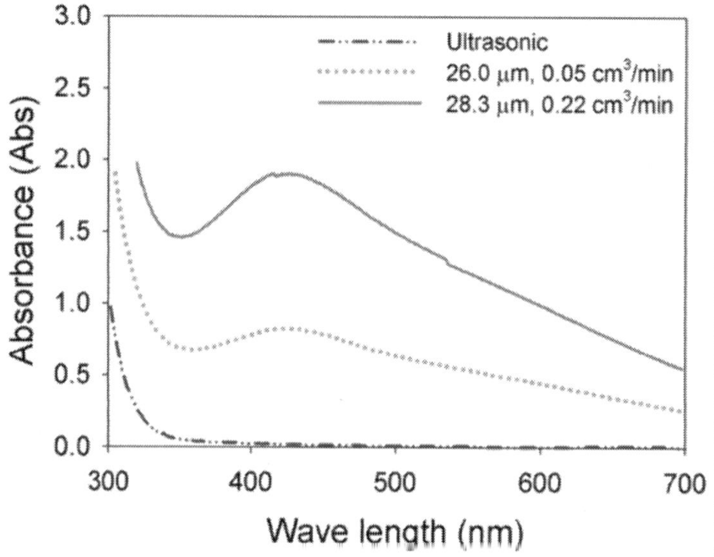

Figure 2. UV-vis Absorption Spectra of the Reaction Solutions after 40-min of Ultrasonic Irradiation.

and the peak strength was increased with an increase in the volume density of the microbubbles.

Figure 3 shows the absorbance at about 400 nm wavelength as a function of the ultrasonic irradiation time. The absorbance was increased almost linearly with the irradiation time, and the slope for the cases with microbubble addition was much higher than that without microbubble addition; the slope was about 45 times larger for the case when the microbubbles of the median diameter of 28.3 μm with the rate of 0.22 cm^3/min were added compared with the case without microbubbles.

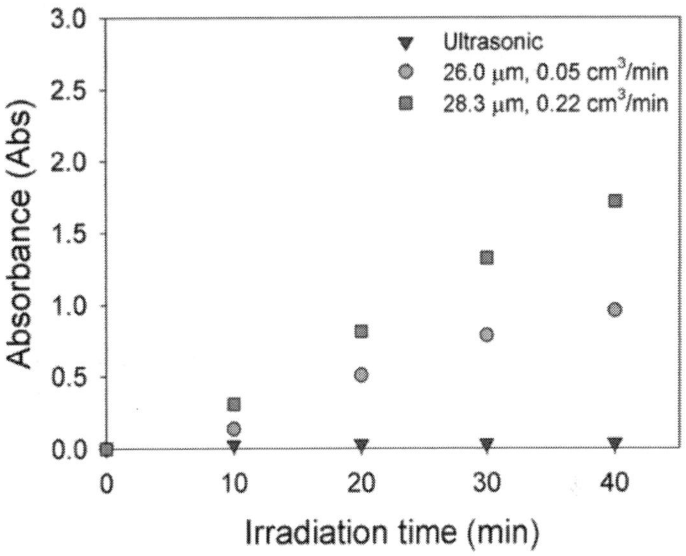

Figure 3. Dependence of the Ultrasonic Irradiation Time for the Absorbance at 400-nm Wavelength.

After a given time of irradiation, the solid particles were observed with TEM. The obtained particles were identified as silver particles with Energy Dispersive X-ray spectra (EDX). The images of the particles produced after 40 min of the irradiation are shown in Figure 4. Particles with the single nanometer size were obtained for all the cases. By an image analysis of several sheets of the TEM images, the average size and the standard deviation of the silver particles are determined. The average sizes of the particles were almost in the same range of 5 to 6 nm, and also the standard deviation was about 1 nm, for all the conditions studied. This results indicate that the size and its distribution of the produced silver particles was almost unaffected by the addition of the microbubbles.

Therefore, the addition of microbubbles with about 25 μm could provide almost similar reaction sites for the silver particles production in the ultrasonic irradiation field.

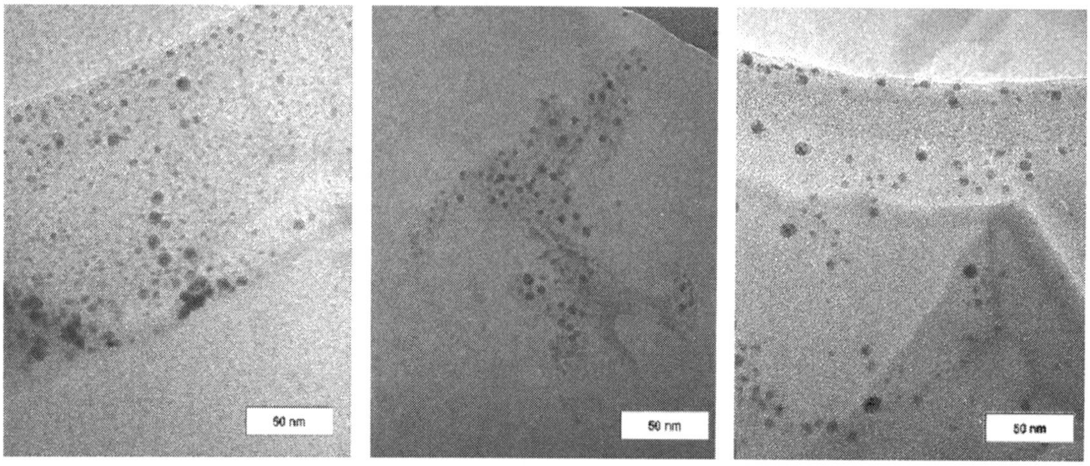

a) b) c)

Figure 4. TEM Images of the Silver Nano Particles
a) without Microbubbles
b) with Microbubbles with 26.0 μm, 0.05 cm^3/min.
c) with Microbubbles with 28.3 μm, 0.22 cm^3/min.

References

1. T.J. Mason and J.P. Lorimer, *Applied Sonochemistry*, Wiley-VCH, Weinheim, **2002**.
   a) W. B. McNamara III, Y. T. Didenko, K. S. Suslick, *Nature (London)*, **1999**, 401, 772. b) W. B. McNamara III, Y. T. Didenko, K. S. Suslick, *J. Phys. Chem. B*, **2003**, 107, 7303.
2. F. Burdin, N.A. Tsochatzidis, P. Guiraud, A.M. Wilhelm, and H. Delmas, *Ultrasonics Sonochemistry*, **1999**, 6, 43.
3. H. Ohnari, *Kagaku Kogaku*, **2007**, 71, 154.

a) K. Okitsu, Y. Mizukoshi, H. Bandou, Y. Maeda, T. Yamamoto, and Y. Nagata, *Ultrasonic Sonochemistry*, **1996**, 3, S249. b) A. Gedanken, *Ultrasonics Sonochemistry*, **2004**, 11, 47.

4. K. Terasaka, Kagaku Kogaku, **2007**, 71, 170.
5. M. Kukizaki, M. Goto, *J. Membrane Sci.*, **2006**, 281, 386.
6. M. Mulder, *Basic Principles of Membrane Technology*, Kluwer, New York, **1991**.
7. S. Link, M. A. El-Sayed, *J. Phys. Chem. B*, **1999**, 103, 8410.

493b Particle Formation Pathways In Antisolvent Precipitation Processes

Jan Sefcik[1], Anna Jawor-Baczynska[1], and Barry D. Moore[2]. (1) Chemical and Process Engineering, University of Strathclyde, James Weir Building, 75 Montrose Street, Glasgow, United Kingdom, (2) Pure and Applied Chemistry, University of Strathclyde, James Weir Building, 75 Montrose Street, Glasgow, United Kingdom

Rapid coprecipitation is a novel method for assembling common pharmaceutical ingredients such as proteins, nanoparticles and nucleic acids onto the surface of crystals [1-3]. The microparticles obtained are typically monodisperse and form free-flowing dry powders that have attractive properties for a range of alternate drug-delivery application. Here we report studies aimed at better determining the mechanism of formation of these microcrystals. The material chosen for this study was D,L-valine (an amino acid) because the kinetics of precipitation of this material were found to be sufficiently slow to allow the process to be monitored in real time. The precipitated microcrystals are generally plate-like with a very high aspect ratio, typically 5-10 microns diameter with a thickness of about 0.2 microns. The basic process involves well controlled rapid mixing of an aqueous solution of valine and an active pharmaceutical ingredient with a water miscible solvent, such as isopropanol, which is a poor solvent for both solutes. This was carried out either in a batch process using a dynamic mixer or in a continuous process using a static mixer, producing particles of very similar morphology. Formation of valine microcrystals and their precursors was monitored in situ using turbidimetry, static and dynamic light scattering over a range of solution compositions at constant temperature [4]. It was found that mixing of isopropanol with an excess aqueous solution resulted in gradual precipitation of stable valine nanoparticles with a constant diameter with a value depending on the solution composition. Subsequently, few large valine crystals grew slowly from these transparent nanoparticle dispersions, while nanoparticle size as detected by dynamic light scattering stayed unchanged. However, mixing of aqueous solutions with an excess isopropanol resulted in liquid-liquid demixing, followed by a sudden appearance of micron-size valine crystals, as observed by fluorescence and small angle static light scattering. In this work we show how careful experimental characterisation of particle formation processes through light scattering, fluorescence and microscopy can be used to identify non-trivial physico-chemical pathways leading to formation of crystals and ordered assemblies of organic molecules, thus providing fundamental understanding crucial for a better design and control of crystallisation processes in pharmaceutical industry.

[1] Kreiner, M.; Moore, B.D.; Parker, M.C. Chem. Comm., 2001, 1096-1097.

[2] Kreiner, M.; Fuglevand, G.; Moore, B.D.; Parker, M.C. Chem. Comm., 2005, 2675-2676.

[3] Murugesan, M.; Cunningham, D.; Martinez-Albertos, J.L.; Vrcelj, R.M.; Moore, B.D. Chem. Comm., 2005, 2677-2679.

[4] Variny M.; Jawor-Baczynska, A., Moore, B.D.; Sefcik, J. J. Disp. Sci. Technol. (in press).

The Effect of Impurities on Borax Decahyrate Crystallization in MSMPR Crystallizer

Birgül BENLİ*, Nusret BULUTÇU

Istanbul Technical University, Chemical Engineering Department, Maslak, Istanbul

In this study, the effect of impurities on the crystal size distribution (CSD) was investigated for industrial crystallization of borax decahydrate. A continuous mixed-suspension, mixed-product removal (MSMPR) crystallizer was used to determine the main characteristics of crystallization using population balance equation. MSMPR experiments on borax decahydrate crystallization showed that large size particles have higher growth rate than small size particles. In addition, increasing on the retention time in the crystallizer increases the particle size of the crystals. On the other hand, there is a significant effect of the impurities on the crystallization of borax decahydrate.

1. Introduction

Boron is one of the most valuable minerals for Turkey in which refined borate production capacity of about 800.000 t/y and borate ore production capacity 1.800.000 t/y (1,2). Borax decahydrate ($Na_2B_4O_7.10H_2O$) is one of the most commercial compounds for borate industry due to their effective usages in many applications such as detergent formulations, metallurgy and high-quality glasses, ceramic materials and enamels, etc. (1,3,4).

Borax decahydrate is refined with vacuum crystallization after several operations. First, it is crushed to about <2.5 cm, and then dissolved in hot recycled borax solution at about 80°C and finally clarification of the solution (5). Crystallization is the most important step in the process. In order to produce crystals with a specified purity and crystal size distribution (CSD) at minimum cost, it is necessary to operate the crystallizer at the optimum of super-saturation within the metastable zone (6). It is known that the growth and dissolution rates are related to the super-saturation and the concentration of the impurities (7). Moreover, the most important parameters affecting the CSD in industrial crystallization are nucleation and growth rate mechanism (8,9,10).

Since the main kinetic parameters such as crystal growth rate (G) and effective rate of nucleation (B_0) influences crystal size distribution (CSD), a continuous mixed-suspension, mixed-product removal (MSMPR) crystallizer is called as an excellent device in pilot scale

(11). In crystallization processes, the population balance equation (PBE), which governs the CSD, is solved together with mass/energy balances and crystallization kinetics (12,13).

In this study, the effect of impurities such as the excess of $NaBO_2$ on the crystal size distribution of borax decahydrate crystallization at steady-state condition using the MSMPR crystallizer was investigated using the population balance equation.

Theory

The particle size distribution of crystals is characterized using population balance theory. Application of the population balance is illustrated in Figure 1. Assumptions for MSMPR crystallizer are given following.

(a) steady state operation,

(b) no crystals in the feed stream,

(c) all crystals have the same shape, and shape factors for particles are not a function of size,

(d) no break-down of crystals by attrition,

(e) crystal growth rate independent of crystal size.

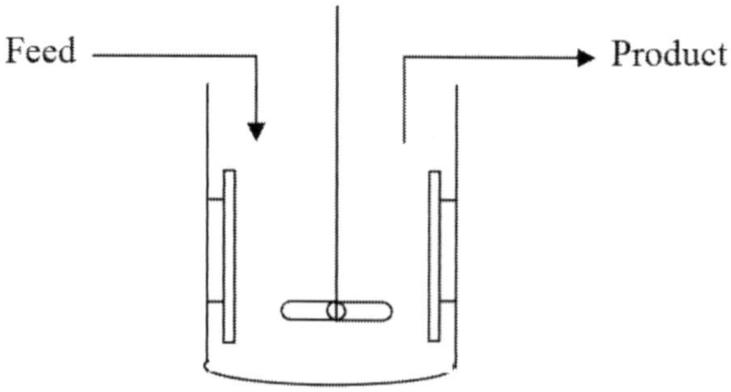

Figure 1. A continuous mixed-suspension, mixed-product removal (MSMPR) crystallizer.

Nomenclature

G	crystal growth rate, dL/dt
L	size of the crystals
ΔL	size range of the crystals
n	population density, number of crystals per unit size per unit volume of the system
$\check{n}$	the average population density
n_0	the population of nuclei density (zero-sized crystals)
Δt	time interval
Q	volumetric feed and discharge rate
V	crystallizer volume
τ	the crystal mean retention time, V_c/ Q

The overview of these assumptions of the crystal population density can be written as

$$\lim \frac{\Delta N}{\Delta L} = \frac{dN}{dL} = n \tag{1}$$

When the value of n depends on the value of L at which the interval dL is taken, the number of crystals in the size range between L_1 and L_2 is given by

$$\Delta N = \int_{L_1}^{L_2} n.dL \tag{2}$$

As shown in Figure 1, the mass transfer or population balance within the crystallizer can be written for the same size range as follows;

$$\text{Input} = \text{Output} \tag{3}$$

Then, population balance in a crystallizer volume V

$$n_1 G_1 V \Delta t = n_2 G_2 V \Delta t + Q\bar{n}\Delta L \Delta t \tag{4}$$

As ΔL approaches to zero,

$$\frac{d(nG)}{dL} = -\frac{Qn}{V} \tag{5}$$

the crystal mean retention time is given by $\tau = V_c / Q$ and assuming no particles in feed stream, and applying Mc Cabe's ΔL law which is assuming that growth is independent of size, $G \neq G(L)$,

$$\frac{dn}{dL} = -\frac{n}{G\tau} \tag{6}$$

which on integration, and if the boundary condition of n_0 representing the population of nuclei density (zero-sized crystals) is employed, $n(0) = n_0$, the result is

$$n = n_0 \exp(-\frac{L}{G\tau}) \tag{7}$$

Equation 7 is the fundamental relationship between crystal size L and population density n characterizing the CSD. Also, this equation can be easily linear in semi logarithmic scale as follows;

$$\log n = \log n_0 - \frac{L}{G\tau} \tag{8}$$

It can be seen from Equation 8 that a plot of log n versus L should give a straight line with an intercept at L=0 equal to n_0 and a slope of $-1/G\tau$. Furthermore, the graphs of this study are arranged according to these equations, in Eq. 7 and 8.

2. Experimental

2.1. Preparation of Borax Decahydrate

Pure borax decahydrate was prepared from industrial grade borax pentahydrate by purifying with two times of recrystallization. A saturated solution of borax pentahydrate was prepared according to the solubility data of Mullin (11). The borax pentahydrate was dissolved in crystallizer with the capacity of 50 liter for purification and production of the borax decahydrate (Figure 2). Then, the solution was heated above the saturation temperature and required Na_2CO_3 was added to the solution in order to precipitate of probertite which is one of the alkaline earth metals in the solution. And the precipitated probertite was removed from the hot solution using a filterpress. Finally purified hot solution was cooled to the room temperature. The crystals obtained from the crystallizer were dried in room temperature.

In the second step of purification, the produced crystals were purified by recrystallization using the same procedure without adding of Na_2CO_3.

Figure 2. Crystallizers used for the production and purification of borax decahydrate.

2.2. MSMPR Setup

All experiments were carried out in a continuous MSMPR crystallizer. A draft tube baffled MSMPR-type crystallizer, made of perspex and glass, with 2 liter of active volume was used in the crystallization experiments. As MSMPR crystallizer and crystallization cell can be seen in Figure 3, no metallic parts were permitted in the crystallization system in order to avoid any decomposition.

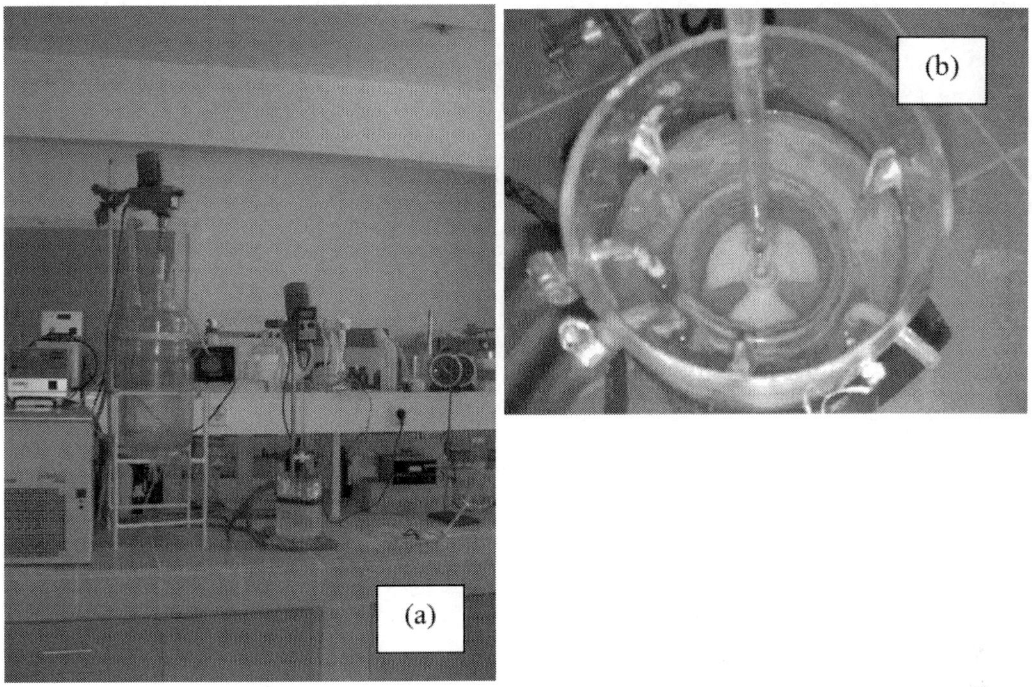

Figure 3. Experimental setup, (a) MSMPR crystallizer, (b) crystallization cell.

MSMPR experiments were performed with and without the excess of $NaBO_2$ ratio which is prepared to the following chemical formulation in Eq. (9).

$$Na_2B_4O_7.10H_2O + 2\ NaOH \rightarrow 4\ NaBO_2 + 11\ H_2O \qquad (9)$$

When the system was reached to the crystallization temperature, 35°C ± 1, the solution velocity and operation parameters such as stirring rate, impeller position, out feed etc, was optimized therefore the crystals were uniformly suspended within the crystallization zone. During the experiments, the samples were taken out with vacuum tube to determine the CSD, $Na_2B_4O_7$ content and bulk density. CSD of the product was determined by sieving, using an ASTM E-11 sieve series in the range of 38-1000 /μm.

3. Results and Discussion

3.1. Effect of Stochiometric Ratio and Retention Time on the Borax Decahydrate Crystallization

The CSD of pure borax decahydrate prepared in the stockiometric ratio using chemical formulation of Eq.(9) is seen in Figure 4 with the different retention times as τ = 0.25h, 0.5h, and 1.0 h. The samples were taken out of the MSMPR at 1.5h, 3h, and 6h, respectively and all experiments were reproduced three times. As shown Figure 4 that the increase on the retention time in the crystallizer, it causes the increase of the particle size of the crystal. The values of both the retention time and mean particle size can be given as $\tau = 0.25, 0.5, 1.0h$ and maximum $L_{mean} = 355, 355$, and 427.5μm, respectively. This result is proof of the theory of crystallization (11).

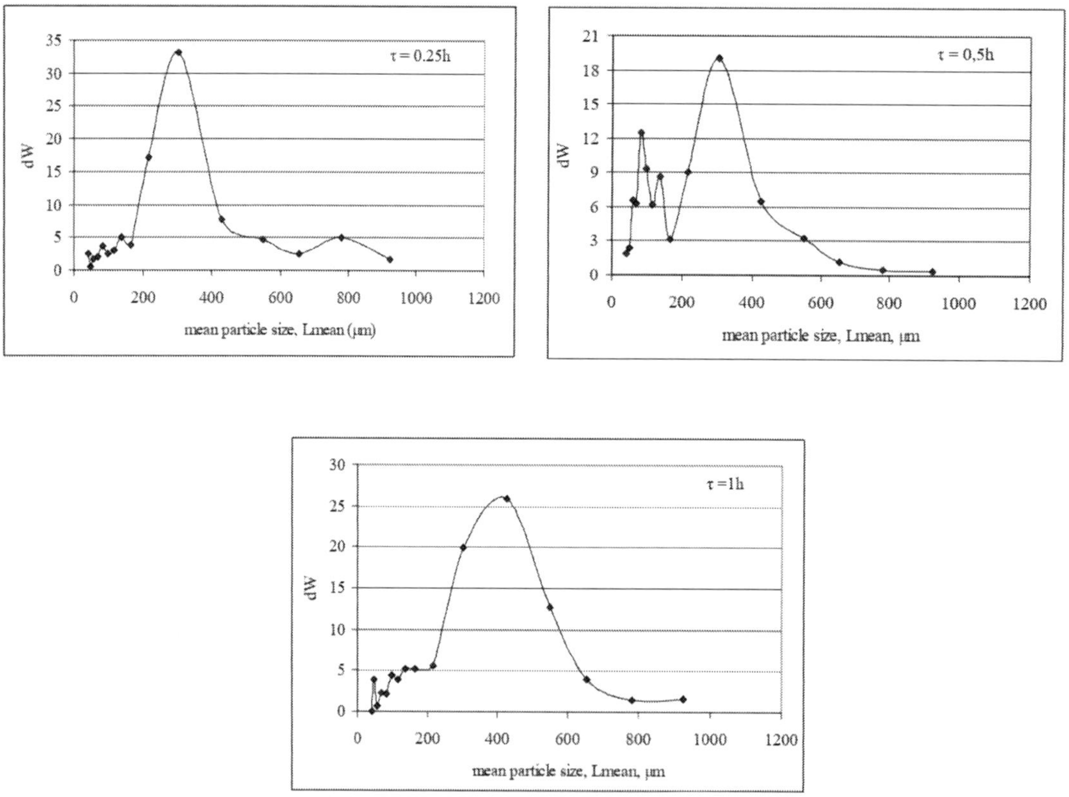

Figure 4. Cumulative sieve analysis results of borax decahydrate as a function of the retention time.

The population balance theory was applied on the CSD results of the data given in Fig.4. The results can be seen in Figure 5. In the calculations of the population balance equation, the density of the borax decahydrate solution was obtained from literature as 1715 kg/m^3 (14), and shape factor was assumed as 1.

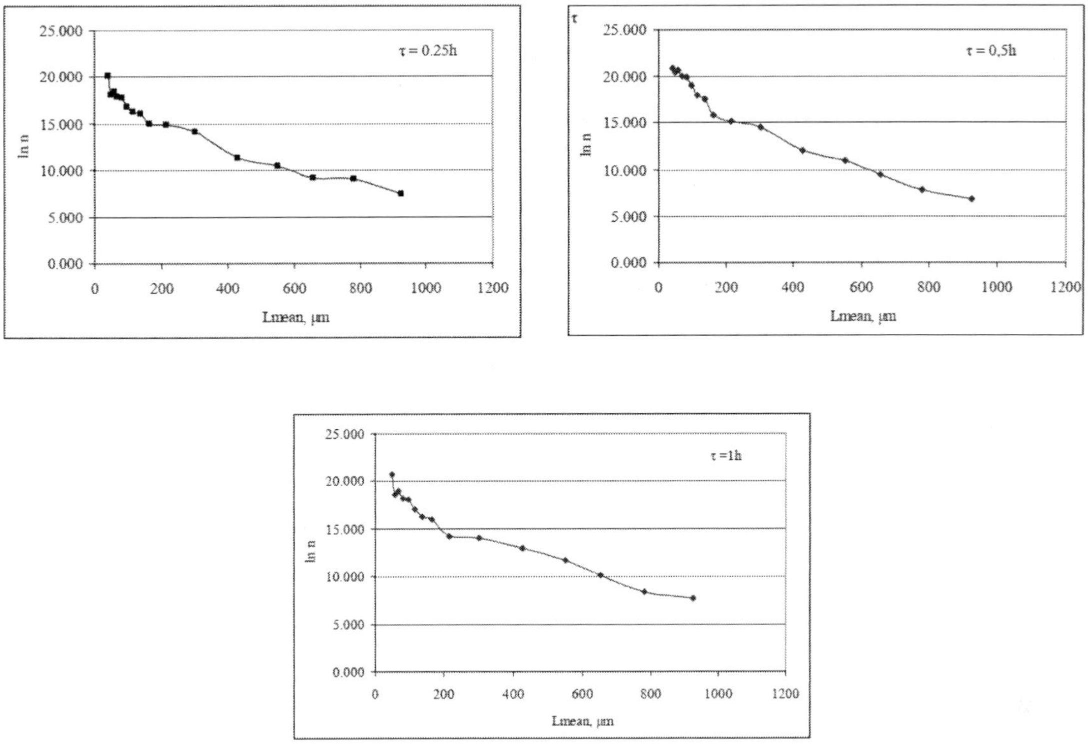

Figure 5. CSD of borax decahydrate using population balance equation.

The graphs in Figure 5 can be divided into two different region; 38μm<Region1<200μm and 200μm<Region2<925μm. Region 1 should be as secondary nucleation and Region 2 should be as attrition fragments and aggregation. It is given in the literature that a plot of ln n versus L_{mean} should give a straight line with an intercept at n_0 and a slope of $-1/G\,\tau$ (11). Using these data (n_0, G), can be calculated nucleation and growth kinetics (1). Therefore, these regions obtained from Fig.5 gives very important information about crystallization of borax decahydrate. Another important point is that there is no data in the literature on the crystal growth and dissolution kinetics of borax decahydrate depending on particle size using MSMPR crystallizer. Other types of cyrstallizers have been used in the previous studies; fluidized-bed crystallizer (1) and rotating disc technique (15).

3.2. Effect of Impurities on the Borax Decahydrate Crystallization

In this part of the study, the effect of impurities on the crystallization was investigated at the retention time τ = 0.5h using MSMPR crystallizer. The excess amounts of $NaBO_2$ (1 % and 3%) from the stochiometric ratio of borax decahydrate chemical formulation which is given in Eq. (9) were used as impurity. CSDs and population balance equations of borax decahydrate crystallization in the excess of $NaBO_2$ are given in Figure 6, (a) and (b).

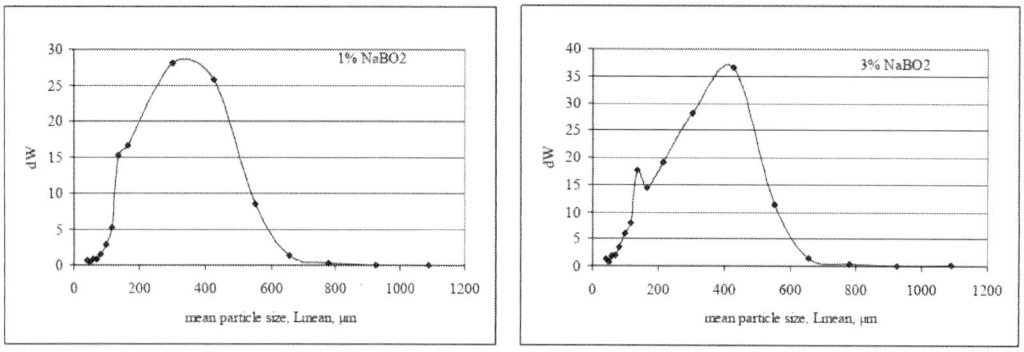

Figure 6 (a). CSDs of borax decahydrate in the excess of $NaBO_2$, 1 - 3%, respectively.

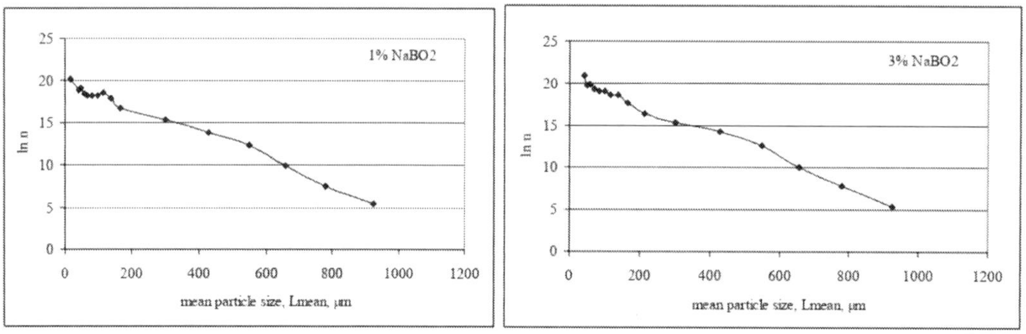

Figure 6 (b). PBEs of borax decahydrate in the excess of $NaBO_2$, 1 - 3%, respectively.

It can be seen from Fig.6 (a) and (b) that the addition of 1% $NaBO_2$ to the solution did not change the mean particle size (305µm). However, the addition of 3% $NaBO_2$ increased the mean particle size to 425µm. The same effect can be seen in Figure 6 (b) that the population balance equation gives a linear relation with ln n versus L_{mean} which is more similar with the general theory (10,12).

The effect of impurities on the shape of particle as well as particle size was investigated with the microscope photographs using a particle image analyzer. The results are shown in Figure 7. As seen from the figures, as the retention times increases, the shape of the crystals are getting smoother.

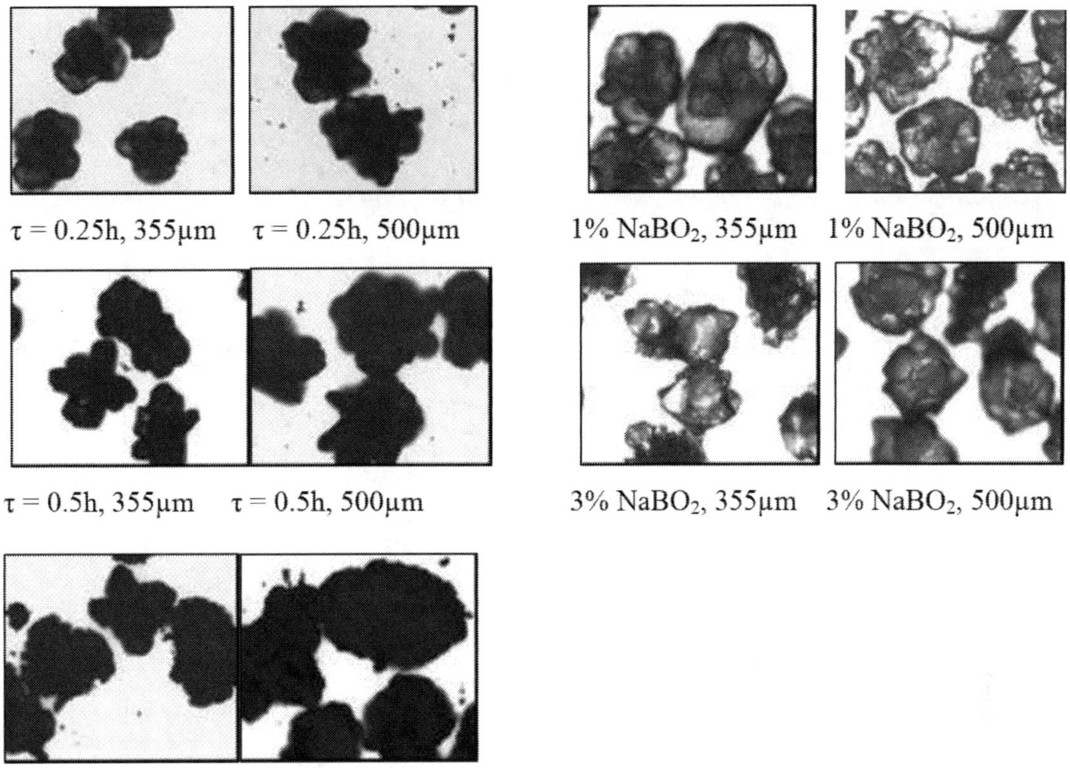

τ = 0.25h, 355μm τ = 0.25h, 500μm 1% $NaBO_2$, 355μm 1% $NaBO_2$, 500μm

τ = 0.5h, 355μm τ = 0.5h, 500μm 3% $NaBO_2$, 355μm 3% $NaBO_2$, 500μm

τ = 1h, 355μm τ = 1h, 500μm

(a) (b)

Figure 7. Borax decahydrate crystals produced (a) the effect of retention time, (b) the effect of impurities.

4. Conclusions

The results from this study can be concluded as follows:

1. The increase on the retention time in the crystallizer increases the particle size of the crystal. The values of both the retention time and mean particle size were obtained as $\tau = 0.25, 0.5, 1.0$h and maximum $L_{mean} = 355, 355$, and 427.5μm.
2. The population balance equation of pure borax decahydrate crystals can be divided into two different region; 38μm<Region1<200μm and 200μm<Region2<925μm.
3. While the addition of 1% $NaBO_2$ did not change the particle size of borax decahydrate, 3% $NaBO_2$ increased the mean particle size to 425μm.
4. This study showed that there is a significant effect of $NaBO_2$ as a function of concentration on the trend of PBEs. The population balance equation of borax decahydrate crystals in excess of $NaBO_2$ shows one linear region.

5. References

1. Ceyhan, A.A., Sahin, O., Bulutcu, A.N., 2007 "Crystallization kinetics of the borax decahydrate", *Journal of Crystal Growth,* 300, pp: 440–447.
2. www.boren.gov.tr
3. Boron, *Kirk-Othmer Encyclopedia of Chemical Technology,* Vol.4, 1992.
4. Crystallization, *Kirk-Othmer Encyclopedia of Chemical Technology,* Vol.4, 1992.
5. Kistler, R.B., Helvaci, C., 1994 "Boron and borates", in: D.D. Carr (Ed.), *Industrial Minerals and Rocks,* 6th Edition, Society of American Mining, Metallurgy and Exploration Inc., Littleton, Co.
6. Gurbuz, H., Ozdemir, B., 2003 "Experimental determination of the metastable zone width of borax decahydrate by ultrasonic velocity measurement", *Journal of Crystal Growth,* 252, pp: 343–349.
7. Al-Jibbouri, S., Ulrich, J., 2001 "The influence of impurities on crystallization kinetics of sodium chloride", *Cryst. Res. Technol.* 36 (12), pp:1365-1375.
8. Sayan, P., Ulrich, J., 2001 "The Effect of İmpurities on the Crystal hardness of NaCl crystals " *Crystal Research and Technology,* 36 (1)11, pp:1253-1262.
9. Al-Jibbouri, S., Ulrich, J., 2004 "Impurity adsorption mechanism of borax for a suspension growth condition: A comparison of models and experimental data", *Cryst. Res. Technol.* 39 (6), pp:540 – 547.
10. Sayan, P., "The effect of the polyelectrolytes on boric acid crystallization", *Doctor of Philosophy Dissertation,* Department of Chemical Engineering, Istanbul Technical University.
11. Mullin, J., W., 1993 "Crystallization", 3th Edition, Butterworth-Heinemann Ltd., Oxford.
12. Choi, B.S., 2005 "Crystallization of NaCl from a concentrated $CaCl_2$-KCl-NaCl solution in a MSMPR crystallizer–observation of crystal size distribution and model validation", *Doctor of Philosophy Dissertation,* Department of Metallurgical Engineering, The University of Utah.
13. Piotrowskia, K., Piotrowskia,J., 2005 "Monte Carlo modelling of continuous adiabatic MSMPR crystallizer", *Chemical Engineering and Processing,* 44, pp:517–529.
14. Snell, D.F., Hilton, C.L., 1968 Eds. *Encylopedia of Industrial Chemical Analysis*, Wiley: New York.
15. Sahin, O., Aslan, F., Ozdemir, M., Durgun, M., 2004 "Determination of kinetic parameters of crystal growth rate of borax in aqueous solution by using the rotating disc technique", *Journal of Crystal Growth,* 270, pp:604–614.

493d Scale-Up Investigation Of Particle Production In Emulsions

Andreas Voigt[1], Michael Fricke[2], and Kai Sundmacher[2]. (1) Process Systems Engineering, University of Magdeburg, Universitaetsplatz 2, Magdeburg, 39106, Germany, (2) PCP, MPI Magdeburg, Sandtorstr. 1, Magdeburg, 39106, Germany

The large scale production of particles with well-defined properties and a narrow property distribution is of great importance for a multitude of technical applications. The particle precipitation in water-in-oil emulsions constitutes a suitable process for the synthesis of particles also in the nanometer range [1,2]. In such a process well-defined water droplets of the emulsion act as reaction vessel with a limited reaction volume. In order to scale this process for industrial applications one will need stringent control of the droplet size distribution in such an emulsion. The first step of a controlled particle production in emulsions is therefore the control of droplet size and distribution [3]. If the production of particles inside droplets is carried out by a diffusion step through the droplet surface, the concentration profile inside the droplets will also be important for the produced particles and their size distribution as well. We will present a combined theoretical and experimental investigation of an emulsion-based particle production process. We present results of population balance modeling as well as molecular modeling which can give significant contributions to improve the understanding of basic phenomena of the process. The simulation results will be compared to our experimental data to validate and support the modeling approach. The conclusions will show interesting and alternative ways towards suitable and efficient particle production technologies.

[1] B. Niemann, F. Rauscher, D. Adityawarman, A. Voigt and K. Sundmacher: Microemulsion-assisted precipitation of particles: Experimental and model based process analysis, Chem. Eng. Process. 45 (10), 2006, p. 917-935.

[2] D. Adityawarman, A. Voigt, P. Veit and K. Sundmacher: Precipitation of BaSO4 nanoparticles in a non-ionic microemulsion: Identification of suitable control parameters, Chem. Eng. Sci. 60 (12), 2005, p. 3373-3381.

[3] M . Fricke, A. Voigt and K. Sundmacher: Droplet size distribution of technical macroemulsions: Variation of process parameters, PARTEC 2007 Conference Proceedings, Nuremberg, 2007.

Web Page: www.mpi-magdeburg.mpg.de/research/projects/1032/1040/MicroemulPre

Membrane Emulsification for PLGA Microparticle Production: Influencing Factors

Gilda Gasparini, Loughborough University, Serguei R. Kosvintsev, Micropore Technologies Ltd, and Richard G. Holdich, Loughborough University.

Introduction

A vibrant field in the pharmaceutical and health care industries is controlled release drug particles. Particles capable of releasing the encapsulated drug are of great interest because:

- they help in maintaining therapeutic levels of a drug
- reduce harmful side effects
- decrease the overall amount of the molecule required
- decrease the number of dosages (increasing patients' compliances)
- facilitate the delivery of drugs with short in vivo half-lives
- protect the incorporated drug by the surrounding aqueous environment [1].

One of the most suitable polymers to produce these particles is PLGA, poly(lactic-glycolic acid). PLGA is a highly biocompatible polymer, it degrades in vivo to lactic ($C_3H_6O_3$) and glycolic ($C_2H_4O_3$) acids which are subsequently eliminated as CO_2 and H_2O via the Krebs cycle [2]. PLGA has been approved by the Food and Drug Administration and it has been used for suture, dental repairs and bone replacement surgery for a long time. It is widely commercially available and the degradation can be controlled. It can entrap the drug either by a matrix encapsulation, where the phase is dispersed through the porous inner morphology of the solid particle or, alternatively in a hollow core-shell encapsulation.

In producing particles for controlled drug release, particle size and particle size distribution are a major issue. Too small particles may encapsulate only a small amount of drug, release it quickly and easily migrate from the place of injection. On the other hand, too big particles are not suitable for some kinds of administration since they may be uncomfortable or dangerous. Polydisperse particles often need to be sieved, wasting valuable material. A narrow size distribution, or deliberately engineered distribution, allows a higher control on release, also reducing the probability of side effects due to overdose [3].

Membrane emulsification is known to produce very monosized droplets and it is easy to tailor the droplet size. Mild production conditions make membrane emulsification a suitable method for working with drugs.

In this work, the production of PLGA particles encapsulating a water soluble model drug is presented. The particles are produced by double emulsification followed by solvent evaporation, hence the particle size is tightly linked to the droplet size.

Emulsification

The emulsification method is a non cross-flow membrane emulsification. The emulsion is obtained by injecting the discontinuous phase through a plate membrane with a regular array of pores into the continuous phase, agitated by a two-blades paddle governed by a motor. It is possible to control the emulsion features by changing operating conditions such as continuous phase agitation speed, discontinuous phase injection rate, membrane pore diameter and pore distance, membrane coating and physical properties of the involved phases.

The studies with a model system were conducted with sunflower oil as discontinuous phase and water and Tween 20 as surfactant were the continuous phase [4]. The droplet diameter can be predict by a simple force balance. When shear is applied, the growing droplets undergo the effect of three forces: Capillary force, F_{Ca}, which acts against the droplet detachment and depends on the interfacial tension, Buoyancy force, F_B, which depends on the density gradient, and the Drag force, F_D, due to the agitation of the continuous face inside the cell [5,6]. F_B is negligible when the droplet diameter is below 300 μm. The Capillary force is expressed in order to consider the neck formed at the late stage of droplet formation, equation 1, just before detachment occurs [6,7]. When the droplet diameter is smaller than twice the pore diameter, the traditional expression for F_{Ca} is to be preferred, Equation 2, since Equation 1 is not valid anymore.

$$F_{Ca} = \pi d_p \gamma \left(1 - \frac{d_p}{d_d}\right) \qquad \text{Eq. 1}$$

$$F_{Ca} = \pi d_p \gamma \qquad \text{Eq. 2}$$

where:
d_p is the membrane pore diameter;
γ is the interfacial tension;
d_d is the droplet diameter.

An expression for the drag force, equation 3, has been seen in [6]. The velocity pattern, equation 4, and the Reynolds number, equation 6, depends on the cell geometry and they are taken from [4]. The shear increases linearly from the centre of the cell, up to a maximum at a distance called transitional radius, r_{trans}, equation 5, from [4]. After this distance it decreases exponentially, see Figure 1.

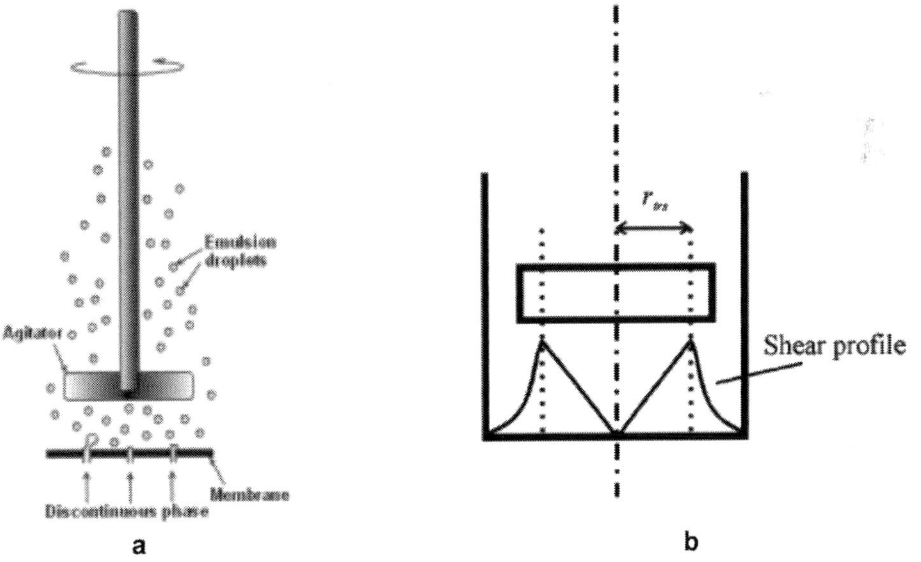

Figure 1 **Schematic diagrams showing a) the stirred cell and b) the shear profile under the stirrer**

The ring area corresponding to the transitional radius is believed to be the most active of the membrane, generating the majority of the droplets [4,8].

$$F_D = 3\pi k_{wl}\eta_{cont} v d_d$$ Eq. 3

$$v = \omega r_{trans}\left(1 - \exp\left(-\frac{d_d}{2\delta}\right)\right)$$ Eq. 4

$$r_{trans} = \frac{D}{2}1.23\left(0.57 + 0.35\frac{D}{T}\right)\left(\frac{b}{T}\right)^{0.036} n_b^{0.116}\frac{\mathrm{Re}}{1000 + 1.43\,\mathrm{Re}}$$ Eq. 5

$$\mathrm{Re} = \frac{\rho_{cont}\omega D^2}{2\pi\eta_{cont}}$$ Eq. 6

where:
k_{wl} is a coefficient to consider the effect of the cell walls, flow and phase viscosity, [9];
η_{cont} is the continuous phase viscosity;
ω is the paddle rotation speed;
δ is the Landau Schwitz constant expressed from [4] as $\delta = \sqrt{\frac{\eta_{cont}}{\rho_{cont}\omega}}$
ρ_{cont} is the density of the continuous phase;
D is the paddle blade size;
T is the cell diameter;
b is the distance between the paddle and the membrane;
n_b is the number of blades.

The predicted droplet diameter is calculated by a force balance between equation 1 and equation 3, which act against each others. It can be only solved numerically. The results of the prediction are compared to the experimental results for the sunflower oil-in-water system in Figure 2 and for the PLGA-in-water system in Figure 3. For this evaluation, when PLGA is used, the continuous phase is water and 1% poly vinyl alcohol while the discontinuous phase is 5%PLGA and dichloromethane. Sunflower oil droplet size and all the following particle diameters were measured by a laser diffraction device, Malvern Mastersizer, while PLGA droplet size, presented in Figure 3, are direct measurements from images taken with an optical microscope.

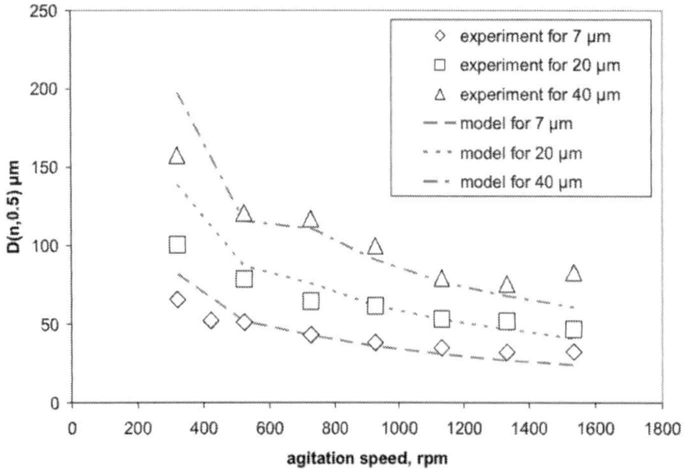

Figure 2 **Droplet diameter model prediction and experiment results (number basis) comparison for different membrane pore size and agitation speed for sunflower oil in water**

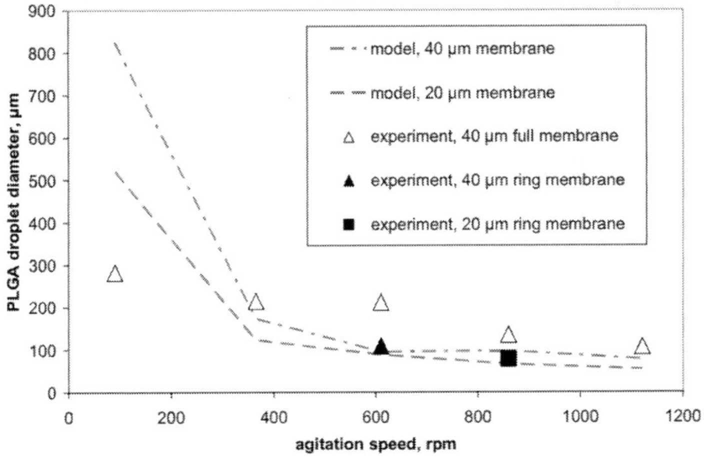

Figure 3 **Droplet diameter model prediction and experimental results comparison for different membrane pore size, different membrane area and agitation speed for PLGA in water system**

It is noticeable that at low agitation speed the model over estimates the droplet size while at agitation rates above 400 rpm the prediction is accurate. PLGA and DCM viscosity is not far different than water and this leads to a wider size distribution when using the full membrane area. To improve this, the membrane working area was modified. All of the membrane was blocked off apart from a ring which corresponds to the region where the shear is maximum so that the droplets face a fixed optimal shear stress. When the full membrane is used, the droplet diameter is less monosized and the model prediction is less accurate.

Because the target is to encapsulate a water soluble drug, a double emulsification was performed. The inner water phase is composed by: reverse osmosis water, blue dextran (model drug) and 40 g/l of salt. 20 ml of this solution was emulsified by a homogenizer with 50 ml of DCM and different concentrations of PLGA. This represents the oil phase which is injected in a water phase made of water, 1% PVA and different salt concentrations via membrane emulsification in the dispersion cell. 10 ml of discontinuous phase are injected in 150 ml of water phase and then the emulsion obtained is solidified to obtain encapsulated PLGA particles.

PLGA Particle Production

Different solidification methods were tried in order to maintain the narrow size distribution and the spherical shape obtained during the emulsification phase. During the solidification process the organic solvent transfers from the drops formed by the emulsification to the external water phase and then to the water-air interface, where it then transfers to the air. Hence changing the water-air evaporation area, or the concentration of DCM in water, it is possible to influence the solvent removal rate. This has been presented in [9]. Briefly, the chosen method is overnight stirring with an evaporation area of 50 cm^2 and a stirring rate of 120 rpm. Figure 4 shows particles obtained by the double emulsion-solvent removal method, encapsulating a water soluble model drug.

The volume shrinkage never represents the actual solvent loss due to the formation of internal porosity and the presence of the inner water phase. The presence of salt and different polymer concentrations can influence the amount of shrinkage. The salt creates osmotic pressure, when the amount of salt in the outer water phase is lower than in the inner water, a inward water flux is created. Hence the solidification shrinkage is lower because of water intake [10].

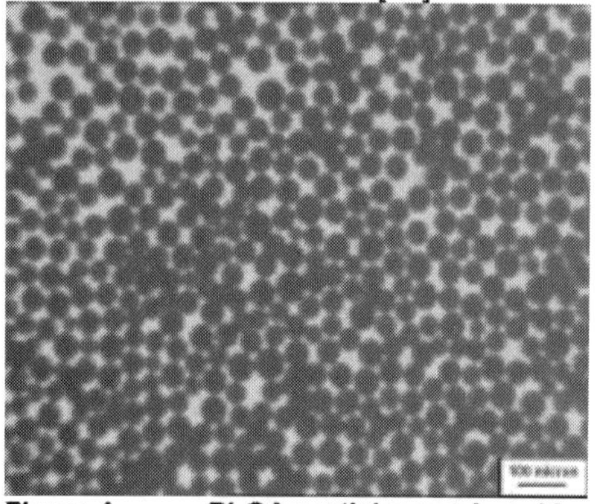

Figure 4 **PLGA particles produced by non-crossflow double membrane emulsification followed by solvent removal, encapsulating a water soluble model drug D(n,0.5)=50 µm Span=0.31**

Changing membrane pore size and other operating conditions it is possible to produce particles with a diameter of 100 µm or 50 µm with a span ranging from 0.3 to 0.5 with a best result of 0.29, hence the particles are very monosized.

Encapsulation Efficiency and Drug Release

At the end of the solidification, a sample of the continuous phase was filtered and analyzed with a UV spectrophotometer at 620 nm to quantify the amount of blue dextran escaped from the particles.

Encapsulation efficiency of less than 100% may be due to two effects: disruption of the primary W/O emulsion which is being emulsified into a W/O/W emulsion, and by the leaching out of material from the inner aqueous phase after the encapsulated particles have been formed. When an osmotic pressure is present, the inward flux creates pores through which the inner water phase can leak leading to a lower encapsulation efficiency. Polymer concentration and particle size affect the solidification rate. Higher PLGA concentrations leads to a quicker solidification period so that the inner water phase has less possibility to move towards the outer phase and also the oil phase is more viscous, so that diffusion through it is more difficult. Particle size has a double effect: bigger size means lower surface/volume rations, hence more difficult diffusion for the inner drug but also for the DCM, hence slower solidification. The resulting total effect of bigger particle size is a higher encapsulation efficiency. Results are shown in Figure 5 and in more detail in [10]. The osmotic pressure is expressed as outer/inner salt ratio. When it is 1 it means that there is the same salt concentration in the inner water phase and in the outer one, hence there is no osmotic pressure due to the salt. On the other hand when it is 1/3, the inner water phase has a salt concentration three times higher than in the outer phase, hence there is a strong osmotic pressure leading the water inside the particles.

Factors influencing the encapsulation efficiency have been highlighted and it is possible to achieve high encapsulations up to 100%.

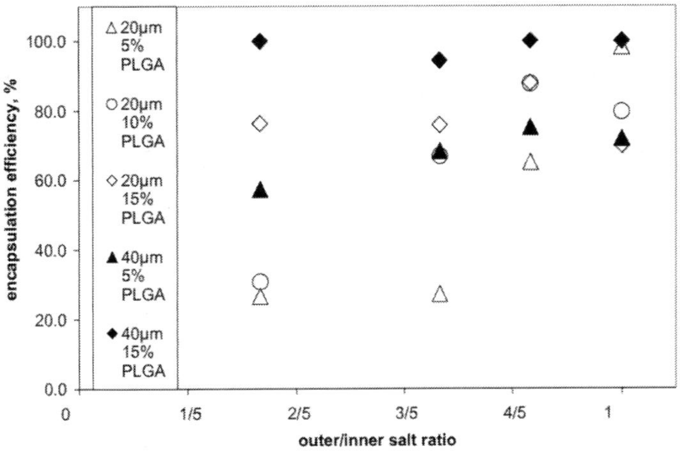

Figure 5 **Encapsulation efficiency dependency on osmotic pressure, PLGA concentration and particle size**

After the particles have solidified and been characterized, they were filtered and washed three times with water. Then they were divided into 10 samples, in each of them 25 ml of phosphate buffer (pH 7.3) was added with 0.1% of sodium azide to prevent bacterial growth. Finally they were placed in a shaking bath at 37°C. At regular intervals, one sample was taken out of the bath, the particles were analyzed to see change in size, size distribution and surface morphology. The external water was filtered and analyzed in the UV spectrophotometer to measure the release. This technique allows a detailed investigation of the change particles undergoing the release.

So far, the effect of two different PLGA concentrations, 5% and 15%, and the effect of two different salt ratios, 1/3 and 1, have been studied. As seen from the microscope and in agreement with the literature, PLGA degradation starts after 20 days (when 15% PLGA is used). During this time, the release is only due to leaking of the inner phase through pores or diffusion. From Figure 6 it is possible to see that before day 20, the particle size is actually increasing due to water intake. After this period the PLGA matrix starts degrading. After the 30th or the 46th day there were not enough particles left to be able to measure them.

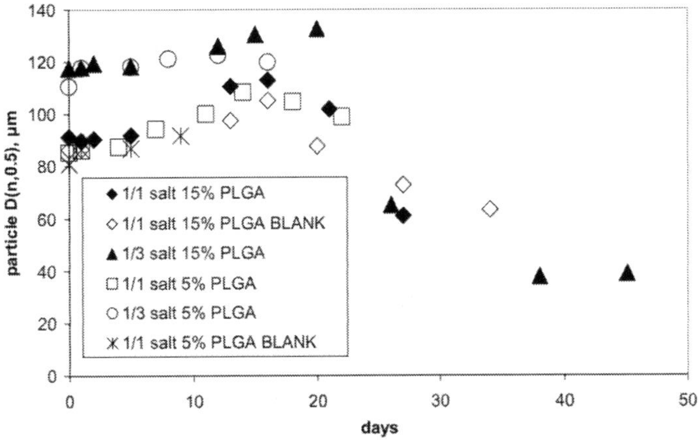

Figure 6 **Particle size changes during the release when different salt and PLGA concentrations are used during particle production**

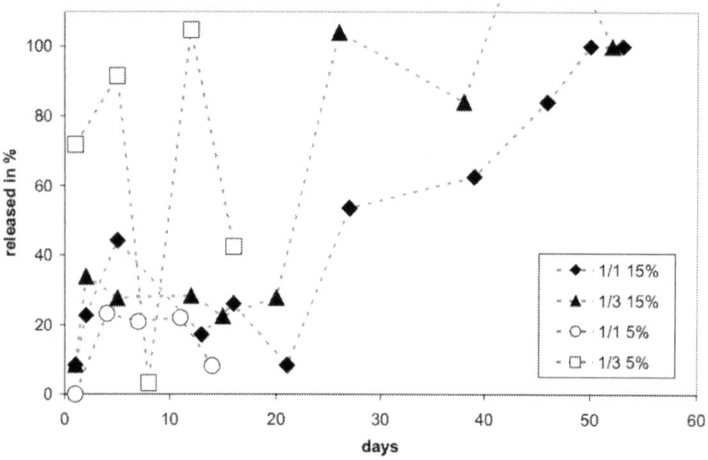

Figure 7 **Blue dextran release from PLGA particles produced with different PLGA concentration and different osmotic pressure during the preparation phase**

Following the release, Figure 7, after an initial burst, a lag time appears lasting until the 20th day. After this period the release continues. Coupling the release data and the size changes, it could be said that the lag time is due to an actual inward water flux which counteracts the release of the inner phase. More extensive studies are in progress to investigate this.

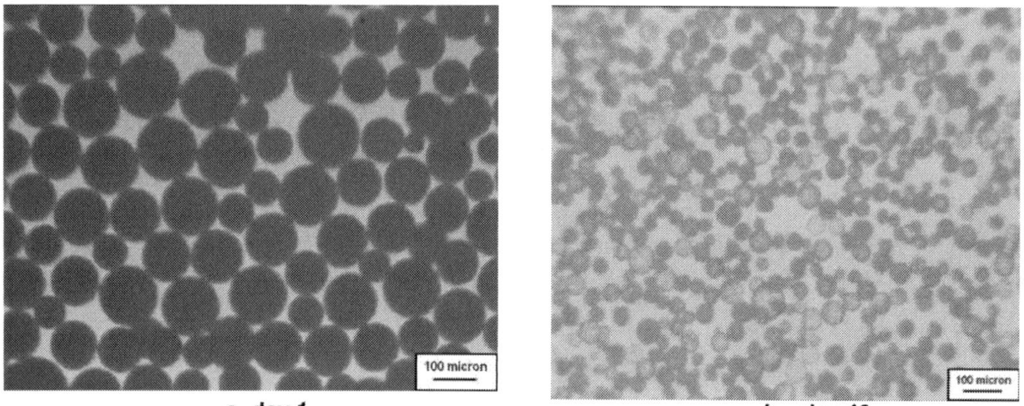

a- day 1 **b – day 46**

Figure 8 **Changes of PLGA particles during release, from a) day 1 to b) day 46: these particles were obtained using 15% PLGA and a outer/inner salt ratio of 1/3**

Conclusions

PLGA particles for encapsulation of controlled release drugs can be successfully produced by the non-crossflow membrane emulsification method presented. The size can be tailored and the particles are monosized. The encapsulation efficiency can be as high as close to 100%, this means that the droplets-particles are not disrupted during production. Operating conditions can be varied in order to obtain the desired target and a model has been proposed to predict the droplet size. Drug release studies are in progress and so far they show an initial swelling of the particles due to water intake. This swelling may act against the release, slowing it down until polymer degradation starts.

References

[1] Kang, F. and Singh, J., 2001. Effect of additives on the release of a model protein from PLGA microspheres. *AAPS PharmSciTech*, 2 (4), article 30.

[2] Dunne, M., Corrigan, O.I. and Ramtoola, Z., 2000. Influence of particle size and dissolution conditions on the degradation properties of polylactide-co-glycolide particles. *Biomaterials*, 21, 1659-1668.

[3] Berkland, C., Kim, K.K. and Pack, D.W., 2001. Fabrication of PLG microspheres with precisely controlled and monodisperse size distributions. *Journal of Controlled Release*, 73, 59-74.

[4] Kosvintsev, S.R., Gasparini, G., Holdich, R.G., Cumming, I.W. and Stillwell, M.T., 2005. Liquid-liquid membrane dispersion in a stirred cell with and without controlled shear. *Ind.Eng.Chem.Res.*, 44, 9323-9330.

[5] Van der Graaf, S., Sttegmans, M.L.J., Van der Sman, R.G.M., Schroen, C.G.P.H. and Boom, R.M., 2005. Droplet formation in a T-shaped microchannel junction: A model system for membrane emulsification. *Colloids and Surfaces A: Physicochem. Eng. Aspects*, 266, 106-116.

[6] Xu, J.H., Luo, G.S., Chen, G.G. and Wang, J.D., 2005. Experimental and theoretical approaches on droplet formation from a micrometer screen hole. *Journal of Membrane Science*, 266, 121-131.

[7] Schröder, V., Behrend, O. and Schubert, H., 1998. Effect of dynamic interfacial tension on the emulsification process using microporous, ceramic membranes. *Journal of Colloid and Interface Science*, 202, 334-340.

[8] Stillwell, M.T., Holdich, R.G., Kosvintsev, S.R., Gasparini, G. and Cumming, I.W., 2007. Stirred cell membrane emulsification and factors influencing dispersion drop size and uniformity. *Ind.Eng.Chem.Res.*, 46(3), 965-972.

[9] Keh, H.J. and Chen, P.Y., 2001. Slow motion of a droplet between two parallel plane walls. *Chemical Engineering Science*, 56, 24, 6863-6871.

[10] Gasparini, G., Kosvintsev, S.R., Stillwell, M.T. and Holdich, R.G.. Preparation and characterization of PLGA particles for subcutaneous controlled drug release by membrane emulsification. Accepted for publication in *Colloids and Surfaces B: Biointerfaces*.

493f Droplet-Based Microfluidics For Monodisperse Particle Synthesis

Nick Carroll[1], Christian Holtze[2], Sergio Mendez[1], Gabriel P. Lopez[1], David A. Weitz[2], and Dimiter Petsev[1]. (1) Chemical and Nuclear Engineering, University of New Mexico, Farris Engineering Center MSC01 1120, 1 The University of New Mexico, Albuquerque, NM 87131, (2) Dept. of Physics and DEAS, Harvard University, ESL 227, 40 Oxford St., Cambridge, MA 02138

Emulsion droplet formation in a flow-focusing microfluidic device allows for the creation of highly monodisperse drops whose volumes are in the nano- to picoliter scale. Each droplet can be considered as an individual chemical reactor. The introduction of polymer or particle precursors as the dispersed phase in droplet-based microfluidics presents a unique approach to particle synthesis.

The development of novel non-ionic fluorosurfactants allows for using fluorocarbon oils as a water-free continuous emulsion phase. Fluorocarbon based emulsions allow for a simple, versatile, and straight forward method for the templated synthesis of particles that in conventional systems require multiple processing steps or may not be made at all.

The promises of this approach are exemplified by the suspension polymerization of polyurethanes (PU), in which the liquid precursor is emulsified into droplets in a microfluidic device that are then converted into polymer particles. The droplet stability against coalescence upon removal of the continuous phase by evaporation confirms the formation of solid PU particles. This proves that the water-free environment of fluorocarbon based emulsions allows for high conversion. Monodisperse, cross-linked, and fluorescently labeled PU-latexes with controllable mesh size have been produced through microfluidic emulsification in a simple one-step process.

The ability to control dispersed phase chemicals and the surrounding continuous phase within a flow-focusing microfluidic device facilitates the synthesis of a variety of monodisperse microsphere particles. Potential synthesis includes silica nano-particles, silica gels with highly ordered nano-structures, and materials which must by synthesized in an anhydrous environment.

Simulant Development for Savannah River Site High Level Waste

R. E. Eibling
D.C. Koopman
D.P. Lambert
P.R. Burket
M.E. Stone
Savannah River National Laboratory, Aiken, SC 29808

The Defense Waste Processing Facility (DWPF) at the Savannah River Site vitrifies High Level Waste (HLW) for repository internment. The process consists of three major steps: waste pretreatment, vitrification, and canister decontamination/sealing. The HLW consists of insoluble metal hydroxides (primarily iron, aluminum, magnesium, manganese, and uranium) and soluble sodium salts (carbonate, hydroxide, nitrite, nitrate, and sulfate). The HLW is processed in large batches through DWPF; DWPF has recently completed processing Sludge Batch 3 (SB3) and is currently processing Sludge Batch 4 (SB4). The composition of metal species in SB4 is shown in Table 1 as a function of the ratio of a metal to iron. Simulants remove radioactive species and renormalize the remaining species. Supernate composition is shown in Table 2.

Table 1. Metal species in SB4 as Ratios to Fe

	40-inch Heel Case	**15C, Blend 1**	15C, Batch 1
Al	1.057	0.707	1.874
Ba	0.0088	0.0059	0.0067
Ca	0.0920	0.0919	0.0946
Ce	0.0108	0.0070	0.000
Cr	0.0115	0.0081	0.0109
Cu	0.0041	0.0027	0.00033
Fe	1.000	1.000	1.000
K	0.0953	0.0156	0.0410
La	0.0047	0.0048	0.0075
Mg	0.0433	0.0811	0.0385
Mn	0.249	0.228	0.238
Na	1.170	0.881	1.394
Ni	0.183	0.0666	0.0870
Pb	0.0122	0.00430	0.00033
Si	0.0859	0.104	0.237
Th	0.0027	0.0032	0.0092
Ti	0.0007	0.0011	0.0008
U	0.402	0.349	0.244
Zn	0.0061	0.0043	0.0003
Zr	0.0135	0.0097	0.0117

Table 2. Supernate Species and Solids Content of SB4

	40-inch heel case	**15C, Blend 1**	15C, Batch 1
SpGr, kg/L	1.0585	1.0410	1.0403
Na^+, M	1.3363	0.9180	0.9606
NO_2^-, M	0.5073	0.3970	0.4124
NO_3^-, M	0.2445	0.1910	0.2426
OH^-, M	0.3369	0.1770	0.1389
Cl^-, M	0.0013	0.0010	0.0007
SO_4^{2-}, M	0.0208	0.0220	0.0235
F^-, M	0.0019	0.0040	0.0004
CO_3^{2-}, M	0.0763	0.0510	0.0490
AlO_2^-, M	0.0598	0.0190	0.0178
$C_2O_4^{2-}$, M	0.0033	0.0080	0.0015
PO_4^{3-}, M	0.0012	0.0010	0.0005
K^+, M	0.0034	0.0040	0.0017
Insoluble Solids (wt%)	14.90	12.20	10.29
Total Solids (wt%)	21.82	17.41	15.90

Simulants are utilized to perform tests of the DWPF process. The sludge is non-Newtonian and is typically described as a Bingham Plastic. The yield stress of actual waste is typically higher than the simulants prepared by the current preparation technique. Improvement of the simulant was desired to allow more accurate representations of plant processes to allow optimization studies and to incorporate techniques developed during preparation of simulants for the Hanford vitrification process.

The original simulant preparation process is performed in four steps: 1) Manganese nitrate precipitation by addition of potassium permanganate 2) Ferric and nickel nitrate addition to the MnO_2 slurry followed by precipitation by addition of sodium hydroxide 3) Solids washing to remove soluble sodium and potassium nitrate 4) Trim chemical addition to add remaining metals (Al, Ba, Ca, Cr, Mg, Na, etc.) and anions (nitrite, sulfate, oxalate). Solids washing can be performed by gravity settling, decanting the supernate, and addition of dilution water; repeating as required to hit the targeted nitrate level. Continuous washing using a cross-flow or rotary microfilter can also be utilized. Small batches can be washed using a centrifuge to aid gravity settling.

The simulant testing proceeded in phases, based on available funding and need dates for process simulants. During Phase 1, the original preparation method was compared to revised processes to determine if the process was sensitive to changes and could be tuned to match the actual waste properties. Phase 2 consisted of testing the simulants through the DWPF Chemical Process Cell (CPC) process to ensure that the changes noted persisted through the process and additional refinement of the revised methods. Phase III was performed after scaling issues were identified during Phase II testing. Phase IV involved testing and implementation of a CSTR process for simulant production.

During Phase I, testing was conducted using a SB3 composition (similar to SB4, but with less Al and more Fe). A revised process was developed that added all species as metal nitrates prior to the manganese strike[1]. Trim chemical additions were limited to base-reactive species (silica and titanium oxide). Different washing techniques were evaluated during the development of the revised process, as well as inclusion of post treatment processes such as heat treatment at boiling and high shear mixing[2]. As shown in Figures 1 and 2, the revised recipe resulted in yield stresses above the actual waste samples (SB2/3 sample results). The baseline case for this testing was the precipitation of all metals except aluminum during the hydroxide addition. As shown in Figure 1, addition of aluminum to the precipitation process (Tests 7 and 8) lowered the yield stress while the application of shear during washing (Test 2 and Test 3) increased yield stress. Heat treatment provided a small benefit.

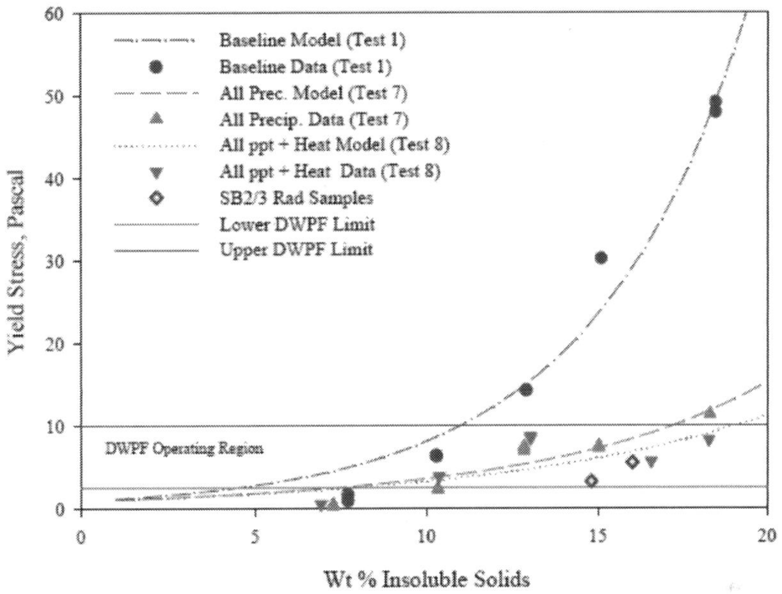

Figure 1. Phase I Test Results for Al Precipitation

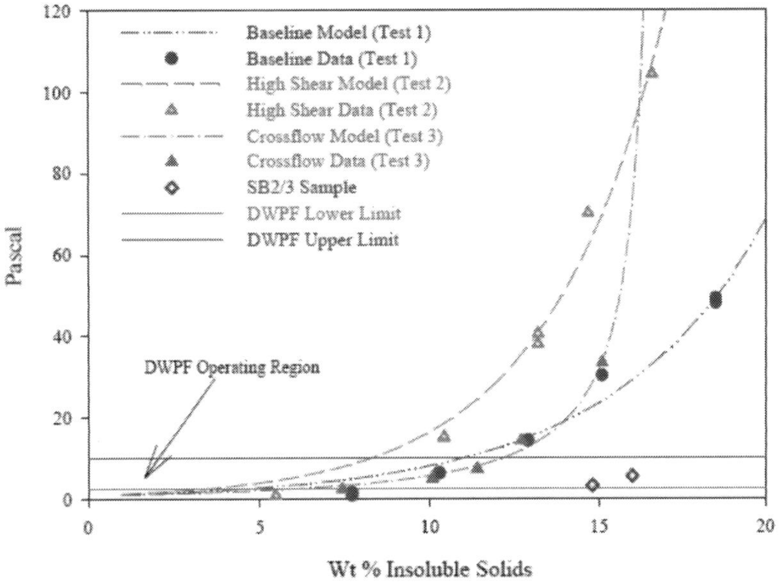

Figure 2. Phase I Test Results: Application of Shear

During Phase II testing, the physical property differences were shown to persist through the DWPF pretreatment process[3]. In addition, irradiation and thermal treatment processes were performed on the simulants to evaluate the impact of these treatments on physical properties. Irradiation did not impact the yield stress, particle size, or crystalline composition of the simulants. The thermal treatment that was performed in an attempt to accelerate any aging processes did impact the yield stress and crystalline makeup of the simulant. The impacts on yield stress were much smaller than the impacts of varying makeup and washing methods, but small benefits were noted as shown in Figure 3.

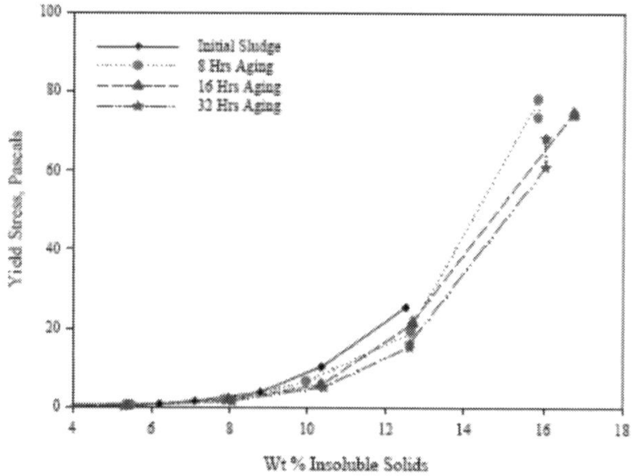

Figure 3. Impact of Thermal Treatment at 90°C on Yield Stress of SB3 Sludge Simulant

During the Phase II testing, a large batch of Sludge Batch 4 simulant was needed. Simulants produced by the original recipe were performing in a marginal manner; therefore, the revised recipe where all species were precipitated together was chosen to produce the needed simulant. The large batch exhibited extremely viscous behavior, likely due to scaling issues with mixing during the precipitation process. The precipitated slurry thickens considerably when the pH reaches ~4, and then thins as the pH exceeds 8. Mixing during the time period with high yield stress was difficult to maintain in the large scale batch.

Phase III testing was performed to evaluate the scaling issues, but efforts were hampered by the requirements to produce batches of SB4 simulant for testing. Modifications were made to the mixing protocols, addition rates of reagents, and the manganese strike prior to metal nitrate additions was reinstated. These changes led to improvements in yield stress as the batch size increased, but all batches greater than 2 liters were significantly higher in yield stress than the 2 liter batches[4,5,6,7], as shown in Figure 4.

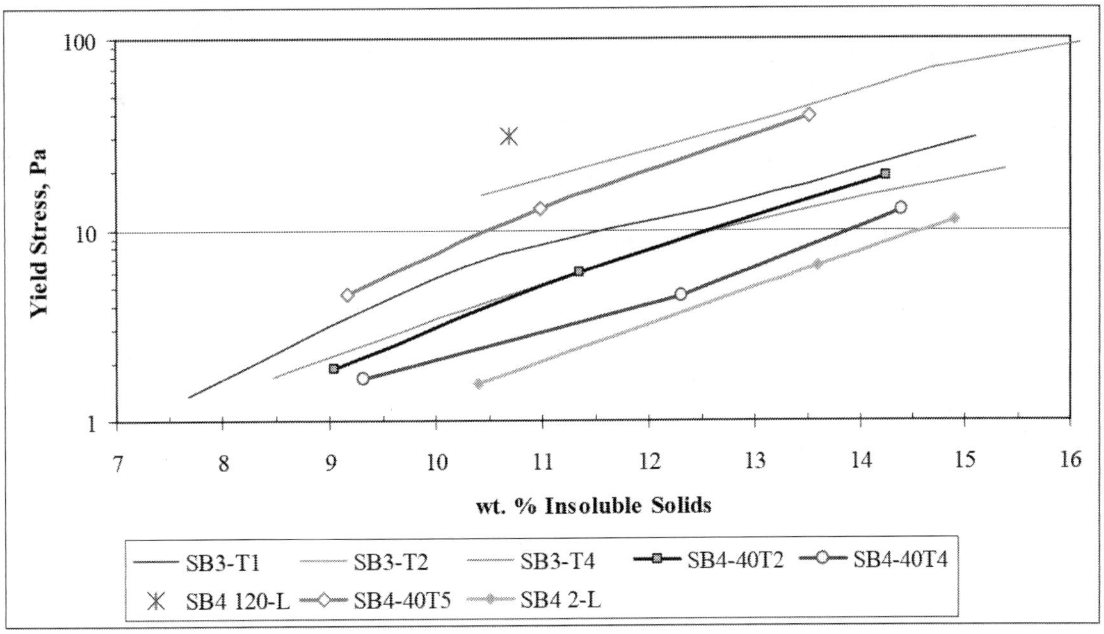

Figure 4. Yield Stress from Batches of SB4 Simulant

A small-scale continuous process was proposed and developed to allow large batches to be produced without the scale-up issues noted during the batch process. The continuous process consisted of a small (2L) CSTR as shown in Figure 5. The automated control system is not shown.

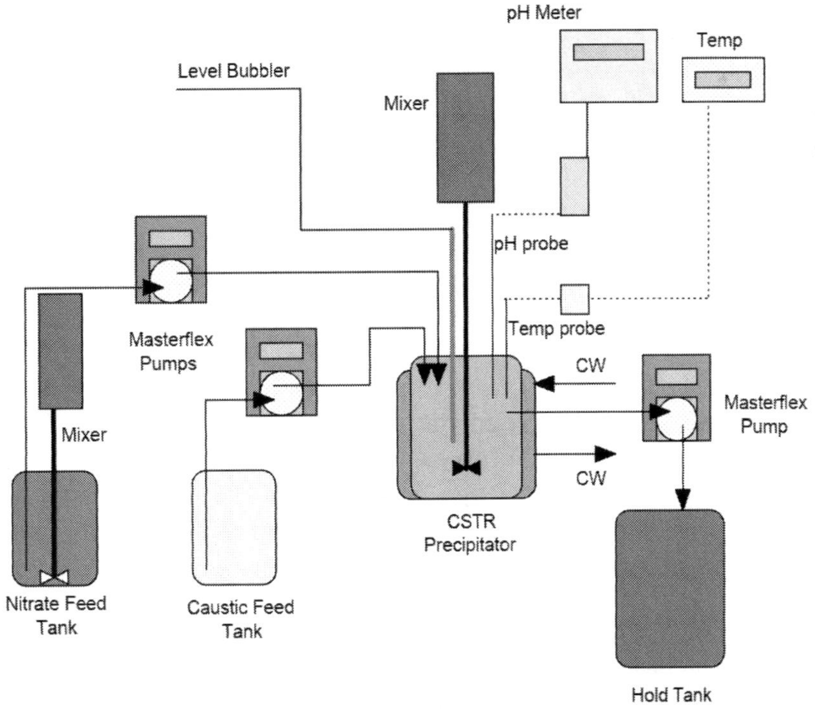

Figure 5. Diagram of CSTR Precipitation Apparatus

The manganese strike is performed in a batch manner in the nitrate feed tank, similar to the original preparation process. Metal nitrates are added to the manganese slurry and a small batch precipitation is then performed and used as the precharge to the CSTR vessel. The remaining metal nitrate slurry is fed to the CSTR concurrently with 50% NaOH to perform the hydroxide precipitation. An automated control system was implemented to maintain a constant level and pH in the CSTR. The CSTR is water-jacketed to maintain a constant temperature. The precipitated slurry is collected in an unmixed container without temperature control.

The CSTR process was utilized to produce large scale batches (20 to 100 liters) with yield stresses typical of those seen with the lab-scale batches (2 liters)[8], as shown in Figure 6 (line SB4-15C-MS2). The process has been successfully utilized to prepare simulants of SB4 and SB5 composition and has been adopted as the baseline process.

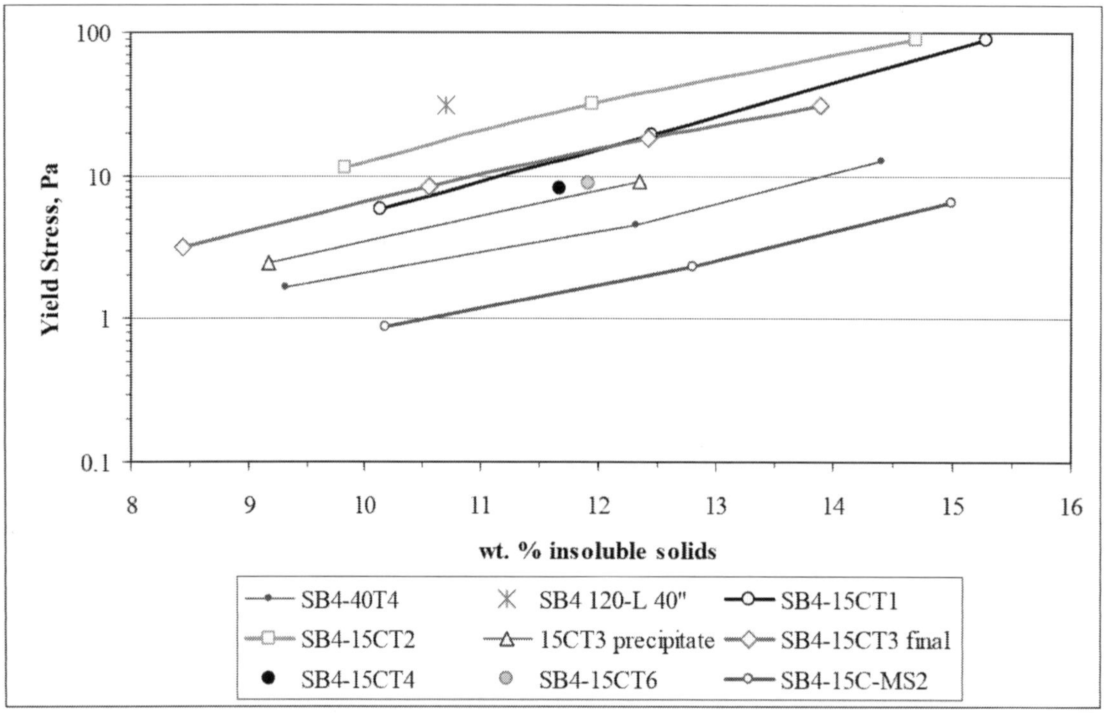

Figure 6. Yield Stress Results from CSTR Testing

Particle size measurements were made throughout the testing. General trends were seen in the data that shearing produced more fines while heat treatment removed some of the fines. The particle size distribution of the simulants was generally similar, with the exception of the CSTR runs. As shown in Figure 7, the particle size distribution was narrower for the CSTR than the batch process. Particle size measurements of actual waste indicate a peak at 3 microns, few fines below 0.8 microns, and a maximum particle size of 50 microns.

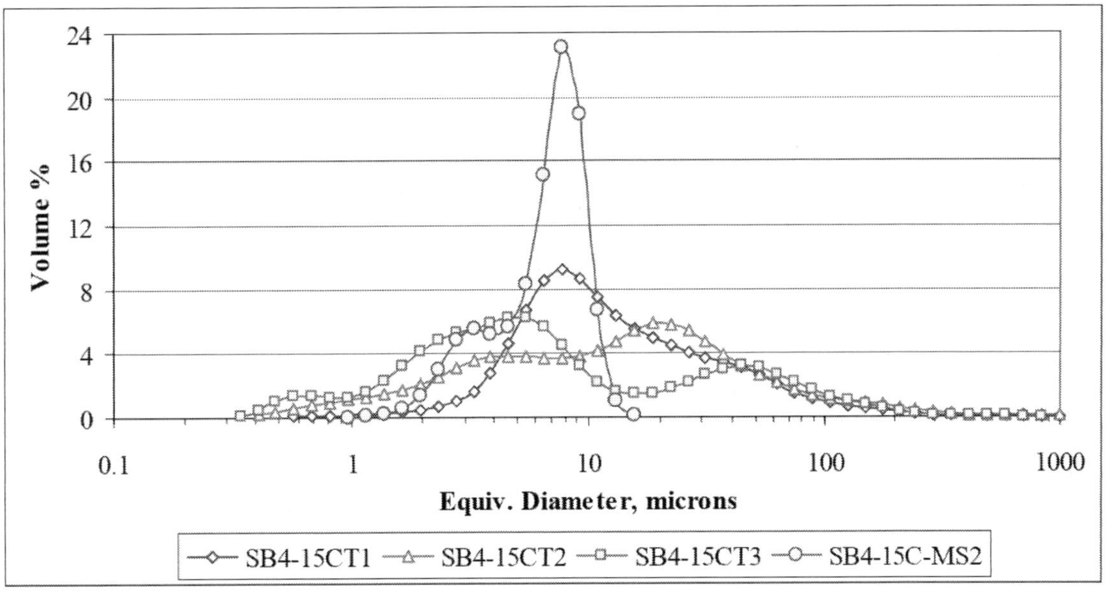

Figure 7. Impact of CSTR Process on Simulant Particle Size

Future testing will prepare a matrix of sludge compositions to determine the impact of varying the sludge composition on the simulant preparation process as well as the DWPF pretreatment and vitrification processes. Additional testing to refine the parameters for the CSTR operation and to scale-up the process to a 20L vessel is under consideration.

References

---

[1] Eibling, R. E., 2004,"Impact of Simulant Production Methods on the Physical Properties of DWPF Sludge Batch 3 Simulant", WSRC-TR-2004-00578, Washington Savannah River Company, Aiken, SC.

[2] Eibling, R. E. and M. E. Stone, 2006,"Impact of Irradiation and Thermal Aging on DWPF Simulated Sludge Properties", WSRC-TR-2005-00543, Washington Savannah River Company, Aiken, SC.

[3] Lambert, D. P., M. E. Stone, and Eibling, R. E., 2005,"Impact of Sludge Production Methods on SRAT Product", WSRC-TR-2005-00294, Washington Savannah River Company, Aiken, SC.

[4] Koopman, D. C. and Eibling, R. E., 2005,"Preparation and Heat Treatment of DWPF Simulants with and without Co-Precipitated Noble Metals", WSRC-TR-2005-00285, Washington Savannah River Company, Aiken, SC.

[5] Koopman, D. C., Lambert, D. P. , Best, D. R., Barnes, M. J., 2006,"Rheology Improvements during Preparation of 40-inch Heel Case Simulants for Sludge Batch 4", WSRC-STI-2006-00067, Washington Savannah River Company, Aiken, SC.

[6] Koopman, D. C., Lambert, D. P., Best, D. R., Barnes, M. J., 2006,"Impact of Preparation Method and Scale Factors on Sludge Batch 4 Simulant Properties", WSRC-STI-2006-00088, Washington Savannah River Company, Aiken, SC.

[7] Koopman, D. C., Lambert, D. P., Barnes, M. J., 2006,"Preparation of Sludge Batch 4 Qualification Simulants for DWPF Process Simulations", WSRC-STI-2006-00242, Washington Savannah River Company, Aiken, SC.

[8] Lambert, D. P., 2007,"SB5A Sludge Simulant: Basis, Preparation, and Chemical Processing", SRNL-PSE-2007-00130, Washington Savannah River Company, Aiken, SC.

AIChE 2007 Annual Meeting

Abstract

Authors (in alphabetical order)
San Kiang, Chiajen Lai*, Thomas LaPorte, Fernando Quiroz, Lori Spangler, Jean Tom, Zhongmin Xu

Seeding—Friend or Foe?

Seeding is used in the crystallization process to control supersaturation and particle size. Although some systems nucleate spontaneously, it is generally a good practice to seed. Seeding provides positive control of the crystallization process, and improves process robustness especially when scaling up.

During crystallization of a pharmaceutical compound, we found significant morphological differences between materials crystallized from seeded and unseeded processes. Crystals from seeded batches form "rosette" structures (agglomerated with flake-like building blocks), while those from unseeded batches appear to be primary particles (non-agglomerated) with significantly less surface area. The "rosette" crystals break up readily in an agitated filter-dryer and this leads to poor filtration. The cake heel consisted mostly of the fine, broken pieces of the rosette. Additional indication of the fragility of the rosette crystals comes from dry powder particle size data collected at different dispersive pressures.

The solubility diagram and metastable zone width (MSZW) were established for this compound as part of a process robustness study. It was found that the final particle size is relatively insensitive to variations in initial concentration and solvent composition at the point of seeding. Morphological differences and product particle size are mainly determined by the amount and size of the seeds. Large quantities of small seeds created small rosettes, while small quantities of large seeds produce large primary crystals. Our discussion will be focused on the effectiveness of seeding, and strategies that can be used to control the product surface area.

*presenting author

551b A High Intensity Emulsification Method For The Production Of Nanoparticles

Eric Jayjock, M. Silvina Tomasone, and Fernando J. Muzzio. Chemical and Biochemical Engineering, Rutgers University, PIscataway, NJ 08817

Recent advances in the ability to structure material on the nano-meter size scale offer the pharmaceutical industry a new tool set: (i) high potency drugs can be made into small controlled release spheres which can survive in the body for a long period and slowly release there payload keeping a steady safe concentration in the body (ii) low solubility (lipophilic) compounds can be made into nano-sized spheres which will have a very high surface area to volume ratio for improved solubility (iii) nano-sized drug carriers can be surface modified with specific proteins to target specific areas of uptake and greatly reduce the occurrence of side effects. While creating these functionalities is already a difficult task at the laboratory scale, scaling-up these techniques to the manufacturing scale poses a whole new set of challenges.

In this talk we focus on a method for creating nano sized particles at a rate suitable for full-sclae manufacturing. The basic mechanism of the Hydrodynamic Formation pathway is the creation of a nano-emulsion of which the droplets can be hardened to form nano-particles. The droplets of the emulsion are either the therapeutic agent themselves or the therapeutic agent and an encapsulation material. Initially the emulsion is rough and the droplets are on the micron size scale. The emulsion is passed through a static mixer at a high rate. The shear inside the mixer causes the micron-sized drops to be reduced down into the 50-1000 nano-meter range. After the size reduction the droplets are hardened by a phase shift brought about by a rapid change in temperature or solubility conditions, or by extracting a solvent.

This work focuses on identifying the critical parameters for control the width and size of the particles produced by the high intensity process. Through the directed design of the static mixers particular attention is payed to the role of cavitation vs shear on the resulting particle sized distribution.

551c Precipitation of Micro/nanoparticles by Enhanced Energy Dissipating Impinging Jets in Stirred-Tank Reactors

Giuseppe Di Benedetto[1], Ewa Sukcik[1], Piero M. Armenante[1], and Somenath Mitra[2]. (1) Otto H. York Department of Chemical Engineering, New Jersey Institute of Technology, 323 King Blvd., Newark, NJ 07102-1982, (2) Department of Chemistry and Environmental Science, New Jersey Institute of Technology, 323 King Blvd., Newark, NJ 07102-1982

Crystallization is a key unit operation of the pharmaceutical industry. Over 90% of all pharmaceutical products contain drug substances in particle form. In the pharmaceutical industry, there is a significant need to control the particle size distribution and the purification of a drug substance. This is accomplished through the control of crystallization. Unfortunately, the pharmaceutical industry works with new drug substances that are made up of complex organic compounds. These chemical compounds are typically hydrophobic. The chemical complexity of these entities creates difficulties for crystallization. The solutions to these difficulties are sought through the various equipment and techniques used to perform crystallization in the pharmaceutical industry.

In this study, an impinging jet mixer was utilized as a crystallizer within a stirred-tank reactor. The impinging jet mixer was composed of two jet nozzles arranged diametrically opposed to each other with the outlet tips of the nozzles directed to face each other. The jet nozzles were organized in a submerged arrangement within the stirred-tank reactor. A solvent/anti-solvent crystallization method was utilized for an active pharmaceutical ingredient (API), more specifically griseofulvin. The two temperature regulated solutions were pumped through the nozzles at a high linear velocity until they met each other head-on at the impinging point or plane. In order to achieve the goal of producing nanoparticles, a submerged impinging jet system coupled with sonication was utilized.

Such a configuration has been shown to produce micron-sized and nano-sized particles. The particles produced showed greater uniformity in the crystal size distribution, which is a very important parameter of crystallization studies. The impingement collision must occur while undergoing sonication to achieve nanoparticles.

Development of a Particle Size Analysis Method for Dry and Wet Granulation

Jennifer Wang, Niranjan Kottala, Robert Jerzewski, Emil Friedman

Laser diffraction particle size analysis technique has been used in a wide range of industries. This technique is known for its repeatability, ease of verification, in addition to speed of measurement. In the pharmaceutical industry, it has been used for particle size analysis of drug substances and also some unique specialty pharmaceutical finished products such as microsphere and nanoparticle dosage forms, aerosols, and inhalers. However, this technique has not yet been widely used for measurements of solid dosage forms despite the importance of particle size analysis in drug development and manufacturing of these products. The objective of this study was to develop a particle size analysis method for dry and wet granulation of solid dosage forms using laser diffraction technique.

For method development during this study, a Sympatec HELOS laser diffraction particle size analyzer was used. A typical milled and dried granulation made by high shear wet granulation was chosen as representative granules for the study. Samples were prepared by riffling into 5 g or 10 g portions using a spinning riffler prior to measurements. A 3^2 factorial design of experiments to include sample sizes of 5 g, 10 g, and 20 g as well as vibration levels at 18%, 22%, and 27% was conducted. Upon completion of the experiments, the RSD values of ten measurements of d10, d50, and d90 for all of the analytical conditions were evaluated. The results showed that a sample size of 10 g and vibration level of 27% provided the optimal testing conditions for dry granulation. The results also showed that vibration level had a larger effect than sample size on the RSD values of measurements. In addition, it was determined that the RSD values were sensitive to vibration levels for 5 g and 20 g sample sizes but not for 10 g sample size while 10 g sample size led to lower RSD at all vibration levels. The method developed in this study showed satisfactory robustness in terms of analytical conditions, repeatability, and reproducibility. During the study, the sample size varied from 8 g to 12 g with dispersion vibration level range of 22-32%. In terms of repeatability (day-to-day variation) by one analyst, the RSD values for d10, d50, d90 measured in three different days were 6.5, 1.6, and 0.7%, respectively. In case of reproducibility (analyst-to-analyst variation), the RSD values for d10, d50, and d90 measured by two different analysts were 2.6%, 3.4%, and 8.3%, respectively.

The method described in this study was successfully developed. It has been so far also proven to be suitable for particle size analysis of solid dosage forms such as wet granulation made with high shear wet granulation, dry granulation made with roller compaction, and direct blending. This study has shown that the pioneering laser diffraction approach can effectively measure the particle growth and agglomeration as well as breakage and attrition dynamics in high shear wet granulation modeling.

551e Computational Modelling Of The Anderson Cascade Impactor

Eric Jayjock, Justin P. Lacombe, and **Fernando J. Muzzio**. Chemical and Biochemical Engineering, Rutgers University, PIscataway, NJ 08817

The Delivery of active ingredients through the pulmonary pathway has received a considerable amount of attention in the past 10 years. The large amount of surface area and relatively quick uptake time make the lungs a good pathway for drug delivery. In recent years researchers have focused on delivering dry powder aerosol particles to the lungs. It is well understood that in this mode of drug delivery the particle size distribution of the active aerosol determines the success or failure of the drug delivery. This creates the need for tools to accurately and quickly test the particle size distribution being emitted from a dosation device. Currently the most commonly used device to try to determine aerosol size distribution is an Anderson Cascade Impactor.

Anderson Cascade Impactors is designed to impinge air onto a metal collection plate at increasing velocities as the air travels through the impactor. In the initial stage of the impactor, where the velocity is the lowest only larger particles with more thermal momentum impact on the collection plates. As the velocity of the air impinging on the surface increases smaller and smaller particles are collected. The plates can then be weighed for mass of impacted particle or sent for HPL analysis.

Inspite of these impactors serving a relatively important roll in dry powder delivery research little work has been done to evaluate the ability of these impactors to differentiate between different particle size distributions of similar size. The goal of this work is the creation of a model to predict the response of an Anderson Cascade Impactor to various dry powder particle size distribution.

In brief, the impactor is broken down into three area's of interest: 1. The entry where the Dry Powder is introduced referred to as the throat 2. the jet holes of each stage of the impactor where impaction is taking place and 3. the area around the edge of the impactor where the air is directed after impaction. Fluent is then used to simulate the flow of the in each area of interest as it would be during an impaction test. Finally, massed particles are tracked over the through the sections. When particles impact on the impaction surface they are counted as being stuck to the plate. After running many particles over all of the separated geometries a statistical model is generated for impaction in the real world ACI. The results of this model match well to data from the literature.

551f Controlled Nanoparticle Agglomeration As A Low-Density Dry Powder Aerosol Formulation Strategy

Lianjun Shi[1], Laura J. Peek[2], Carl Plumley[3], Mark Bailey[3], and **Cory Berkland**[2]. (1) Department of Chemical and Petroleum Engineering, University of Kansas, 2030 Becker Drive, Lawrence, KS 66047, (2) University of Kansas, 2030 Becker Drive, Lawrence, KS 66047, (3) Chemical and Petroleum Engineering, University of Kansas, 2030 Becker Drive, Lawrence, KS 66047

Nanoparticle agglomeration in solution represents a new strategy for formulating dry powder aerosols as an alternative to conventional spray drying, spray freeze drying, lyophilization, etc. Dry powders were formed by obtaining a suspension of drug nanoparticles or nanoencapsulated drug and destabilizing the colloid to form agglomerates of a desirable size in solution. Nanoparticle agglomerates were then filtered or concentrated and dried to form a free-flowing powder. A variety of API and excipients such as nifedipine, ciprofloxacin, insulin, stearic acid, poly(dl-lactic-co-glycolic acid), etc., were formulated as low density "nanocluster" powders (~0.1 g/cm3) exhibiting appropriate aerodynamic diameter (~1-2 µm) for pulmonary administration. Excellent control over agglomerate size and density was achieved as well by controlling shear, particle concentration, particle zeta potential, and the concentration of destabilizer (e.g. polymer or salt). Finally, nanocluster particles possessed desirable, irregular surface microstructure for facilitating powder dispersion, decreasing agglomeration, and decreasing adhesion to surfaces. Nanoparticle agglomerates or "clusters" represent a unique and simple method for utilizing nanoparticles in a dry powder formulation.

Process engineering aspects of plasmid-based biopharmaceuticals production: Tackling the threatening vaccine shortages to prevent global pandemics

Michael K. Danquah*, Gareth M. Forde

BEL (Bio Engineering Laboratory), Department of Chemical Engineering, Monash University, Wellington Road, Melbourne, 3800, Australia

Abstract

Infectious diseases such as SARS, influenza and bird flu have the potential to cause global pandemics; a key intervention will be vaccination. Hence, it is imperative to have in place the capacity to create vaccines against new diseases in the shortest time possible. In 2004, The Institute of Medicine asserted that the world is tottering on the verge of a colossal influenza outbreak. The institute stated that, inadequate production system for influenza vaccines is a major obstruction in the preparation towards influenza outbreaks. Because of production issues, the vaccine industry is facing financial and technological bottlenecks: In October 2004, the FDA was caught off guard by the shortage of flu vaccine, caused by a contamination in a US-based plant (Chiron Corporation), one of the only two suppliers of US flu vaccine. Due to difficulties in production and long processing times, the bulk of the world's vaccine production comes from very small number of companies compared to the number of companies producing drugs.

Conventional vaccines are made of attenuated or modified forms of viruses. Relatively high and continuous doses are administered when a non-viable vaccine is used and the overall protective immunity obtained is ephemeral. The safety concerns of viral vaccines have propelled interest in creating a viable replacement that would be more effective and safer to use.

Nucleic acid molecules are heralding a new generation of reverse engineered biopharmaceuticals. One example is plasmid DNA (pDNA) which has an excellent safety profile as it is free from safety concerns associated with viral vectors, displays no toxicity and is simpler to develop. Hence pDNA is considered a viable alternative to viral vectors. Reverse designed pDNA vaccines have the potential to revolutionalize production time and methods, aside being a safer way to invoke immune responses. The utilization of pDNA as a vector for the introduction of genes into mammalian cells via traditional immuno-response methods is widely reported. The increasing number of pre-clinical and clinical trials utilizing pDNA has resulted in pressure to make more in less time. Considerable attention has been given to the potential of pDNA vaccines to mitigate and prevent a number of infections but substantially less examination has been given to the practical challenges of producing large quantities to meet the current rising demand. Plasmid DNA is produced from fermentation processes and the success of this fermentation process hinges on the interactions between the host organism harboring the recombinant plasmid vector and the growth environment. A major advantage of fermentation is that conditions that influence cell growth, plasmid yield, quality and stability can be examined and controlled. These include media composition, temperature, pH, dissolved oxygen and build-up of waste metabolites. The growth medium formulation dramatically affects the performance and productivity of microbial processes, especially bacterial fermentation. Media containing yeast extract and hydrolyzed protein are often used because they are relatively simple to prepare and generates high cell densities. Meat extracts are also rich sources of nutrients for fermentation, but there is the risk of contamination with animal viruses. Commercially available growth media are usually employed for the fermentation of E. coli for pDNA production but most of these on-sale media are directed toward cell proliferation and/or protein expression and not pDNA replication. Due to the simplicity of use associated with these off-the-shelf media, a number of lab-scale pDNA vaccine production schemes rely on the use of un-optimized small-scale processes employing these complex formulations. Studies on the quantitative investigation of comparative effects of different media on cell growth and pDNA yield kinetics as well as economic consideration of medium formulation are limited. This limitation makes cultivation medium selection more of a "trial and error" process without any cost and yield analysis prior to fermentation. This constitutes generally to the low pDNA yield that is encountered even for expensive medium formulations and high copy number plasmids. Studies on pDNA yield kinetics of a particular medium can also dictate when fermentation can be halted for optimum performance rather than the normal random stoppage which constantly produces small quantities of pDNA.

Molecular biologists have developed efficient lab-scale protocols for pDNA purification which employ operations such as ultracentrifugation and solvent extraction. Although routinely used for obtaining pDNA for research applications, these protocols are not scalable or use reagents such as phenol, ethidium bromide, RNase and CsCl that are expensive or not compatible with regulatory guidelines. Further disadvantages include long processing times, limited capacity, low recovery and high cost per dose. The creation of a commercially viable purification process for pDNA requires a scalable technique employing optimized stationary adsorption phase(s) in a minimum number of processing steps without the use of expensive and toxic chemicals, avian and bovine derived enzymes. It is in this light that a chromatographic technique for rapid pDNA purification is essential.

The isolation and purification of pDNA is hampered by the low performance of conventional chromatographic supports with a small particle pore diameter. Most of these chromatographic supports are made for the high adsorption capacity of proteins with size less than 10 nm. In columns packed with such supports, large molecules such as pDNA with sizes greater than 100 nm adsorb predominantly at the particle outer surface. Consequently capacities are in the order of tenths of mg pDNA/mL support compared to hundreds of mg/mL for proteins.

A monolith is a continuous phase consisting of a single piece of a highly porous organic or inorganic solid material. The most important feature of this support is that all the mobile phase is forced to flow through the large pores of the monolith. As a consequence, mass transport is enhanced by convection; dramatically reducing the long diffusion time required by particle-based chromatographic supports. Therefore, the chromatographic separation process on monoliths is practically not diffusion-limited. The large pores of these monoliths allow penetration of large pDNA molecules to the internal surface area at high flow rate with low pressure drop. Different types of monolithic supports are currently available based on preparation and chemistry. These are silica-based, polyacrylamide-based and gel-based monolithic resins. Polymethacrylate monolithic support is an optimal adsorbent for pDNA separation. These adsorbents have large pore diameters, thus no significant hindrance to convective mass transport. They are resistant to pH, non-toxic, economically favorable to synthesize and can be easily modified by functionalizing with anion-exchange, hydrophobic interaction or affinity ligand. The flexibility and the ease to tailor their pore and surface characteristics to the target pDNA molecule through alteration in synthesis conditions make them more attractive.

In this current study, different standard media (LB, TB and SOC) for culturing recombinant E. coliDH5alpha harboring pUC19 were compared to a medium optimized for pDNA production. Lab scale, shake flask fermentations using the standard media showed that the highest pDNA volumetric and specific yields were for TB (11.4 mg/l and 6.3 mg/g dry cell mass respectively) and the lowest was for LB (2.8 mg/l and 3.3 mg/g dry cell mass respectively). A fourth medium, PDMR, designed by modifying a stoichiometrically-formulated medium with an optimized carbon source concentration and carbon to nitrogen ratio displayed pDNA volumetric and specific yields of 23.8 mg/l and 11.2 mg/g dry cell mass respectively. However, it is the economic advantage of the optimized medium that makes it so attractive. Keeping all variables constant except growth medium and using LB as a base scenario (100 MC units / mg pDNA), the optimized PDMR medium yielded pDNA at a cost of only 27 MC units / mg pDNA. These results show that greater amounts of pDNA can be obtained more economically with minimal extra effort simply by using a medium optimized for pDNA production. Results from pilot-scale fermentations endorsed the suitability of the PDMR medium for plasmid production on the commercial level.

A monolithic sorbent was synthesized via free radical copolymerization of ethylene glycol dimethacrylate (EDMA) and glycidyl methacrylate (GMA) with pore diameter tailored specifically for plasmid binding, retention and elution. The polymer was functionalized with 2-Chloro-N,N-diethylethylamine hydrochloride for anion-exchange direct isolation of pDNA from clarified lysate obtained from E. coli DH5alpha-pUC19 culture in a ribonuclease/protease-free environment. Characterization of the monolithic resin showed a porous material with 68 % of the pores existing in the matrix having diameters above 300 nm. The final product isolated from a single-stage rapid purification was a pure and homogenous supercoiled pDNA with no endotoxins, gDNA, RNA and protein contaminations as confirmed with EtBr-AGE, enzyme restriction analysis, SDS-PAGE and limulus amoebocyte lysate.

The extent of exothermicity associated with the construction of large-volume methacrylate monolithic columns has somewhat obstructed the realisation of large-scale rapid plasmid purification. A novel synthesis technique via a heat expulsion mechanism was employed to prepare a 40 mL methacrylate monolith with a homogeneous radial pore structure along its thickness. Radial temperature gradient was recorded to be only 1.8 °C. Maximum radial temperature recorded at the centre of the monolith was 62.3 °C, which was only 2.3 °C higher than the water bath temperature. Pore characterisation of the monolithic polymer showed uniform pore size distributions at different radial positions. Chromatographic characterisation of the polymer after functionalisation with amino groups displayed a persistent dynamic binding capacity of 15.5 mg/mL. The maximum pressure drop recorded was only 0.12 MPa at a flow rate of 10 mL/min. The polymer demonstrated rapid separation ability by fractionating E. coliDH5alpha-pUC19 clarified lysate in only 3 minutes after loading. The plasmid sample collected from the fast purification process was tested to be a homogenous supercoiled plasmid with DNA electrophoresis and restriction analysis. This cost effective and non-toxic technique is cGMP compatible and commercially-viable for rapid production of pDNA.

553b Continuous Thermal Epimerization in High Temperature Tubular Reactor

Amy Jines, Chemical Product Research and Development, Eli Lilly and Company, Inc., 1400 W. Raymond Street, Indianapolis, IN 46221

A thermal epimerization reaction was conducted in a continuous thermal tube reactor. A research scale reactor was utilized to rapidly screen reaction temperatures in the range 100 to 320 degrees C and residence times in the range 10 to 60 minutes. This proved to be a powerful tool for defining reaction kinetics. Optimum reaction conditions were 300 degrees C and 15 minutes residence time. Subsequently the process was rapidly scaled up to 200 g/h rate by increasing tube dimensions. This continuous process offers significant advantages over the previous batch process, including safety, speed to scale up, and operating temperatures and pressures outside normal bounds for typical batch reactors.

553c Continuous Newman Kwart Rearrangement In Thermal Tube Reactor

Martin Johnson, Chemical Process Research & Development, Eli Lilly &Co., Inc., 1400 W. Raymond St., Indianapolis, IN 46221

A continuous Newman Kwart rearrangement was conducted in a thermal tube reactor in supercritical DME at 300 degrees C and 1000 psig. This process offered significant safety advantages compared to the previous batch process, speed to scale up, selectivity, and yield. Compound was produced at a rate of 50 g/h for early phase delivery.

553d Controlling The Key Response Variables Through Scaling Rules During The Scale-Up Of A Fluidized Bed Granulation Process

Ecevit Bilgili[1], Jeff Ko[1], Adam Chen[1], Edward Smith[1], Pavol Rajniak[1], Kyle Fliszar[2], Gordon Wong[3], Larry Rosen[1], Daniel Grubb[1], and Dafni Bika[1]. (1) Pharmaceutical Commercialization Technology, Merck Manufacturing Division, Merck & Co., Inc., WP78-110, P.O. Box 4, West Point, PA 19486, (2) Analytical Chemistry and Development, Merck & Co., Inc., WP78-110, P.O. Box 4, West Point, PA 19486, (3) Center for Materials Science and Engineering, Merck & Co., Inc., WP78-110, P.O. Box 4, West Point, PA 19486

This paper presents recent work to develop a fluidized bed process for co-granulation of two highly water-soluble active pharmaceutical ingredients (APIs). The granulations were produced at three scales corresponding to batch sizes of 2 kg, 13 kg, and 130 kg. The effects of binder concentration, binder addition rate, and inlet air temperature were investigated at the 2 kg scale. Based on the process knowledge generated at this scale, four granulations were produced at the 13 kg scale to determine the effects of binder addition rate and moisture level on the process, while using the same superficial air velocity. A Malvern Insitec RTSizer was used to determine the droplet size distribution of binder solutions, which was then used as a guide in our scale-up procedure. The scale up from 13 kg to 130 kg was based on maintaining similar excess air velocity, moisture level, and droplet size distribution. The above-produced granulations were subjected to various analyses: particle size distribution, porosity, SEM imaging, sieve assay of the APIs, bulk and tap density.

The granule growth was monotonic, and the granule size distributions exhibited self-similar character. A simple one-parameter population balance model explained the growth behavior. In general, the granules were very porous, relatively fragile, and free-flowing. Downstream processing of the granules in compression led to acceptable tablet attributes. At higher moisture levels, a drop in granulation yield was observed due to an increased loss into the filters and column surfaces. This, along with other process observations, points to an optimum moisture level for fluid bed granulation.

The paper will highlight both the practical issues faced during the development and the applicability of various proposed scaling rules. Excess air velocity, moisture level in the powder bed, and the droplet size distribution appear to be the key response variables, which are governed by various equipment, process, and formulation parameters. The presentation demonstrates the success of scaling rules for maintaining the key response variables, and thus the granule size distribution upon scale-up.

553e Development And Scale-Up Of Efficient Api Drying Using Fluid Bed Dryer

Yubo Yang, Jenny Lau, Ilaria Popolla, Gerrie Liaw, Melissa Ostberg, and Stephen Tyler. Abbott Laboratories, 1401 Sheridan Road, North Chicago, IL 60064-6292

The API wet cake, isolated from a mixture of water, ethanol, and small percentage of heptanes and other solvents, has a LOD approximately 50% and is dried with an agitated drier under vacuum and heat. The dried API from the agitated drier in production typically produces small "ball" shaped materials, which makes the drying operation less efficient and takes over 2 days to dry a batch. The demand for the API has increased significantly in recent years, thus, additional drying capacity and alternative drying technology is required to meet the increased demand. This led us to investigate an alternative and more efficient drying technology to dry the wet cake, and to seek an alternative milling technique to consistently produce the desired particle size. Fluid bed dryer (FBD) was chosen because it allows for faster drying and produces a consistent pre-milled particle size distribution.

We will present the detailed development and scale-up of FBD drying technology to dry the API wet cake, from lab scale to pilot scale and finally to production scale. Many obstacles have been overcome successfully, such as potential degradation of API by hot FBD air, flowability of wet cake for delumping and automatic feeding, efficient removal of residual heptanes to minimize environmental emissions, and handling of the hygroscopic dry API. The outcome of this project is that an efficient drying procedure using fluid bed dryer has been developed and scaled-up successfully to commercial production with a dramatic reduction of drying cycle time (to approximately 3 hours per FBD cart), producing a consistent particle size distribution with desired physical properties and minimum environmental emission. The efficient removal of heptanes prior to FBD drying will be also discussed. This work significantly increases the throughput of API drying in commercial production.

553f Quality Risk Analysis of a Tablet Coating Process

Steve S. Y. Wang, Steven H. Chan, and San Kiang. Engineering Technology, Bristol-Myers Squibb Company, One Squibb Drive, New Brunswick, NJ 08903

In this presentation, we will demonstrate how the Failure Mode and Effect Analysis (FMEA) technique can be adopted as a tool for the management of quality risk associated with a tablet coating process. The drug product process chosen for this analysis, involves the coating of an Active Pharmaceutical Ingredient (API) onto core tablets. Content uniformity is a critical quality attribute. The FMEA analysis examines all the processing parameters that can impact the content uniformity in the final drug product. Other critical quality attributes are also identified, with their corresponding systems and subsystems. The risk analysis, using the FMEA approach, quantifies each failure mode, and ranks them according to an overall risk priority number (RPN). PAT applications, mechanistic model development, design of experiments (DOE), and other specific engineering controls that can be used to mitigate quality risks with high RPN's will also be discussed.

553g Novel Scaling of Rotary Drum Film Coating Developed from Commercialization of Pharmaceutical Tablets

Alan R. Silverman[1], Jeffrey Givand[1], Brad Holstine[1], and Michael Gentzler[2]. (1) Pharmaceutical Commercialization Technology, Merck & Co., Inc., PO Box 4 Sumneytown Pike, Mailstop WP78-110, West Point, PA 19486, (2) Center for Material Science and Engineering, Merck & Co., Inc., PO Box 4 Sumneytown Pike, Mailstop WP75B-210, West Point, PA 19486

The scale-up of pharmaceutical tablet film coating is examined within the context of maintaining the tablet elegance observed at the pilot scale. Tablet elegance was defined by several physical characteristics: surface finish, tablet damage, and color uniformity. Successful bridging from 24" (pilot) to 64" (commercial) diameter pans required a novel scaling rationale based upon a constant spray flux, bed temperature, and tablet coating mass variance. This new method incorporates tablet variables and was demonstrated across an eight fold range of core tablet mass. In contrast, scale-up based on literature recommendations of controlling process parameters - outlet temperature or relative humidity, spray/air ratio, and peripheral drum speed was relatively unsuccessful.

References:

Turton, R., Cheng, X. X. (2005). "The scale-up of spray coating processes for granular solids and tablets." Powder Technology, 150, 78-85

Levin, M, ed. (2002). "Pharmaceutical Process Scale-up." Drugs and the Pharmaceutical Sciences, vol. 118, Marcel Dekker, Inc.

565a Oral Absorption of Weakly Acidic Compounds

Monica L. Adams, James Ormes, and Ajit Thakur. Biopharmaceutics Research and Development, Bristol-Myers Squibb, 1 Squibb Drive, New Brunswick, NJ 08903

Dosage form design begins with suitable understanding of the biopharmaceutical characteristics of the therapeutic agent under consideration. While chemical and physical properties govern biological performance of a specific compound, it is often possible to modulate oral absorption through a variety of techniques to maximize the dissolution characteristics of the drug in the formulation. Following oral administration, drug substances are exposed to a range of pH conditions which impact the ionization state of drug molecules as they pass through the gastrointestinal tract. For poorly water soluble drugs, particle size control is important to enhance dissolution from the solid dosage form. Poorly soluble drugs containing weakly acidic chemical functionalities are subject to limited aqueous solubility under gastric pH conditions (i.e., pH < pKa). General strategies for overcoming limitations related to oral absorption will be highlighted. As examples, two weakly acidic compounds with relatively high pKa values (BMS-A and BMS-B) will be discussed. Both particle size reduction and salt formation were explored in order to achieve suitable blood levels. The in vitro dissolution characteristics of the various forms as well as the rate and extent of oral absorption were evaluated in a canine model. While dissolution studies could often be used to rank-order in vivo performance in terms of oral exposure, the relative exposure of salt forms versus free acid forms could not be adequately predicted by simple in vitro dissolution experiments.

565b Oral Delivery Of Insulin-Transferrin Bioconjugates Using Intelligent Complexation Hydrogels

Justin P. Shofner and Nicholas A. Peppas. Department of Chemical Engineering, University of Texas at Austin, 1 University Station C0400, Austin, TX 78712

Complexation graft copolymer of poly(methacrylic acid-g-ethylene glycol), designated as P(MAA-g-EG), has been shown to be effective in oral delivery of theraputic proteins. Their hydrogen bonding complexation/decomplexation characteristics render these responsive hydrogels able to protect the insulin in the harsh, acidic environment of the stomach before releasing the bioactive agent in the small intestine. Further, it can inhibit the activity of Ca2+ dependent proteolytic enzymes, increase the residence time of the drug in the small intestine by mucoadhesion, and reversibly open the tight junctions between the intestinal cells. Transport studies using Caco-2 cells, a widely used in vitro model for intestinal absorption of drugs, have proven the efficacy of the polymeric system. Further enhancement in the insulin bioavailability may be obtained by conjugating it to molecules that can recognize specific receptors on the epithelial cells and get transported across the intestinal epithelium. This approach eliminates the side effects associated with prolonged opening of the tight junctions. Since the conjugate complex crosses the cell barrier by the transcellular mechanism and is specific towards the protein of interest, unspecific transport of toxic compounds is avoided. Researchers have investigated receptor-mediated transcytosis of the insulin-transferrin conjugate in Caco-2 cells. Transferrin is a naturally occurring protein that is involved in the uptake of iron by the cells. The iron-bound transferrin (holo-transferrin) binds to the specific receptor on the epithelial cells and the complex is then endocytosed into the cells. The use of a complexation hydrogel as a delivery vehicle for the insulin conjugates is under investigation. The transferrin molecule (~80 kDa) is significantly bigger than the insulin (~5.8kDa). Hence, as a preliminary step towards developing the delivery system, the loading and release profile of the transferrin from the swollen and deswollen polymer microparticles was investigated by spectroscopic methods. Loading of the microparticles near the iso-electric point of the protein resulted into high loading efficiencies. Understanding the loading and release of transferrin from the complexation hydrogels is important in developing a delivery system for the insulin-transferrin conjugates. The loading and release profile of the insulin-transferrin conjugates from the polymer microparticles was also investigated. It was found that altering the polymer synthesis parameters such as crosslinker, solvent to monomer ratio, particle size, and specific copolymer had an effect on the release profile of the conjugates. Based on these release profiles, the synthesis parameters were adjusted to provide the maximal release of the insulin-transferrin conjugates from the complexation hydrogels.

This work was supported by a NIH grant (EB 000246) and a NSF/IGERT Fellowship to JS

565c Development Of A Computer Program Predicting The Influence Of Cyclodextrins On Oral Bioavailability Of Insoluble Drugs: From Modeling To In Vivo Comparison

Ece D. Gamsiz, Deparment of Chemical Engineering, Northeastern University, 342 Snell Engineering Center, Boston, MA 02115, Lee A. Miller, Pfizer Global Research and Development, Ann Arbor, MI, Avinash G. Thombre, Pfizer, Inc., Eastern Point Road, Groton, CT 06340, and Rebecca L. Carrier, Department of Chemical Engineering, Northeastern University, 342 Snell Engineering Center, Boston, MA 02115.

Drug delivery technologies such as cyclodextrins, lipid-based systems, and polymers have been used to deliver drugs with low oral bioavailability. Low oral bioavailability is often caused by poor absorption in the gastrointestinal (GI) tract. Dissolution and permeation of the drug through the intestinal membrane are the key processes affecting drug absorption in the GI tract environment. Predicting the influence of a drug delivery technology on oral bioavailability of an insoluble drug would be highly beneficial in selection of drug delivery technology and device design. Cyclodextrins (CD) are cyclic oligosaccharides which form inclusion complexes with a large variety of compounds including drugs. There are many studies in the literature showing that complexation between cyclodextrin and drug enhances oral bioavailability and some demonstrating failure of CD in bioavailability enhancement, but relatively little guidance regarding when CD can be used to enhance bioavailability.

Four models were developed considering oral delivery of insoluble drug (either neutral form or saltform) in the presence of CD in different dosing schemes (physical mixture vs. pre-formed complex). These models describe: 1. dosing neutral compound as a physical mixture with cyclodextrin (NCPM), 2. dosing a saltform of a drug as a physical mixture with cyclodextrin (SCPM), 3. dosing a neutral compound as a pre-formed complex (NCC) and 4. dosing a saltform of a compound as a pre-formed complex (SCC). In these models, dissolution is described by the Noyes-Whitney equation and absorption is considered as a first-order process. Complexation between drug and CD is considered to be in pseudo-equilibrium. Precipitation is described by first-order kinetics proportional to supersaturation. Simulations were run for different ranges of parameter values reflecting the physical and chemical properties of drug, CD and the GI tract environment. Simulations results were analyzed using statistical software to understand which parameter combinations are effective in enhancing drug absorption. In order to predict the oral bioavailability of insoluble drugs, a pharmacokinetic (PK) model was incorporated into each model.

In order to test the ability of the models to predict the oral absorption of insoluble drugs, in vitro tests (dissolution, precipitation and absorption) were performed for each key process as well as combinations of the processes. In the experiments, insoluble drugs such as Naproxen, Nifedipine and Spironolactone were tested either in neutral form or saltform depending on the model being validated. For example, for NCC model validation pre-formed complex of Naproxen with B-CD was prepared, and dissolution, precipitation and absorption in simulated intestinal fluids were tested. Simulation results accurately predicted the in vitro experimental results for all four models.

Simulations were run to compare predictions with in vivo experimental results taken from the literature to assess the ability of the models to predict oral bioavailability of insoluble drugs. For example, in vivo experimental results for Carbamazepine delivered in the presence of HP-B-CD in rats were compared with NCPM/PK model predictions. Carbamazepine was orally administered as a physical mixture with HP-B-CD to rats[1]. NCPM/PK model results agreed with in vivo experimental results for Carbamazepine in both the presence and absence of CD. The results showed no influence of CD on drug bioavailability. Depending on the physical and chemical properties of drug, CD and biological environments, the models indicate the positive, negative or lack of effect of the drug delivery technology on oral bioavailability of insoluble drugs. The information obtained from the models would be highly useful in the process of formulation design.

1. Choudhury S, Nelson KF 1992. Improvement of oral bioavailability of carbamazepine by inclusion in 2-hydroxypropyl-b-cyclodextrin. International Journal of Pharmaceutics 85(1-3):175-180.

USING PHYSICALLY-BASED DISSOLUTION SIMULATIONS TO DEVELOP IDEAL PARTICLE DISTRIBUTIONS IN TABLETS

DANIEL BRAIDO, ALBERTO CUITINO

ABSTRACT. The purpose of this investigation is to develop modeling and simulation tools to conduct in-silico design of delivery systems for pharmaceuticals, in particular tablets. A basic component is the systematic correlation of end product quantities (e.g. dissolution profile) with the physical characteristics of the tablet (e.g. spatial distribution of particles). In this paper we describe a methodology to determine optimal active distribution for a prescribed dissolution profile. This methodology integrates physically-based solvers that track the effects of fluid penetration with nonlinear optimization techniques. This methodology opens the possibility of aiding in the design process of tablets with desired or required performance.

The computational tool initially simulates tablets with a predefined distribution of active nano-particles on a three dimensional fixed grid. A steady-state concentration is used to develop a localized flux for each of the nano-particles, and determine the amount of time it would take for each of the nano-particles to dissolve completely. All of the nano-particle volumes are reduced accordingly in steps and particles below a critical volume are removed from the simulation, until all of the nano-particles have dissolved completely. The outputs of these simulations are then compared to the desired output and the differences in composition to determine a rate of change. This rate of change is then used to develop a new particle distribution whose output is closer to the desired output. This process is repeated until the desired output is reached to within an acceptable tolerance.

The simulation produces profiles for active release, water concentration and surface fluxes at every time step. The program is easily adaptable to multiple grid sizes, active nano-particle distribution concentrations, and diffusion properties through several global constants. The backsolving algorithm is capable of dealing with particle distribution profiles of varying complexity.

This program provides a good basis for future extensions and demonstrates feasibility of the method. The process can be adapted into larger simulations incorporating a moving fluid boundary layer, variable tablet geometry, and composite particle systems. The backsolving algorithm is removable from the dissolution program and is easily adapted to future dissolution simulations.

INTRODUCTION

As the pharmaceutical industry works towards achieving more online monitoring of its manufacturing processes, there is an increasing need to correlate properties of tablet structure with tablet performance. Furthermore, in understanding which aspects of tablet structure are alterable, it is possible to engineer tablet formulations to achieve a desired performance. To this end we have developed a physics based simulation which incorporates several levels of interaction to create a meaningful evaluation of tablet properties. The simulation deals with surface erosion, solvent penetration and particle diffusion(all of which are modeled using the same cubic mesh) to develop a release profile for the active ingredient. The surface erosion is handled using a level set method. The current version of the simulation involves a spheroidal tablet in a motionless solvent which acts as a perfect sink. This is represented in the surface boundary conditions, which treat everything outside the tablet as having an active concentration of zero. Once the solvent moves past the surface, its progress into the tablet is tracked based on Fick's Second Law. The increasing levels of solvent inside the tablet affect the rate at which the internal particles dissolve. The active particles are individually tracked inside the tablet. Each particle is represented as its own node in the internal mesh. The particle diffuse based on the same Fick's law diffusion formula as the solvent penetration. The simulation steps forward through time until all of the particles have dissolved. This simulation is the basis for a backwards solving algorithm. The algorithm takes as input tablet properties and a desired release profile. The algorithm utilizes a bracketing method to determine alternative active particle distributions which produce release profiles increasingly closer to the input desired release profile.

LEVEL SETS

A large focus of this work is the proper implementation of level sets for the outer surface of the tablet. Level sets are well suited to this problem due to the simplicity with which the method can be utilized and parallelized while tracking complex changes in three dimensions. Level sets can operate on the same square grid matrix as all the other calculations. This reduces the mathematical complexity of the problem without sacrificing higher level resolution. Such simplification will be even more important when the method is moved to a more efficient language and parallelization is possible. Parallelization is vital to any computationally intensive modeling as it will always be easier to add more processors to a computer than to make a processor run that many times faster. The simplicity of parallelization of the level set formulation hinges on the single mesh representation of the volume being considered. Level set value, tablet concentration, excipient concentration, active concentration; all necessary parameters can be included in the same grid matrix. In addition, the level set need only be updated near surfaces of the tablet, and the calculations necessary are based only on grid points near these surfaces. The main value of interest of a level set is 0(zero). The surface being tracked is located at the zero level set value. By only calculating changes near the surface the number of operations being performed can be greatly reduced, thus reducing the computational cost of the algorithm. As the formulation is being developed on an evenly spaced cubic lattice, the lattice can be broken down into several smaller cubes with a small amount of overlap and sent to multiple processors for simultaneous solving. Level sets not only track the propagation of the initial surfaces, but also register the formation of any new surfaces in or on the tablet. This is useful in identifying the formation of any cracks in the surface or fissures developing inside of tablets. Formation of new surfaces will lead to increased dissolution activity and can vastly affect the release profile of a tablet. Essentially the rate at which the level set changes is based on how fast the surface is moving as well as the current slope of the level set. The inclusion of the slope of the level set allows for appropriate changes to be made only near the tablet surfaces without compromising the meaning of the remaining cells in the level set. In the example, a level set of first order space convex is implemented. The main equation in operation is:

$$\phi_{ijk}^{n+1} = \phi_{ijk}^{n} - \Delta t \left[max\left(F_{ijk}, 0\right) \bigtriangledown^{+} + min\left(F_{ijk}, 0\right) \bigtriangledown^{-} \right] \quad (0.0.1)$$

Where:
ϕ_{ijk} represents the value of the level set
Δt is the timestep
F_{ijk} represents the overall magnitude of the surface velocity
$\bigtriangledown$are the positive and negative directions of the surface velocity

First the gradient of the level set is calculated using a central differencing technique. Then the concentration gradient is calculated again utilizing a central differencing scheme. The concentration gradient is then used to create a normalization factor. The individual gradients of the level set are then divided by this normalization factor to produce pointer matrices. The pointer matrices represent each dimensions contribution to the overall velocity of the surface as a function of each associated node's position in reference to that surface. The concentration gradients are then multiplied by their corresponding pointer matrix to produce each dimension's velocity matrix. This velocity matrix represents the rate at which the surface is moving between the given nodes in their given direction. The three velocities are then combined into a single matrix, F, in order to be inserted into the main equation. Then the most positive and most negative slopes of the level set at each point are calculated, these are the Deltap and Deltam matrices. The dL is calculated using the overall corrected velocity matrix, F, and the associated direction matrices. This rate of change is multiplied by the time elapsed and then subtracted from the current values of the level set to produce the level set values at the current timestep.

Solvent Penetration

After the surface effects which are handled by the level set formulation, the next phase of the simulation deals with solvent penetration. The present iteration models solvent penetration via a centrally based finite differencing method. The method is based on Fick's second law:

(0.0.2)
$$\frac{\partial C}{\partial t} = \alpha \frac{\partial C}{\partial V}$$

This equation uses the current solvent concentration values of the six surrounding nodes and a prescribed coefficient to determine the rate of change of the solvent concentration with respect to time. While both the porosity and hydrophillicity of the tablet are described by the diffusion coefficient, this value is not necessarily the same at all points inside the tablet. Since the coefficient is stored in matrix form, local differences in solvent absorption affinity are represented by differences in their respective coefficient. In terms of boundary and initial conditions, it is assumed that the concentration of solvent outside the surface of the tablet is always '1'(pure solvent), and that initially the internal concentration is '0'(no solvent). As time progresses the internal concentration gradually increases until it reaches a value of '1'. This equation is the most frequently iterated as it has, by far, the smallest critical timestep. Initially, for approximately every 1000 iterations of this equation, the other two main equations are updated once. As the shock conditions soften, the timestep becomes progressively closer to the critical timesteps for the particle diffusion and level set operations.

Particle Diffusion

Once the solvent has begun to penetrate the surface of the tablet, the API begins interacting and eventually dissolving. The internal concentration is developed using a steady state approximation. Assuming there is no change in the solvent concentration or particle size, the internal concentration matrix is relaxed to represent a constant concentration distribution. The specific solvent concentration used for a particle diffusion based timestep is determined by the average of each cell's current solvent concentration and the solvent concentration at the previous timestep. The steady state concentration is then used in conjunction with each particle's surface area to determine their individual volumetric rate of change, $\frac{dV}{dt}$.

(0.0.3)
$$\frac{dV}{dt} = 4\pi M C_s A$$

Where:
M is the diffusion coefficient

C_s is the solvent concentration
A is the surface area of a particle
The matrix containing particle radii is then updated, and if the radius of any particle has dropped below a threshold value, that particle is eliminated from the particle matrix.

Release Profile. The release profile of the tablet is determined using the particle radii matrix. At each particle diffusion timestep the total volume of the active particles remaining is calculated. This value is then subtracted from the initial volume of the active particles, divided by the initial total volume and then multiplied by 100 to give the percentage of total active released. The below figure shows some sample release profiles generated by the simulation.

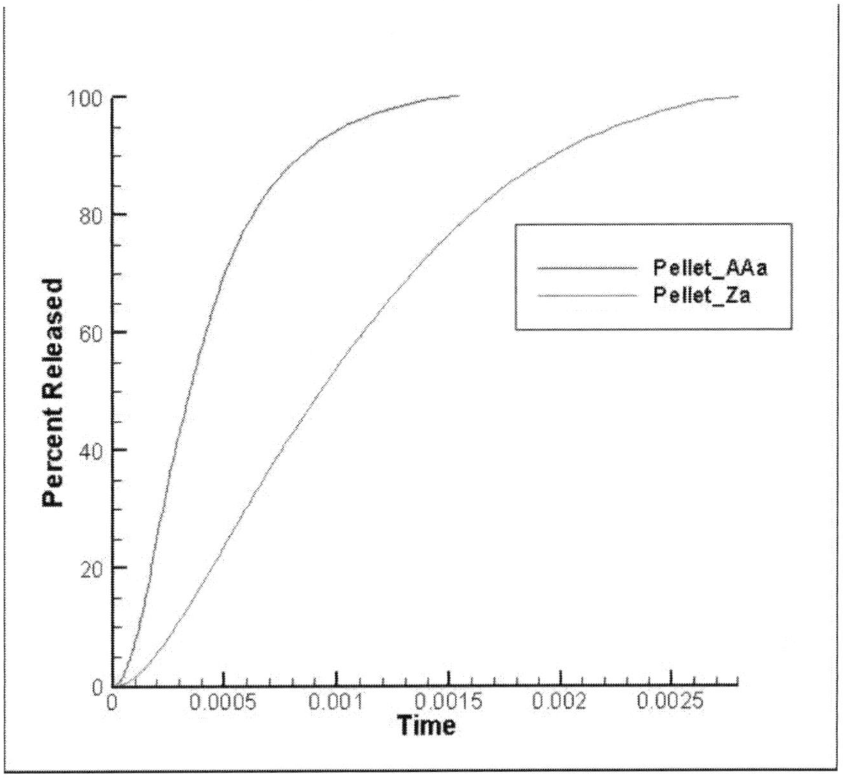

FIGURE 1. Samples of typical release profiles produced by the simulation

SUBJECT CONSTRUCTION

As in the physical world, in order to examine a tablet numerically, it is first necessary to build that tablet. The tablets to be simulated need to be defined on several levels. First, the surfaces are defined using R-functions. In most cases, a basic shape is defined using standard geometric equations. Several basic shapes can be added or subtracted from one another to form more complex shapes. Such is the case with the most prevalent shape used in the simulations, which was made by adding the volumes of a sphere and a disk. The surfaces are defined by a cell value of zero. This makes it easy to replace the given values with a smoother function ranging from negative one at the center of the tablet to one at the edge of the defined cell, with the surface again being equal to zero. Thus the initial value of the level set is defined. Next the tablet must be filled with particles of API. Since each particle is represented individually at a specific node, the number of particles is defined as a percentage of the total number of nodes

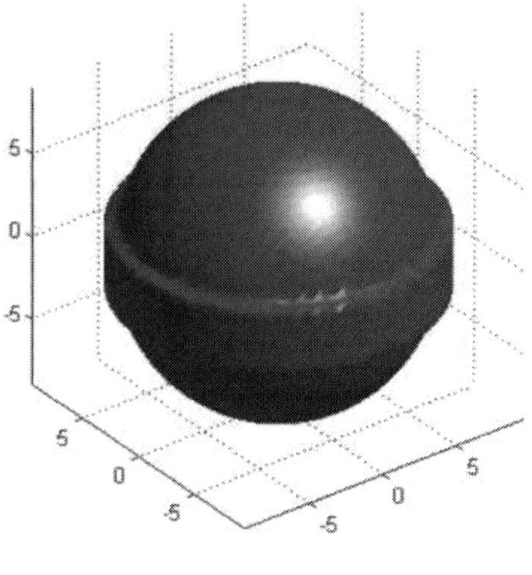

FIGURE 2. Outer surface of example tablet as represented by level set

inside the tablet. The particles are then placed in a semi-random manner about the tablet. In most cases the tablet has been divided into a series of volumetrically equivalent shells and an equal number of mono-sized particles are placed in each shell at random locations. This particle location matrix is then associated with a set of matrices representing the radius (and therefore volume and surface area as all the particles are assumed to be perfect spheres) of the particles and the solvent penetration coefficient. All of this initial data is then stored in a file to be accessed by the simulation.

BACKSOLVING ALGORITHM

Once a generally acceptable model of tablet dissolution has been created, the next step is to use this numerical simulation to solve the problem of tablet formation. That is to say given a desired release profile, what is the desired active concentration distribution of a tablet necessary to achieve this profile. One such back-solving method involves an iterative tablet creation/dissolution algorithm. The user first inputs a desired release profile as a function of time, as well as any specific requirements for the tablet such as total number of particles of a given size, diffusion coefficients, initial particle distribution and any other conditions necessary for the tablet. Once an initial guess tablet has been constructed, it is dissolved via the dissolution simulation. Concurrently, an additional tablet slightly different from this initial guess, a "perturbation" tablet is also dissolved via simulation. A simple example using a linear concentration distribution of uniformly sized active particles was performed to test the viability of such a method. The distribution for this example was modeled by the equation:

$$C = \alpha_0 + \alpha_1 L \tag{0.0.4}$$

Where:
α_0 is the baseline concentration
α_1 is the slope of the deviation from this concentration
L is a function of radial location in the tablet based on the level set

In this instance the tablet is divided radially into shell regions. These shells exist in between prescribed radii, similar to the multiple layers of a golf ball or jawbreaker, and all are of the same volume. The dissolution performance of the "perturbation" tablet is compared to the desired dissolution profile as a measure of goodness. Localized differences are focused upon so that the next iteration will produce a curve that not only matches the overall output desired, but conforms to desired shape of the output profile. This is achieved via a Riemann summing technique. The absolute difference at every timestep stored is summed up in an effort to reduce the likelihood of producing a curve which behaves counter to the desired output. That is to say the method in place would discourage the production of an output curve which spent most of the release time either above or below the desired curve producing the same overall output, as opposed to one which may produce a slightly different overall output, but followed the shape of the desired curve more precisely. The calculated difference between the profiles is then correlated with the changes made to the active concentration distributions in the tablet. This ratio is then used to determine a new guess for active concentration distribution in the tablet. With the baseline concentration representing a constant concentration for the middle shell, particles are either moved from the more inner shells to the more outer, or vice versa in accordance with the effects of the initial perturbation. This new tablet is then simulated, and the sequence repeated until the desired release profile is achieved to within an allowable tolerance. This method allows for automated determination of active concentration distribution in a tablet, which could prove a highly valuable tool to drug developers.

CONCLUSIONS

Physics based simulations will become increasingly useful as the cost of computing decreases. The simulation described here is able to produce a large variety of release profiles while only manipulating a few variables of tablet structure. As the simulation grows in complexity, so will the value of its output. In the sample configurations examined by this simulation changing the active concentration distribution had similar effects on the shape of the release profile to changes in particle size.

Acknowledgment. This work was completed as part of my Master's Degree study as an NSF IGERT fellow in the program for Nano-Pharmaceutical Engineering at Rutgers University.

565e A New Hypromellose Excipient For Direct Compression Controlled Release

Karl Jacob[1], Kacee Ender[2], Tim Cabelka[2], Dan Denomme[2], and Dave Wallick[2]. (1) The Dow Chemical Company, 1319 Bldg, Midland, MI 48667, (2) Water Soluble Polymers R&D, The Dow Chemical Company, Larkin Lab, Midland, MI 48674

A common issue with controlled release direct compression excipients is their poor flowability. The goal of this project was to develop a new hypromellose excipient for direct compression (DC) controlled release formulations and to evaluate the performance of the new direct compression hypromellose in a direct compression tabletting process by comparing it to a standard METHOCEL(TM) K4M Premium controlled release grade in the same application. By appropriate application of particle engineering techniques, a better flowing excipient was prepared with improved flow as measured by reduced mean time to avalanche in an Aeroflow tester. In addition, the new excipient (in a metoprolol tartrate formulation) flowed freely through the tablet press hopper without the application of any external vibration. The variation in tablet weight was reduced from 31 mg to 4 mg (target tablet weight of 400 mg) with the improved excipient. The dissolution profiles of the new DC grade excipient and the standard controlled release grade nearly overlay (62-63% release after 5 hours) indicating comparable controlled release. The new DC hypromellose excipient improved the flowability of the directly compressed metoprolol tartrate formulation resulting in reduced tablet-to-tablet variability, while maintaining sufficient controlled drug release performance and tablet hardness.

565f Sublingual Drug Delivery; Will It Always Give 'Fast Onset'? A Product Enhancement Perspective

Jeremy A. Bartlett, Oral Products Center of Emphasis, Pfizer Inc., Eastern Point Road MS#8156-042, Groton, CT 06340

In general, absorption of drugs from the oral cavity via solutions, sublingual, fast dissolving, or buccal dosage forms into the mucosal tissues is typically a fast event, while transport of drug from the mucosal tissue into the systemic circulation can be rate limiting to overall systemic bioavailability. While it is common practice to consider administration via the oral mucosa (e.g. sublingual) when a short Tmax or rapid onset is desired, the approach may not always be appropriate depending on the chemical entity.

This presentation will briefly discuss the clinical models available for studying drug transport via the oral mucosa and highlight the similarities and differences to drug absorption via the GI tract. This fundamental understanding of drug transport via the oral cavity is needed to help explain and model pharmacokinetic data that may be generated following administration to the oral cavity. A review of clinical Tmax values for drugs administered via the oral cavity was completed. This data review has allowed for further modeling and correlations to be drawn between chemical structure and Tmax, which may offer the ability to predict a priori certain PK parameters following oral cavity administration based on physiochemical properties alone. If the goal in a drug development program is rapid PK onset, such a model may allow for better prediction of which drug candidates have a higher probability of achieving this goal, and which candidates are likely to be less successful based on physiochemical properties alone.

616a Conversion Of Chord Length Distribution To Particle Size Distribution In Needle Systems

Sze Wing Wong and Christos Georgakis. Department of Chemical and Biological Engineering, Tufts University, Science and Technology Center, 4 Colby Street, Medford, MA 02155

Conversion of Chord Length Distribution to Particle Size Distribution in Needle Systems

Information regarding the time-evolving Particle Size Distribution (PSD) of the product is of crucial importance in any crystallization processes. The real-time measurement of the PSD along with the use of population balance equation provides valuable information about the nucleation and growth kinetics. By modeling the complex crystallization kinetics, the final PSD can then be accurately predicted and the process controlled to achieve the desired one. The final characteristics of the PSD are critical since it directly affects the ease of solids handling in downstream operations. It also affects the product quality such as the dissolution rate of an Active Pharmaceutical Ingredient (API). Although the available on-line analytical tools, such as Focus Beam Reflectance Measurement (FBRM), provide real time measurement capability, additional steps are required to convert the raw Chord Length Distribution (CLD) data into useful information such as PSD. This presentation addresses a new methodology developed to convert Chord Length Distribution (CLD) obtained from the FBRM measurement into PSD. We are also comparing the results to two other existing methods.

In this presentation, the forward CLD-to-PSD modeling method is proposed to estimate PSD from the measured CLD with FBRM. It is different form the exiting inverse modeling methods[1], [2] in that it does not attempt to invert the ill-conditioned matrix that characterized the PSD-to-CLD information. It assumes a probability density function of the particles with 2 or 3 unknown parameters. It also assumes a constant aspect ratio as all other methods have done. It estimates the 2 or 3 parameters probability density function by minimizing error between the measured CLD and the calculated one. Furthermore, the forward CLD-to-PSD modeling method is capable of estimating PSDs of crystals with two different crystal shapes. This is particularly important in cases where the crystallization system has undergone secondary nucleation, polymorphic transformation, or another diastereomer has appeared in the slurry.

The advantage of the forward CLD-to-PSD modeling method is that it does not invert the conversion matrix that is ill-conditioned in nature. As a result, the solution will not have oscillation or sometime result in negative number of particle size. In order to quantify the conversion accuracy, the proposed method was compared with two inverse modeling methods, the regularization method[1] and the Project Over Convex Sets (POCS) method[2], in eight different simulated cases and with real FBRM measurements from one crystallization experiments. The simulated cases were necessary to evaluate the relative merits of all three approaches. The experimental cases, even though they are affected by measurement errors, provide a cross-examination of the approaches discussed.

In all eight simulated test cases, we focused on a needle crystal system and we considered different realistic examples of particle size distributions (PSD). Knowing the shape of the crystals, one can calculate the CLD of a single particle or of an assumed PSD. The calculated CLD is compared to the measured CLD and the probability density function parameters are set to minimize the difference between the two CLDs. In the experimental case, the CLDs measured by the FBRM instrument in a crystallization experiment were used. It should be noted that the estimated PSDs from all methods are compared with the PSDs obtained via image analysis. The PSD differences were due to three causes, CLD measurement errors, conversion errors, and PSD measurement errors.

The forward CLD-to-PSD modeling method can easily be applied on-line for process modeling and control purposes. It requires minimal computation time of less than 30 seconds. With the implementation of the forward CLD-to-PSD model on-line, the PSD can then be controlled on-line within specification provided the real-time measurements of PSD and solute concentration are made available.

---

[1] Hukkanen, E. J. and R. D. Braatz (2003). "Measurement of particle size distribution in suspension polymerization using in situ laser backscattering." Sensors and Actuators B **96**: 451-459.

[2] Worlitschek, J., T. Hocker, et al. (2005). "Restoration of PSD from Chord Length Disribution Data using the Method of Projections onto Convex Sets." Particle & Particle Systems Characterization **22**(2): 81-98.

616b Observation Of Polymorphic Change Through Analysis Of Fbrm Data: Applications

Stephanie Barthe, Martha Gallivan, and Ronald W. Rousseau. School of Chemical & Biomolecular Engineering, Georgia Institute of Technology, Atlanta, GA 30332-0100

Polymorphism is an important aspect of crystallization processes used in the pharmaceutical industry, as many solid drugs exist in different crystalline forms. It is widely recognized that product properties, some of which may be related directly to the utility of the drug, and downstream processes, such as tableting, are influenced by crystal morphology, size, and shape. For these reasons and others, being able to detect a polymorphic transformation in situ would have great value.

Focused-beam reflectance measurement (FBRM) is among the process analytical technologies (PAT) that hold promise for enhanced monitoring of pharmaceutical crystallization. It is based on scattering of laser light and provides a methodology for in-line monitoring of a representation of the crystal population in either batch or continuous crystallization systems. If properly installed, it allows in-line determination of the chord-length density, which is a complex function of crystal geometry. In batch crystallizations, which are especially important in pharmaceutical processing, FBRM instrumentation, offers in situ, real-time information regarding the evolution of crystal size distribution.

In the present research, batch cooling crystallization of paracetamol from ethanol solutions is used as a model system for exploring the utility of FBRM data in detection of the formation and transformations of polymorphs. Paracetamol is known to exist under three different polymorphic forms: a stable Form I (monoclinic lattice), a metastable Form II (orthorhombic lattice) and an unstable Form III. Form II is preferred to Form I because of superior tablet-forming properties due to a higher compressibility. Unfortunately, since Form I is more stable, it is the dominant product form.

Our experiments found that both Forms I and II were nucleated upon batch cooling of a paracetamol-ethanol solution. However, Form II disappeared shortly after nucleation as it underwent a solution-mediated transformation and left an end-product that was uniformly Form I. By slightly modifying the experimental conditions, Form II was nucleated in larger amounts and was retained for a longer period of time. The two forms have different crystal habits; Form I crystals are octahedral, Form II crystals are needle-like. Such different shapes generate different chord length densities thus a transition from one form to the other can be detected through an efficient use of the FBRM.

By varying the conditions at which crystallization was induced, it is possible to induce nearly exclusive nucleation of Form II. Tracking the chord-length distribution of the resulting crystal population makes it possible to monitor the transition between Form II and Form I.

616c Relationship between Particle Size Distribution and Chord Length Distribution

Pui Shan Chow, Zai Qun Yu, and Reginald B.H. Tan. Crystallization and Particle Science, Institute of Chemical & Engineering Sciences, 1 Pesek Road, Jurong Island, Singapore, Singapore

Focused Beam Reflectance Measurement (FBRM) has become increasingly popular in pharmaceutical production and research as an in situ particle sizing tool. The interpretation of FBRM data, however, is far from certain and consistent. The objective of this study is to investigate the dependence of chord length counts on the number of particles and particle size, and to elucidate the relationship between particle size distribution (PSD) and chord length distribution (CLD). A series of particle samples with known sizes were prepared. They were added to a crystallizer equipped with a FBRM probe in such a sequence and proportion that from the perspective of FBRM, crystallization events such as nucleation and crystal growth were taking place separately or simultaneously. FBRM data, i.e. counts of chord length and chord length distribution, could be scrutinized in light of the true particle size distribution and the number of particles since they were known at each moment. Perfect spheres as well as real particles commonly used in crystallization research were employed. Three scenarios in crystallization were simulated by blending particle samples in a stirred vessel, that is, pure nucleation, pure crystal growth, simultaneous nucleation and crystal growth. With mono-sized particles, there was a linear relationship between the counts of chord lengths and the number of particles. At the same time, the counts of chord lengths were affected by particle size apart from particle counts. The statistics of CLD such as weighted or un-weighted mean chord length and mode could not be linearly correlated with the statistics of PSD when both particle counts and PSD were changing. The CLD of a mono-sized population deviated significantly from theoretical values as obtained in first principles methods to restore PSD from CLD. FBRM could be used to follow crystallization processes, but it was extremely difficult to extract two critical process variables, i.e., the number of particles and PSD, either by empirical or first principles methods without supplementary information from other process analytical technologies.

616d In Situ Psd Monitoring At High Solids Concentration Using Video Imaging And Statistical Estimation

Paul A. Larsen and James B. Rawlings. Department of Chemical and Biological Engineering, University of Wisconsin-Madison, 1415 Engineering Drive, Madison, WI 53706

Advanced control of crystal shape, size distribution, and polymorphic form (internal lattice structure) in suspension crystallization processes has been hindered by the limitations of available on-line sensors. High-speed, in situ video microscopy is a promising technology for measuring these critical solid-phase properties. Extracting the size and shape distributions from the video images requires not only suitable image analysis but also statistical estimation to correct for sampling biases inherent in imaging-based measurement. We have developed advanced image analysis algorithms that are effective for in situ images of needle-like particles[1] and particles with complicated shapes that can be modeled geometrically using a wireframe model[2]. The former algorithm has been licensed by an industrial manufacturer of in situ video imaging probes and is expected to be packaged with their next-generation model this fall. We also have derived a maximum likelihood estimator (MLE) of the particle size distribution (PSD) that corrects sampling bias due to the finite size of the imaging window, enabling unbiased estimation of the PSD as well as quantification of measurement uncertainty[3].

The derived MLE provides a necessary and useful starting point for imaging-based PSD estimation, but is based on the assumption that image analysis is perfect, i.e. that the image analyzer identifies every single particle appearing completely inside the imaging window. In reality, however, image analyzers often miss particles that are overlapping or occluded. In this talk, we present a method for estimating the number density of monodisperse particle populations at high solids concentrations with heavy particle overlap and imperfect image analysis. The method is based on a first-principles, probabilistic model that describes the degree of particle overlap as a function of particle geometry and number density. The effectiveness of the method is demonstrated using data obtained from a complicated image analysis algorithm applied to images simulated under a variety of conditions.

Reconstructing the PSD using imaging requires a large number of samples (i.e. images), but the amount of time available to acquire samples is limited by crystallization kinetics. By applying our image analysis and estimation tools to images from a simulated batch crystallization of an industrial photochemical, we demonstrate the feasibility of on-line, imaging-based, high-resolution PSD measurement given the time constraints imposed by a typical crystallization process.

[1] Paul A. Larsen, James B. Rawlings, and Nicola J. Ferrier. An algorithm for analyzing noisy, in situ images of high-aspect-aspect ratio crystals to monitor particle size distribution. *Chem. Eng. Sci.*, 61 (16):5236--5248, 2006.

[2] Paul A. Larsen, James B. Rawlings, and Nicola J. Ferrier. Model-based object recognition to measure crystal size and shape distributions from in situ video images. *Chem. Eng. Sci.*, 62:1430--1441,

616e Real-Time Particle Size And Shape Characterization Using In-Process Video Imaging Process Analytical Technology (Pat) In Dynamic Processes

Eric J. Hukkanen, Real-Time Analytics, Mettler-Toledo AutoChem, Inc., 14833 NE 87th Street, Redmond, WA 98052 and Terry P. Redman, Particle Systems Characterization, METTLER TOLEDO, Redmond, WA 98052.

The Mettler-Toledo Lasentec in-line Particle Vision and Measurement (PVM) in-process imaging system have been successfully implemented in the pharmaceutical world to monitor and characterize particle growth at the mesoscale and the particle size distribution at the macroscale. The PVM is a unique in-process imaging system capable of providing high-resolution images at most solids concentrations. With resolution below 2 microns and the ability to be inserted directly in a process line or reactor, the in-process video microscope provides an unsurpassed qualitative understanding of the change to particles and particle structures during dynamic processes. The PVM allows the user to characterize or visually confirm surface morphology and/or microstructures. As a quantitative tool, advanced image analysis techniques are implemented, in which particle properties and characteristics specific to the process (i.e., crystallization, granulation, polymorphic shift, etc.) are reported (e.g., length, width, aspect ratio, particle projected area and perimeter). Several examples from the pharmaceutical and petrochemical industry are provided to demonstrate the application of in-process video imaging and image analysis to characterize particle shape and size, as well as dynamic process changes.

Web Page: www.mt.com/pvm

616f Automated Control Of Batch Cooling Crystallization Using Fbrm

Pui Shan Chow, Jia Wei Chew, and Reginald B.H. Tan. Crystallization and Particle Science, Institute of Chemical & Engineering Sciences, 1 Pesek Road, Jurong Island, Singapore, Singapore

Accurate in-line sensors that are robust enough to be used in production environment have become available in recent years. This opens up the possibility of using such measurements to control crystallizations interactively in order to improve crystal product consistency. Closed-loop concentration-control (C-control) using Attentuated Total Reflectance – Fourier Transform Infrared (ATR-FTIR) has been the most commonly used feedback control method in which supersaturation is controlled at a constant level. Although C-control has been proven to be robust, a recent study by Chew et al.[1] found that C-control was unable to ensure product consistency in unseeded crystallization systems due to the random nature of primary nucleation.

Focused Beam Reflectance Measurement (FBRM) has emerged as a widely used technique for the in-situ characterization of crystallization systems by monitoring the chord length distributions. It has been used to develop and optimize crystallization processes, track and trouble-shoot crystallizer systems, and to monitor polymorphic forms. Yet, despite the proven useful applicability of FBRM in crystallization, there has not been any published work of automated closed loop strategies using FBRM. In view of the potential difficulty in implementing ATR-FTIR in the production systems due to calibration complexities, it would be advantageous to be able to rely solely on FBRM for reproducibility in product crystals.

In this work, we made use of FBRM in-situ in a feedback loop to improve crystal product quality in a batch crystallizer, without any external loop. The method consists of automated FBRM detection of nucleation, followed by controlled internal seed generation. The technique has been successfully implemented for the cooling crystallization of glycine and paracetamol in water.[2] Results showed that the crystal product consistency using our technique was comparable to that of externally seeded crystallizations.

1. Chew, J. W.; Black, S. N.; Chow, P. S.; Tan, R. B. H., Comparison between open-loop temperature control and closed-loop supersaturation control for cooling crystallization of glycine. *Industrial & Engineering Chemistry Research* **2007**, 46, (3), 830-838.

2. Chew, J. W. Experimental investigation of the applicability of FBRM in the control of batch cooling crystallization. MEng, National University of Singapore, Singapore, 2006.

660a Pharmaceutical Engineering as An Emerging Academic Field - Challenges and Opportunities

Piero M. Armenante, Otto H. York Department of Chemical Engineering, New Jersey Institute of Technology, 323 King Blvd., Newark, NJ 07102-1982

Pharmaceutical engineering is still a new but rapidly growing engineering field in academia. Therefore, it is important to be able to define its range of activities, since any academic program devoted to it must operate within these boundaries. Because of their novelty, pharmaceutical engineering curricula and the courses that constitute them need to be identified, and their contents must be defined, without the benefit of relying on significant previous experience in this area.

This presentation examines the current status of pharmaceutical engineering education in general, and the educational programs currently available in this area in particular, with specific reference to the Pharmaceutical Engineering program which has was established at NJIT in 2002.

The discussion of what constitutes the common knowledge in the evolving area of pharmaceutical engineering in an academic curriculum is an ongoing debate, and constitutes an additional focus of this contribution. An examination of the approach that should be taken to determine the range and extent of the core knowledge and activities in the rapidly expanding pharmaceutical engineering field will be presented and discussed.

660b Technology Roadmap for Pharmaceutical Development and Manufacturing

Gintaras V. Reklaitis, School of Chemical Engineering, Purdue University, 480 Stadium Mall Drive, West Lafayette, IN 47907

Text Not Available.

660c NSF Engineering Research Center for Structured Organic Particulate Systems

Fernando J. Muzzio, Chemical and Biochemical Engineering, Rutgers University, 98 Brett Road, Piscataway, NJ 08854

Text to come from Dr. Muzzio.

AUTHOR INDEX

Abbas, A. 65
Abhinava, k. 142
Achenie, L. E. K. 245
Acquah, C. 245
Adams, M. L. 330
Alexander, A. W. 93
Allen, B. R. 208
Amenitsch, H. 233
Anand, A. 30
Arastoopour, H. 36, 201
Arendt, P. D. 234
Armaou, A. 200
Armenante, P. M. 95, 208, 316, 349
Arntz, M. M. H. D. 34
Aviles, M. O. 47
Bailey, M. 319
Balani, P. N. 16
Barresi, A. 168
Barrow, R. 29
Barthe, S. 344
Bartlett, J. A. 341
Bekker, A. V. 53
Benli, B. 287
Berkland, C. 319
Berry, M. R. 138
Berthiaux, H. 215
Bertrand, F. 28
Bhamidi, V. 259
Bika, D. 326
Bilgili, E. 140, 209, 326
Borchert, C. 51
Bordawekar, S. 278
Boudouvis, A. G. 189
Braatz, R. D. 26, 234, 271
Braido, D. 334
Breting, J. 112
Briels, W. J. 34
Brower, K. P. 46
Bulutcu, N. 287
Burcham, C. L. 37, 155
Burket, P. R. 307
Busch, F. 138
Cabelka, T. 340
Cameron, I, T. 83
Cameron, I. T. 170
Caramellino, M. 95
Carrier, R. L. 332
Carroll, N. 306
Carvajal, M. T. 214
Chaduvula, A.I.R. 20, 157
Chan, S. H. 103, 328
Chan, S.Y. 16
Chang, S-Y. 103
Chaouki, J. 28
Chappidi, V.S. 20
Chattopadhyay, S. 233
Chaudhuri, B. 14
Chen, A. 326
Chen, J-S. 203
Chen, W. 103
Chern, R. 1, 94, 97, 140
Chew, J.W. 348
Chow, P.S. 345, 348
Chow, Pui Shan 244
Coles, A. 156
Colton, C. K. 46
Cooney, C. 224
Cuitino, A. 143, 334
Curtis, J. S. 30
Danquah, M. K. 320
Dave, R. 29
Davis, R. J. 202

AUTHOR INDEX

Davis, R. M. 48
Den Otter, W. K. 34
Denomme, D. 340
Descamps, N. 67
Dhanasekharan, K. 1
Di Benedetto, G. 208, 316
Diemer, R. B. 12
Ding, Y. 12
Dirion, J.L. 215
Doherty, M. F. 242
Doucet, J. 28
Doyle II, F. J. 170
Doyle III, F. J. 83
Ebner, A. D. 47
Eder, R. 204
Ehrl, L. 63
Eibling, R. E. 307
Ende, D. J. A. 207
Ende, M. T. A. 138
Ender, K. 340
Erdemir, D. 233
Ermoline, A. 29
Fan, R. 1
Farber, L. 94
Fichthorn, K. A. 269
Figueroa, C.V. 32
Fisher, R. J. 46
Fliszar, K. 326
Foddi, O. 65
Forde, G. M. 320
Foster, N. R. 255, 257
Fricke, M. 297
Friedman, E. 317
Fukumori, Y. 246
Furcas, F. 168
Gallagher, K. 97
Gallivan, M. 344
Gamsiz, E. D. 332
Gao, Qi 232
Gao, Y. 113
Garcia, T. P. 138
Gascon, J. 245
Gasparini, G. 298
Gatumel, C. 215
Gentzler, M. 209, 329
Georgakis, C. 342
Ghadiri, M. 12
Gillanders, A. 104
Givand, J. 329
Glaser, T. 83, 170
Glasser, B. 13
Glasser, D. 13
Glassey, J. 113
Goh, L.M. 271
Griffiths, K. 142
Grossmann, I. E. 24
Grosso, M. 65
Grubb, D. 326
Guo, L. 233
Hancock, B. C. 30
Hardin, M. T. 260
Hausberger, B. 13
He, G. 259
Heidel, S. J. 209
Higgins, J. P. 154, 223
Hildebrandt, D. 13
Hill, B. 94
Holdich, R. G. 298
Holstine, B. 329
Holtze, C. 306
Hsu, S-H. 210
Hukkanen, E. J. 347

AUTHOR INDEX

Ichikawa, H. 246
Ierapetritou, M. G. 212
Ikeda, C. 97
Ikeda, M. 246
Ilavsky, J. 233
Immanuel, C. D. 83, 170
Irizarry, R. 190
Jacob, K. 340
Jacobson-Andrews, G. B. 258
Jain, S. 202
Jawor-Baczynska, A. 286
Jayjock, E. 315, 318
Jeanquartier, C. 204
Jerzewski, R. 317
Jin, M. 256
Jines, A. 324
Joglekar, G. 210
Johnson, M. 325
Jones, C. W. 202
Josyula, S. B. 20
Karunanithi, A. T. 245
Kataria, A. 38
Kavousanakis, M. E. 189
Kee, N. 234
Kenis, P. J. A. 259
Kenis, P.J.A. 270
Khalili, S. 200
Khan, M. A. 123
Khinast, J. 203
Khinast, J. G. 204
Khumalo, N. 13
Kiang, S. 103, 314, 328
Kienle, A. 160
Kientzler, D. 103
Kipling, K. 113
Kiyono, F. 279
Ko, J. 326
Ko, J. S. 154, 223
Kokitkar, P. B. 96
Koopman, D. C. 307
Koparde, V. 269
Kosvintsev, S. R. 298
Kottala, N. 317
Koynov, A. A. 143
Krasnyk, M. 160
Kumagai, K. 279
Kumar, V. 39
Lacombe, J. P. 318
Lai, C. 314
Lam, U.T. 255
Lambert, D. P. 307
Lamm, M. 64
LaPack, M. A. 158
LaPorte, T. 314
Larsen, P. A. 346
Lattuada, M. 63
Lau, J. 327
Lenhoff, A. 260
Li, Y-E. 156
Liaw, G. 156, 327
Lin, S. 159, 206
Litster, J. D. 260
Litster, J. D. 27, 83, 170
Livk, I. 53
Lopez, G. P. 306
Lowinger, M. 154, 223
Lyngberg, O. 103
Macias, K. 214
Mammucari, R. 255
Mancinelli, C. 94
Mangold, M. 160
Mangual, J. O. 47

AUTHOR INDEX

Mantzaris, N. V. 189
Marchisio, D. 168
Marchut, A. J. 103
Marek, J. 159, 206
Matsoukas, T. 200
McCarthy, J. J. 31
McCready, C. 152
McCurdy, V. E. 138
McGarvey, B. 142
Mckelvey, C. A. 154, 223
Mehta, C. M. 260
Mendez, S. 306
Mendez-Roman, R. 33
Metzger, M. J. 13
Meyer, R. F. 154, 223
Miles, W. 48
Miller, L. A. 332
Miller, M. T. 46
Mitchell, C. 159, 206
Mitchell, J. D. 142
Mitra, S. 316
Montague, G. 113
Moore, B. D. 286
Moore, C.M.V. 22, 25
Morbidelli, M. 63
Morris, K. R. 23
Moser, J. 97
Mukherjee, S. 112, 278
Mustakis, J. 207
Muzzio, F. J. 33, 212, 315, 318, 351
Myerson, A. S. 233
Nahar, A. Abu 15
Nalla, K.N. 20
Nere, N. K. 51
Ng, W.K. 16
Nikku, M.Y. 20
Olson, J. H. 12
Olusanmi, D. 12
Ormes, J. 330
Ostberg, M. 327
Palzer, S. 67
Pandey, M. K. 46
Parker, W. L. 232
Peek, L. J. 319
Peppas, N. A. 331
Pernenkil, L. 224
Perry, S.L. 270
Petsev, D. 306
Plumley, C. 319
Poon, J. M. H. 83
Poon, J. M.H. 170
Poornachary, S. 244
Popolla, I. 327
Portillo, P. M. 212
Pranami, G. 64
Prud'homme, R. K. 39
Quan, C. 258
Quiñones, L.G.O. 32
Quiroz, F. 314
Radosz, M. 49
Rajniak, P. 1, 94, 97, 140, 326
Ramachandran, R. 83, 170
Ramasamy, M. 209
Ramkrishna, D. 51
Rawlings, J. B. 346
Realpe, A. 75
Redman, T. P. 347
Reklaitis, G. V. 210, 350
Richter, S. 159
Ritter, J. A. 47
Roberts, G. W. 270
Roberts, K. J. 12

AUTHOR INDEX

Romagnoli, J. A. 65
Rosen, L. 326
Rousseau, R. W. 344
Rudeen, B. L. 154, 209, 223
Sanders, C. F. W. 83, 170
Schenck, L. 1, 154, 223
Schitter, G. 204
Schreyer, E. 67
Schroer, J. G. 205
Scicolone, J. V. 29
Sefcik, J. 286
Segre, C. U. 233
Seibert, K. 142
Seibert, K. D. 139
Sethuraman, S. 139, 142
Shi, D. 31
Shi, L. 319
Shofner, J. P. 331
Sih, R. 257
Silverman, A. R. 329
Sithambaram, S. 245
Sizemore, J. P. 242
Smith, E. 326
Snyder, R. C. 242
Soos, M. 63
Spangler, L. 314
Stepanek, F. 94, 170
Stone, M. 307
Strumendo, M. 201
Suib, S. 245
Sukcik, E. 316
Sundmacher, K. 51, 297
Szilágyi, E. M. 258
Tajima, H. 279
Takeda, S. 279
Talreja, S. 259
Tan, R. B.H. 16, 234, 244, 345, 348
Tawakkul, M. 123
Temphuwapat, W. 246
Thakur, A. 330
Thombre, A. G. 332
Tice, J.D. 270
Titchener-Hooker, N. 113
Tolle, J. 278
Tom, J. 314
Tomasone, M. S. 315
Troup, G. M. 154, 223
Tsutsumi, A. 246
Turner, C. 258
Tüzün, U 15
Tyagi, R. 46
Tyler, S. 156, 327
Tyrrell, Z. 49
Uppuluri, K.B. 157
Vargas, W. 31
Velazquez, C. 33, 75
Venkatasubramanian, V. 35, 210
Vicenzi, J. T. 155
Vigil, R. D. 64
Viswanath, S. 112, 159, 206
Voigt, A. 51, 297
Wallick, D. 340
Wang, F-Y. 83, 170
Wang, J. 317
Wang, S. S. Y. 328
Wang, Z. 159
Wassgren, C. R. 30
Watterson, A. C. 46
Weck, M. 202
Weitz, D. A. 306
Wertman, J. 102
White, E. T. 260

AUTHOR INDEX

Whittaker, J. 224

Willcox, D. 153

Willis, M. 113

Winoto, W. 49

Woelfler, H. 204

Wolfe, C. 139

Wong, G. 326

Wong, S,W. 342

Wu, H. 123

Xu, Z. 314

Yamasaki, A. 279

Yanagisawa, Y. 279

Yang, Y. 327

Yazdi, F. J. 224

Yedluri, I. 20

Yu, Z.Q. 345

Zare, R. N. 258

Zharov, I. 50

Zheng, X. 202

Zhong, Q. 256

Zhou, J. 46

Zhou, Y. 113

Zucca, A. 168

Zukoski, C. F. 259